Physics Behind Music

This engaging undergraduate text uses the performance, recording, and enjoyment of music to present basic principles of physics. The narrative lays out specific results from physics, as well as some of the methodology, thought processes, and "interconnectedness" of physics concepts, results, and ideas. Short chapters start with basic definitions and everyday observations and ultimately work through standard topics, including vibrations, waves, acoustics, and electronics applications. Each chapter includes problems, some of which are suited for longer-term projects, and suggestions for extra reading that guide students toward a deeper understanding of the physics behind music applications. To aid teaching, additional review questions, audio and video clips, and suggestions for class activities are provided online for instructors.

Bryan H. Suits has been a professor in the Department of Physics at Michigan Technological University (MTU) since 1985. An award-winning teacher, he has taught physics courses at all levels. An accomplished amateur musician with decades of experience, the author has combined his enthusiasm for physics with that for music to find ways to improve physics literacy among nonphysicists.

"This textbook is written with a palpable passion for physics, music, and communicating science to others.

Clearly a result of years of teaching the topic, it covers the basics in both physics and music comprehensively. While not requiring prior knowledge, the physics principles covered here are advanced, from the concepts of superposition and normal vibrational modes to waves and uncertainty principle. All are considered with a good depth despite using barely any mathematics.

What makes this stand out from other physics texts is the deep connection with music both in terms of using physics to explain principles underpinning music, but also in terms of capturing the philosophy of both fields. The book does not shy away from the human perception side of the story, which is critical in music and rarely, if ever, considered in physics.

I believe it will be valuable for teaching to interdisciplinary audiences or when introducing more advanced physics topics to A-level students. I also believe music students would benefit from looking at least at the parts describing physical aspects of various musical instruments and hall acoustics.

I particularly like that the text is written as a textbook with exercises, and I hope it will encourage more universities to teach physics in an interdisciplinary context and in relation to music. Once it is published, I will use it in the reading list for the Science of Music module that I teach."
Oksana Trushkevych, University of Warwick

"I highly recommend this book. The text is readable, and suitable for a broad range of student levels. The end-of-chapter problems connect well with the chapter content, and provide a reasonable balance of beginning and advanced questions. The progression of topics is logical, the range is wide, and the content intriguing."
Ananda Shastri, Minnesota State University Moorhead

Physics Behind Music
An Introduction

Bryan H. Suits
Michigan Technological University

CAMBRIDGE
UNIVERSITY PRESS

Shaftesbury Road, Cambridge CB2 8EA, United Kingdom

One Liberty Plaza, 20th Floor, New York, NY 10006, USA

477 Williamstown Road, Port Melbourne, VIC 3207, Australia

314–321, 3rd Floor, Plot 3, Splendor Forum, Jasola District Centre, New Delhi – 110025, India

103 Penang Road, #05–06/07, Visioncrest Commercial, Singapore 238467

Cambridge University Press is part of Cambridge University Press & Assessment, a department of the University of Cambridge.

We share the University's mission to contribute to society through the pursuit of education, learning and research at the highest international levels of excellence.

www.cambridge.org
Information on this title: www.cambridge.org/highereducation/isbn/9781108844659

DOI: 10.1017/9781108953153

First published 2023

A catalogue record for this publication is available from the British Library.

A Cataloging-in-Publication data record for this book is available from the Library of Congress

ISBN 978-1-108-84465-9 Hardback
ISBN 978-1-108-94870-8 Paperback

Additional resources for this publication at www.cambridge.org/suits.

Contents

Preface   *page* xi

1 Introduction   1

   What Is Music?   4
   What Is Physics?   5
   Describing Quantities   7
   Prefixes and Scientific Notation   8
   Summary   12
   Additional Reading   13
   Problems   13

2 Frequency and Rates   15

   Frequency   15
   Rates   17
   Musical Intervals   20
   The Notes on a Musical Keyboard   21
   Logarithms and the Log Scale   23
   Summary   25
   Additional Reading   26
   Problems   26

3 The Notes We Use   28

   Scales   28
   Chords   31
   Pythagoras's Monochord   32
   Beats   33
   Temperaments   37
   Nice Chords   41
   Summary   42
   Additional Reading   42
   Problems   43

4 The Frequency Domain and Pitch   45

   The Fourier Series   46

Spectra   49
How a Fourier Transform Works (Optional)   57
Spectral Resolution and Range   59
Pitch   59
Summary   65
Additional Reading   65
Problems   65

5 Harmonic Oscillators and Resonance   67

Newton's Laws and Gravity   67
Gravity Near the Surface of the Earth   70
Hooke's Law for Springs   71
The Harmonic Oscillator   73
The Driven Harmonic Oscillator – Resonance   81
Quality Factor   85
Sympathetic Resonances   87
Fictitious Forces and Physics Demonstrations (Optional)   89
Summary   91
Additional Reading   92
Problems   92

6 String Theory   94

The Frequency of Vibration for a String Under Tension   94
String Resonances   98
Standing Waves   101
The Plucked String   104
End Conditions   106
Quality Factor   109
Summary   110
Additional Reading   111
Problems   111

7 Normal Modes   114

Indexing the Modes   114
Chimes   117
Vibrations in Two Dimensions   127
Degeneracies   131
Bells and Other Shapes   137
Torsional Modes   138
Visualizing Node Lines   139
Summary   140

Additional Reading   140
Problems   141

8 Traveling Waves   143

Motion of a Pulse   144
Mathematical Description of Traveling Waves   145
Traveling and Standing Waves   151
Polarization   154
Traveling Waves Transport Energy   155
Dispersion   156
Summary   157
Additional Reading   158
Problems   158

9 The Uncertainty Principle   161

Wave Pulses   162
Attack and Release   165
ADSR – a Model   166
Spectrographs   166
Phase Shifts and Transients   167
Summary   168
Additional Reading   169
Problems   169

10 Nonlinear Physics   171

Linear Problems   171
What Makes a Problem Nonlinear?   172
Friction   174
The Stick–Slip Mechanism   176
Back to the Swing   178
Reed Instruments Simplified   180
Playing in Tune   184
Mode Locking   185
Chaos (Optional)   186
Summary   188
Additional Reading   188
Problems   188

11 Classical Gases   190

The Elements   191
Isotopes   194

Molecules 196
States of Matter 197
Air 198
Temperature and Kinetic Energy 199
Impulse 201
Pressure 203
Ideal Gas Law Examples 206
Pressure for Wind Instruments 208
Summary 209
Additional Reading 210
Problems 210

12　The Speed of Sound in a Gas 212

Measured Values 212
Adiabatic Processes 214
The "Springiness" of a Gas 217
The Adiabatic Exponent 221
Speed of Sound 223
Summary 228
Additional Reading 228
Problems 228

13　Sounds We Hear 230

Fluid Mechanics 230
Sound Waves Are Longitudinal 232
The Amplitude of Sound 233
Combining Signals 238
Summary 240
Additional Reading 240
Problems 241

14　Sound in Pipes 242

Impedance 243
Sound in Air-Filled Pipes 245
Pipes with Holes 254
More Complicated Pipe Resonances 259
Losses 265
Derivation of R and T (Optional) 265
Summary 266
Additional Reading 267
Problems 267

15 Sound in Three Dimensions 269

Isotropic and Point Sources 269
More Complicated Sources 270
The Inverse Square Law 278
Echoes and Reverberation 280
Room Features of Note 287
The Cocktail Party Effect 289
Summary 292
Additional Reading 293
Problems 293

16 Interference, Diffraction, and Diffusion 295

Interference 295
Diffraction 299
Diffusion 301
Summary 307
Additional Reading 308
Problems 308

17 Faraday's Laws of Induction 310

Properties of Magnets 310
The Magnetic Field 312
Magnetic Materials 314
Nonmagnetic Materials 318
Electromagnets 319
The Electric Guitar Pickup 324
Transformers (Optional) 326
Summary 327
Additional Reading 328
Problems 328

18 RC Time Constants 330

Electric Fields 330
Conductors and Insulators 333
Electrostatic Discharge (Optional) 334
Capacitors 335
Resistors 338
Circuit Diagrams 339
RC Time Constants 339
Condenser and Electret Microphones 344

Microphone Sensitivity (Optional)   345
Summary   347
Additional Reading   348
Problems   348

19 Physics and Recording Technology   350

Physical Media   351
Magnetic Recording   355
Optical Media   361
Digital Storage and Playback   367
Summary   374
Additional Reading   375
Problems   375

20 Electronics and Music   376

Electronic Effects   376
Making Music with Electronics   380
Electronic Synthesis   384
Interfaces   386
Concluding Remarks   387
Summary   387
Additional Material   388
Problems   388

Appendices   390
Appendix A   Mathematics   390
            I.   Exponents and Standard International (SI) Unit
                 Prefixes   390
            II. Some Mathematical Properties   392
            III. The Summation Symbol   397
Appendix B   Greek Alphabet   398
Appendix C   Note Frequencies   399
Appendix D   Answers to Selected Problems   401
Index   403

Preface

This book is about the many areas of basic physics, and the application and practice of physics, behind the production, performance, recording, and enjoyment of music. The emphasis and the organization follow from the development of those basic physics topics from a physicist's point of view. The expectation here is that the reader has a math background up to but not including calculus. That includes some algebra, plane geometry, and familiarity with the use of simple graphs and functions. Those who may have learned those skills at one time but have not used them in a while must be willing to relearn them.

My parents were both strong in the physical sciences as well as active amateur musicians. It is not surprising that some of that rubbed off on me. I started to get serious about making the connection between the two over 20 years ago, and I was surprised to see how much information was available. Around the same time, I gained an increased interest in improving physics literacy, especially among nonphysicists. While there were already textbooks available that were very nontechnical and often math-less, as well as highly technical references works, there did not seem to be much in the middle. I felt it appropriate to develop some teaching materials that do require some math skills but are not overly technical, certainly less than might be expected of a graduate student or specialist.

The materials in this book were first developed to meet the needs of a short summer course, to develop some related web pages on the subject, and then for a semester-long course. The audience for that semester-long course is students studying technical disciplines, although not usually in physics. During the development of this book, I found myself asking, in the context of the application to music, "What would I, a physicist, like a well-educated nonphysicist to know about physics and the practice of physics?" The answer includes some specific results from physics but also some of the methodology; some of the thought processes; and some of the "interconnectedness" of physics concepts, results, and ideas. The specifics are, admittedly, my personal choices, and to be sure, not all physicists will agree with those choices or the order of their presentation.

I have tried to develop the book with a variety of readers in mind. That includes those who use it as a primary or secondary text for a course on the subject, as well as individuals who would like to learn more about the subject on their own. It is recognized that this is a wide audience that shares an enthusiasm for both science and music but, individually, may have a wide range of experience, expertise, motivation, and other interests. To keep the material available to as wide an audience as possible, I have assumed that the reader, although technically inclined, has no particular musical skills and little, if any, prior experience with physics as a discipline, with apologies to those who have either or both.

The sequence of topics in the book is an attempt to initially define and introduce the subject using material already known more than 2,000 years ago, long before the development of the formal mathematics of Descartes, Leibniz, Fourier, and many others, and before the advances in physics by Galileo, Newton, and Einstein, to name a few. It should be possible to at least start the discussion using the simpler tools and understandings from long ago. This is not to say mathematics is avoided. Mathematics is the language of physics, and ultimately, physics results are normally expressed in that form. The level of mathematics is then pushed a bit in some of the later chapters. A review of some relevant mathematical material is included in the appendices for the convenience of those who need some review.

To make it easier to adapt the material for a variety of circumstances, it is presented using a large number of shorter chapters. If a particular topic is not of interest, it can be skipped. Some of the more technical material that is, perhaps, a bit less musical can be omitted if desired and is noted as optional.

The 20 chapters, each of which can be covered in one or two class periods, represent material that can be spread over a somewhat intense semester-long course. The first group of four chapters discusses musical notes, frequency and pitch, chords, and ultimately, the description in terms of spectra. The next five chapters look at oscillating mechanical systems, starting with the harmonic oscillator, vibrating strings generalized to plates and chimes, and then simple traveling waves in one dimension and the uncertainty principle. A brief discussion of the very difficult, but essential and often omitted, topic of nonlinear physics follows. The next six chapters develop sound and acoustics, starting from atoms and molecules and ultimately leading to the wave behavior of sound in three dimensions. The final four chapters look at electricity, magnetism, and electronics in musically relevant situations. In each case, the emphasis is on the basic physics of the situation, including, where possible, how that same physics may show up in other situations. The discussion generally does not delve into, for example, the details (i.e., the engineering) of any particular musical instrument. There are other texts specifically about the physics of musical instruments that can be consulted for such details.

There are a few themes that span multiple chapters. Those include exponentials and logs, beats in various contexts, the interplay of restoring forces and inertia for mechanical oscillations and waves, and the idea that the physics seen here is also applicable in many other circumstances.

Suggestions for extra reading are included at the end of each chapter. These include references specifically mentioned in the text, as well as some others that simply may be interesting for additional exploration. An exception is the very last chapter, where the suggested resources also include readily available audiovisual materials. These lists are not intended to constitute a complete bibliography.

A set of problems can be found at the end of each chapter. Some of these are intended to be straightforward and follow the examples given in the chapter text. When solved in the examples, where possible, the solution is presented using an approach based on understanding the physics rather than "finding a formula and plugging in values." That latter approach is a strategy often used by beginning students and should be strongly discouraged. The later, more difficult end-of-chapter problems may work best with small-group discussions (e.g., in class), and some may require additional investigation beyond what is in this text. Some might even be best suited for longer-term projects.

It would be expected that many demonstrations, hands-on activities, and possibly projects will also be carried out in class. The details will depend on what is available and of interest to the instructor, and suggestions for each chapter are available as an ancillary resource accompanying this book. Instructors are especially encouraged to use real equipment, rather than simulations, whenever possible, at least to get started. While it is tempting to use simulations, and many are readily available, they are not real, and in practice, not all students understand what simulations are showing. If real apparatus is not available live, then there are likely many recorded versions that can be used.

There are supplementary chapter review questions, in a multiple-choice format, that are available to instructors from the publisher's website (www.cambridge.org/suits). These questions are supplied in a format that can be easily imported into most learning management systems (LMSs). The questions test for knowledge of basic definitions and concepts from the chapter and are intended as an important complement to the end-of-chapter problems. Some additional ancillary materials are also available at that site, including audio and video files that may help in the classroom, as well as some supplemental material for selected topics. Instructors are especially encouraged to visit that site.

As an added feature, figures identified in the figure caption with a "★" are available in an alternate, animated format that can be found in the e-book version and at the publisher's website (www.cambridge.org/suits).

1 Introduction

A very, very long time ago, the earliest modern humans already knew that unsupported objects would fall to the ground, that the sun and moon would rise in the east, and that objects could be hot or cold. They undoubtedly looked at the stars with a sense of wonderment. And as best as our current science can determine, those earliest humans also had music. There are drawings of musicians playing various flutes and stringed instruments on ancient Grecian pottery dating back to more than 2,500 years ago. Within the ancient Egyptian ruins, dating back to more than 4,000 years ago, there are pictures of musicians playing various stringed instruments. Relatively recently, a collection of 9,000-year-old flutes was found in China, one of which was reported as still playable. In Europe, bone flutes have been discovered that date back to about 35,000 years ago, a time when Neanderthals were still to be found (Figure 1.1).

Those ancient humans, being human, likely wondered about how things worked, how it all fit together – the philosophy of it all. That was the beginning of what is now called the natural sciences, one branch of which is now identified as physics. As time progressed and human study of the universe continued, it was only natural that music would have been considered as much a part of the universe as anything else. After all, as far as humans were concerned, music had been around as long as the wind, the earth, the sun, and the stars. They all came as a package, and so it follows that music and nature were not treated as separate areas of study.

One of the earliest written records of the scientific study of music dates back to the Pythagoreans of ancient Greece. The Pythagoreans were followers of the Greek philosopher, mathematician, and cult leader known as Pythagoras of Samos. Pythagoras lived in the sixth century BCE, although much of what we know about him and his discoveries comes from the writings of his followers. Those writings came centuries after his death and so may or may not be entirely reliable. Nevertheless, many who have studied even a little geometry are undoubtedly aware of the theorem credited to Pythagoras that relates the lengths of the sides of right triangles – the Pythagorean theorem. Fewer may be aware of his, or

(a) (b)

(c) (d)

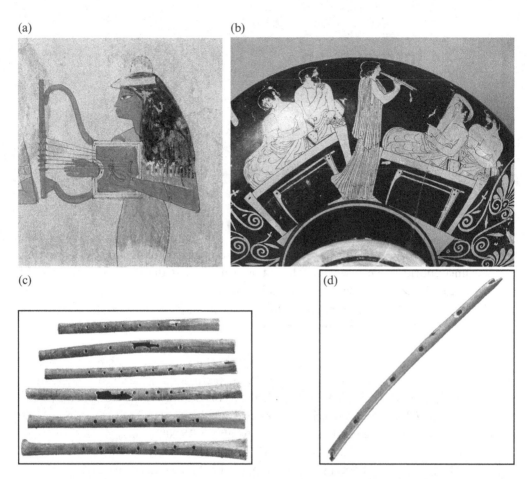

Figure 1.1 Examples of archeological evidence of the presence of ancient human music, including (a) ancient Egyptian wall painting, ca. 1400 BCE (photo by Werner Forman, Universal Images Group, Getty Images); (b) antique Roman cup, fifth century BCE (photo by DEA/G. Dagli Orti/De Agostini, Getty Images); (c) set of bone flutes found at the Jiahu site, China, as much as 9,000 years old (reprinted by permission from Springer Nature Customer Service Centre GmbH: Springer Nature, *Nature*, Oldest playable musical instruments found at Jiahu early Neolithic site in China, Zhang, J., Harbottle, G., Wang, C., et al., © [1991]); and (d) flute made from vulture bone, found in a Stone Age cave, Germany, at least 35,000 years old (reprinted by permission from Springer Nature Customer Service Centre GmbH: Springer Nature, *Nature*, The earliest musical tradition, Adler, D., © [2009]).

his followers', studies of the behavior of musical instruments and their subsequent contribution to the development of modern music theory.

A philosopher in ancient times was someone who studied anything and everything and tried to unify observations and thoughts in a way that would lead to a better understanding. That included unifying the natural sciences, music, and also religious beliefs. It would have been natural for the Pythagoreans to seek to justify astronomical observations using the harmonies of music in a way that seemed appropriate for their religious beliefs. One such description was the "music of the spheres." After all, since there were seven known astronomical bodies that were distinct from the background of the stars (five planets, the moon, and the sun) and seven notes in their musical scale, there surely had to be a connection. The idea was that each astronomical body was affixed to a rotating sphere, and as the sphere moved, it emitted a tone. The assignment of the tones may, today, seem to be somewhat arbitrary, although presumably it made sense at the time. Most importantly, the combination of all the tones was alleged to be one large harmonious chord. It was said that the fact that humans could not hear these tones was because the tones are so ever present in the background, and have been since our birth, that our minds ignore them. Of course, the great Pythagoras was thought to have been able to hear them. Many of these Pythagorean ideas about music theory and its relation to the universe were picked up by Plato and others and have lived on for many centuries. Interestingly, one of the Pythagoreans' other ideas, that of a heliocentric (sun-centered) universe, lasted only a short time, only to be rediscovered many centuries later amid great controversy.

Pythagoras is not the only well-known scientist from history who has looked at science and music. Sir Isaac Newton (1643–1727), perhaps one of the most well-known physicists of all time and also a devotee of the ancient philosophers, studied the colors that can be derived from white light and that give the colors of the rainbow. He identified seven colors, having made what seemed to be a very natural connection between colors and the seven notes of the musical scale. After all, such a connection made things very harmonious.

By the nineteenth century, however, any belief in a direct connection between music and the harmony of the universe had waned. In fact, the German physicist Hermann von Helmholtz (1821–1894), well known among physicists for his discoveries related to electromagnetism and acoustics, stated in the introduction to his book *On the Sensations of Tone* that he considered the study of physical science and that of musical science to have been (in translation) "hitherto practically distinct," thus denying any previous connection between the two.

There certainly are modern scientists of great repute who enjoy music and make it a significant part of their lives. Sir William Herschel (1738–1822), who gained most of his present-day fame as the astronomer who discovered infrared light and who discovered that Uranus was a planet, spent much of his early career as a musician and composer. Herschel has two dozen symphonies and numerous other compositions to his credit. There are many photos of Albert Einstein (1879–1955) playing his violin, which he reportedly used as a way to relax. Richard Feynman (1918–1988), who made many significant contributions to physics and won the Nobel Prize for his advanced theoretical work, also proudly spent time beating out rhythms on the bongos. However, in modern times, music and the natural sciences are considered very disparate disciplines. If a modern scientist, no matter how musically adept, were to attempt to make any direct connection between musical harmony and, say, the orbits of the planets, that scientist certainly would not be taken seriously.

So, what do the advanced studies from the forefront of physics, such as the pursuit of dark energy or the search for the Higgs boson, have to do with a symphony? What is the connection between extremely mathematically complex work in quantum electrodynamics and, for example, jazz improvisation? The modern physicist might say, "Not much, directly, but underlying each, if you look deep enough, is a common set of rules governing how the universe works." If you understand the rules for one situation, you also understand the rules for others. Hence, if you enjoy music, why not use it as a vehicle when learning about the universe? That is what will be attempted here – to learn a bit about how the universe works using music as a backdrop.

An important first step for any study is to define, as best as is possible, the terms that are used to describe the area to be studied. It is quite difficult to give a precise definition of what music is. To define the field of physics seems a bit more straightforward.

What Is Music?

Music is part of our everyday lives. Music is an aspect of many ceremonies, and it is used for courtship, storytelling, and "rousing the troops." Music is also used as part of advertisements, as background sounds in stores, and as an important part of all forms of video entertainment. In addition, music can be created as an object of "pure art" simply to be experienced. Music is so endemic to being human that when a National Aeronautics and Space Administration (NASA) committee, chaired by astronomer Carl Sagan (1934–1996), was asked to create a data set describing humanity to anyone, or anything, who might later discover the two *Voyager* spacecraft destined to leave our solar system, an ample supply of music was included.

A definition for music can be found in any dictionary, although the entries vary considerably from dictionary to dictionary. Typical definitions resemble these:

- Patterns of sound varying in pitch and time produced for emotional, social, cultural, and cognitive purposes.
- The art of producing pleasing, expressive, or intelligible combinations of tones.
- The science or art of ordering tones or sounds in succession, in combination, and in temporal relationships to produce a composition having unity and harmony.
- Vocal, instrumental, or mechanical sounds having rhythm, melody, or harmony.

All these definitions are reasonable while at the same time are imprecise. Clearly, music involves sound, but do these definitions truly distinguish music from other sounds or noises? The first three definitions, though not the last, all suggest that in order to be music, the sounds should have been constructed with a purpose in mind. Should two identical sounds, one that appeared naturally and one produced for "emotional, social, cultural, and cognitive purposes," be classified differently? And certainly, there are many examples of music where rhythm, melody, and/or harmony is apparently lacking and yet most would still classify it as music. It is not hard, in fact, to find examples of sounds accepted by most, at least, as music but that would not be classified as music using one or more of these definitions.

For most of us, **we know music when we hear it**, even if we cannot put a definition into a concise wording suitable for a dictionary. However, there is often disagreement. If you play a section of a Bach organ prelude, rarely (if ever) will you find anyone who would not call that music. But is the call of a bird also music? How about a selection of bird sounds edited together in a manner designed to evoke an emotional impact, such as can be found in Alfred Hitchcock's famous 1963 movie *The Birds*? Is that music? And when a two-year-old child happily bangs out successive, although apparently random, notes on a piano to both express their emotional state and evoke a response from their parents, is that music? Of course, there are also compositions by various modern composers that clearly push our ideas about what constitutes music. A particular example that stands out as one that challenges all our definitions is John Cage's composition *4'33"*, during which the musician(s), sitting in front of an audience, makes no sound at all for just over four and a half minutes. There has been considerable debate whether or not Cage's piece, which is certainly an auditory experience that inevitably evokes a response from the audience as well as the musicians, should also be classified as music.

It appears that there is no simple quantitative assessment that can be performed where the results can be used to determine whether or not something is music, at least not for every case. **Music is an art**, and in the arts, it is normal to have some disagreement, it is normal to push the boundaries, and it is just fine if "the exception proves the rule." The boundaries are fuzzy, and the context is important, but we know music when we hear it. Usually.

What Is Physics?

Science, on the other hand, seeks a universal agreement about how nature works. **Modern scientific methods attempt to find agreement through information gathering, analysis, testing, and verification**. While scientists often disagree, such disagreements, no matter how heated, must be considered temporary while more information is being collected and analyzed. There is an underlying belief that there is one right answer, even if that answer is not yet known. Along the way, candidate correct answers – theories – are thoroughly tested and retested to look for any exceptions. For science, a single exception, even a seemingly minor one, can disprove a rule.

Physics is the branch of science that deals with physical objects, their interactions, and the consequences of those interactions. These are all things that are directly measurable. Someone who practices physics, a physicist, will strive to understand the general relationships between measured quantities and, at the same time, to be able to make precise predictions of those quantities. Physical objects include the most elementary of particles, atoms, molecules, clusters of molecules, solids, liquids, plasmas, planets, stars, galaxies, the universe, and perhaps even more beyond that. The basic interactions are referred to as *forces*, although physicists will often use the related concept of energy to understand and deal with them. Since physics deals with quantities and their relationships, the language of mathematics is a particularly useful tool.

Thus, the parts of music that can be studied using physics will be those parts of music and music production that involve measurable physical objects and/or their interactions. Such studies may include sound and how it behaves, how sounds are produced, when we hear different sounds, what is actually different about them, electromagnetism used for sound production and sound processing, and other related topics. Physics will probably never tell us whether a sound is or is not music.

There is certainly an art to the practice of physics. In this case, the word *art* is referring to "a skill obtained through experience." To understand and describe a situation, the physicist will reduce a problem to what is essential in order to understand the central idea or concept at hand at an adequate level. In particular, the physicist will separate problems into different levels of detail and develop models and approximations to get to the essence of what is going on in each case. For example, when talking about the trajectories of tennis balls lofted into the air, the physicist will use a simplified model for the gravitational pull of the earth and will start by ignoring the presence of the air and even the size and shape of the ball. Such a simplified model may be quite adequate for many purposes. Then, for example, should the effects of air be too large to be ignored, that is, if the air is "nonnegligible," the forces from the air are added, usually using a simplified model. Certainly, if the trajectory of a tennis ball with significant topspin were to be considered, the effect of the air would need to be included from the start. The next level of detail for understanding the physics involved might deal with issues such as what the ball is made of, where the gravitational force comes from, where the simplified model for the forces from air comes from, and so on. Physicists follow a tiered approach to understanding using their skill derived through experience.

In the remainder of this text, that part of music that is physical and can be measured will be considered. To begin, after a few preliminary comments, a fundamental component of music – the musical note – will be examined. After that, how the notes are produced, how their properties can be measured, how they propagate to the listener, and ultimately how electronics may be involved will be presented. However, before starting that process, it is useful to review some of the symbols and language used by physicists when describing quantities that are measured.

Describing Quantities

Quantities are described with a magnitude (a numerical value), units, and possibly other qualifiers, such as a direction. The number describing the magnitude will rarely be exact and will have an implied accuracy. In an equation, such a quantity might be represented symbolically. For example, a single letter might stand in for the magnitude, units, and other possible qualifiers. Calculators work well with numbers, the magnitudes, but generally are poor at keeping track of units, accuracy, and the other properties that are equally important. Additional information must be supplied by the user.

Units

The quantities a physicist will deal with include all things measurable, such as mass, position, velocity, electric current, and many others. In each case, those measurements are a comparison to some established standard unit of measure. If the standard unit is not specified, the quantities can become meaningless. **Units define what is being quantified and the scale being used** – that is, what you are talking about. If you do not know what you are talking about, then talking about it is probably not appropriate. Do not forget to include the units.

When quantities are used in combination, that is, in formulas or equations, the units combine in the same way as the numerical values, thus also telling you about the nature of the computed quantity. This latter point is, on the one hand, an extremely important part of the practice of physics and, on the other, easy to forget since electronic calculators generally deal only with the numerical values. Some examples, which will be presented later, illustrate that the appropriate relationships between certain quantities – "the physics" – can often be deduced based on units alone.

The **Standard International units**,[1] or *SI units* for short, are the current internationally accepted standardized units. Seven SI base units are separately defined. Sometimes referred to as the *metric system*, the four most common base units are the **meter** (m), **kilogram** (kg), **second** (s), and **ampere** (A). The other three are the **kelvin** (K), used for temperature; the **mole** (mol), which defines a quantity of particles; and the **candela** (cd), used for luminosity. The base units

---

[1] Also referred to as the *International System of Units*.

are currently defined in terms of seven measurable constants of nature, thought likely to be the same throughout the universe.

Other units can be defined in terms of these base units. These are called **derived units**. The imperial and US customary systems currently use units defined in terms of SI standards. For example, 1 foot is defined to be exactly 0.3048 meters. Velocities do not have separately named units but are described in meters per second (m/s) or a similar combination of length and time units. The units for many other quantities have their own names, but those units are defined through the base units. For example, the newton (N) is the SI unit of force, where $1\ N = 1\ m\cdot kg/s^2$. Since the meter, kilogram, and second are all defined, a separate definition for the newton is not needed.

Note that an abbreviation for each unit has also been standardized. Those abbreviations are lowercase except when the unit is named for a person (for example, the units newton, ampere, and kelvin are named for scientists named Newton, Ampere, and Kelvin, and the units are abbreviated N, A, and K, respectively).

To add to the confusion, however, it should be pointed out that having a special name for a combination of base units specifies more than just what combination of the basic units to use. For example, the SI unit of energy, the joule (J), is the same as the product of 1 newton and 1 meter – that is, $1\ J = 1\ N\cdot m$. However, 1 N·m can also arise when discussing torques, the forces that cause rotations. In that case, 1 N·m would not be the same as 1 J – a torque is not an energy.[2] Hence, while you can replace 1 J with 1 N·m, you cannot always replace 1 N·m with 1 J – the context is important.

There are other units that have special names even though there are actually no base units involved. These are more like labels than units. Examples that will be seen include the decibel (dB) and the radian (rad). In these cases, the units (or labels) only serve to remind you of what is being discussed. The radian is defined using the ratio of two distances, and so in terms of base units, the units are $m/m = 1$, and the base units cancel. A decibel is also defined using a ratio of two quantities, and so the units cancel. However, decibels and radians are used to describe completely different situations, even though the base units, or lack thereof, are the same.

Prefixes and Scientific Notation

Since the physicist takes the entire universe as subject matter, the quantities considered can vary over a vast range. For example, the diameter of the nucleus of a hydrogen atom is roughly 0.0000000000000017 m, whereas the diameter of the

---

[2] The similarity in base units arises because energy is involved when you apply a force *along* a distance, whereas a torque arises when you apply a force *perpendicular* to a distance. In each case, you have a product of a force and a distance, and so the combination of base units is the same, even though the resulting quantity is completely different.

Table 1.1 Selected prefixes used to describe powers of 10.

Factor	Name	Symbol	Factor	Name	Symbol
10^{-12}	pico-	p	10^3	kilo-	k
10^{-9}	nano-	n	10^6	mega-	M
10^{-6}	micro-	μ	10^9	giga-	G
10^{-3}	milli-	m	10^{12}	tera-	T
10^{-2}	centi-	c	10^{15}	peta-	P

known universe is estimated to be about 8,700,000,000,000,000,000,000,000 m. In each case, there are many digits, but most are zeros that are simply acting as placeholders. **Prefixes and scientific notation are used as tools to deal with the wide range of values encountered.**

Prefixes are often used with all SI units. These prefixes are used as multipliers to scale the units as needed for the measurement at hand. For example, c is used to stand for the prefix *centi-*, meaning "1/100th of." That is, 1 cm = 0.01 m. These prefixes are very common, and a well-educated person should know a good selection of them. A list of the more common prefixes is included in Table 1.1. Note that whether a prefix is upper- or lowercase is very important. Generally, uppercase is used for multipliers larger than 1,000, and lowercase is used for those less than 1,000. Prefixes should not be combined.

In addition to prefixes, powers of 10 can be handled using **scientific notation**. Numbers written in scientific notation are written as a value near 1, with a power of 10 to move the decimal point. For example, the value 35,100 can be written as 3.51×10^4. The superscript "4" in this case can be interpreted to mean "move the decimal point 4 places to the right."[3] Aside from possibly conserving some space, scientific notation can also give you a hint as to the accuracy of a value. For example, 3.51×10^4 implies that the last two zeros of the original number come from rounding the value and are placeholders – that is, only three of the digits are significant, whereas 3.5100×10^4 implies that the value is known to have those zeros present – there are five **significant digits**, and it is thus known more precisely. Such a distinction is important for any measured quantity since nothing is measured to infinite precision. If needed, more examples involving prefixes, powers of 10, and combinations of prefixes and powers of 10 can be found in Appendix A.

---

[3] A negative superscript is used to specify a move to the left. In both cases, leading and trailing zeros are added if necessary.

Example 1.1

The statement that a store is "5 km east of here" describes the location of a store quantitatively. It includes a magnitude (5), units (km = kilometers), and a direction. In this case, the distance would be assumed to have been rounded off to the nearest kilometer.

When rounding off numbers with a higher number of digits to a smaller number of digits, **round to the nearest value**. If the value is exactly halfway between two values, round up. Rounding may be necessary if the result of a calculation includes more digits than is reasonable considering the initial data. Rounding may also be used for convenience when the full accuracy of a known value is not necessary for the purposes at hand.

When writing a value that is smaller than 1, the use of a leading zero, to the left of the decimal point, is considered good practice. Leading zeros are placeholders and are not considered significant, but they help draw attention to the presence of the decimal point.

Example 1.2

The value 4,523,110 rounded to include two significant digits is $4,500,000 = 4.5 \times 10^6$.
The value 7.3845 rounded to four digits is 7.385.
The value 0.002789 rounded to two digits is $0.0028 = 0.28 \times 10^{-2}$.

Symbols and Equations

In modern physics, the relationship between physical quantities is usually expressed using symbols and the language of mathematics. When doing so, it is natural to represent unknown quantities symbolically. Most often a single letter, possibly with a subscript or other symbolic modifier, will be used, although there are occasions when multiple letters are employed. However, when using these relationships, and certainly when memorizing them, it is often more productive to remember the relationship, at least in part, using words rather than abstract symbols.

Example 1.3

Consider the **Pythagorean theorem** for right triangles:

In a right triangle, the sum of the squares of the lengths of the sides adjacent to the right angle is equal to the square of the remaining side.

The same statement can be written in a concise symbolic form using a diagram to define symbols and an equation to show how they are related.

$$a^2 + b^2 = c^2$$

Example 1.4

Consider the following proposition:[4]

The distance traveled in a certain time by a body which begins to fall from rest is half the distance which it would cross in an equal time with a uniform motion having the velocity acquired at the end of the fall.

This can be made symbolic by defining symbols for each quantity, then showing how the quantities are related using mathematics. For example, one alternate way to write this proposition is as follows:

Defining t as the time to fall from rest a distance d, and v as the velocity at d, then $d = vt/2$.

One of the important reasons for thinking in terms of words rather than symbols, at least some of the time, is that **throughout physics, many symbols will have multiple meanings**. A reader of physics is expected to deduce the correct meaning from the context. A frequently occurring example is the symbol for millimeters, mm, where the first m represents the prefix *milli-* and the second m represents the unit, *meters*. Note that the letter m is also commonly used to symbolically represent the mass of an object; at the same time, it may be used as the name of that object, and sometimes the letter m may be used for other purposes entirely.

While the symbols used for prefixes and units should follow well-established standards, the choice of which symbol to use for mathematical variables ("unknowns") is generally up to the writer, with the understanding that it is the writer's obligation to ensure that the symbols used are always well defined for their readers. Those symbols are usually letters drawn from an alphabet (Roman, Greek,[5] etc.), although some are not. In practice, there will be some symbols that are so ensconced in the physicist's lexicon that to substitute another, without good cause, might be considered inappropriate. And yet, to avoid confusion, some other symbol may need to be substituted even then.

There are a few symbols that are used as modifiers for other symbols. One example that is very common in physics is the uppercase Greek delta, Δ. While the delta may appear as a variable by itself, it is often used to mean "a change between a first and second value of." For example, when using the variable t to stand for time, Δt would represent a single value equal to the change in time from some first event to a second event. Similarly, if velocities are symbolized using the variable v, then Δv would represent a change in that velocity between two events, such as starting and ending times or locations.

---

[4] This is proposition V of part II of Christiaan Huygen's *The Pendulum Clock, or, Geometrical Demonstrations Concerning the Motion of Pendula as Applied to Clocks*, originally written in 1673, as translated from Latin by R. J. Blackwell (Iowa State University Press, 1986).

[5] For reference, the Greek alphabet is listed in Appendix B.

Example 1.5

The position of a rolling ball is measured at two different times, as measured with a stopwatch: $t_1 = 2.2$ s and $t_2 = 5.3$ s. The time interval between the measurements is $\Delta t = t_2 - t_1 = 3.1$ s.

In physics, although there may be rare exceptions, **upper- and lowercase symbols should always be considered to be different**. For example, the uppercase M can represent the prefix *mega-* (but not *milli-* or *meters*), or it may be a variable. A mass labeled M would not be the same as a mass labeled m. Part of the art of reading physics is to be able to distinguish the different meanings from the context. Having language elements with multiple meanings, where the correct meaning must be deduced from context, seems to be a normal part of any language. For example, consider the following somewhat contrived, although grammatically correct, English sentence: "With a plant of his foot, Professor Plant said we should plant a plant at the auto plant." For those well versed in English, the different meanings of *plant* are likely clear from the context. For those just learning, these different meanings can be problematic. The same difficulty will be present for the language of the physicist.

With these elements of the physicist's language in mind, we can proceed. Additional mathematical details can be found in Appendix A, for those who wish to, or need to, refer to them.

Summary

The relationship between music and the sciences has changed considerably throughout history. To the Pythagoreans, some 2,500 years ago, there was little separation between the two. In current times, physics and music are considered very different disciplines. Music is somewhat difficult to define and includes subjective elements that are not universally agreed upon. Physics is a science based on the measurement of those things that are physical, and the physicist strives to be as objective as possible.

A value that describes the result of a physical measurement includes a number that may be written using scientific notation, standardized units describing the nature of the quantity that was measured, and some indication of the accuracy of the value. In some cases, there may be additional qualifiers, such as a direction. A standardized prefix may be included with the units to indicate a scale factor that is a power of 10.

Symbols, usually letters of the Roman or Greek alphabet, are often used to stand in for quantities, especially when a specific value is unknown. Symbols may be used for more than one purpose, and their meaning must be determined from the context.

ADDITIONAL READING

Bridgman, P. W. *The Logic of Modern Physics.* MacMillan, 1928.
James, J. *The Music of the Spheres.* Copernicus/Springer, 1995.
Wallin, N. O., B. Merker, and S. Brown, eds. *The Origins of Music.* MIT Press, 2000.

PROBLEMS

1.1 The typical size of an atom is 10^{-1} nm. How many centimeters is that?

1.2 Microwaves are often specified using units of hertz (Hz). Certain microwave signals associated with hydrogen atoms in interstellar space, referred to as *21-cm lines*, are observed at 1,420 MHz. How many GHz is that?

1.3 The distance light travels in one year, a "light-year," is 9,460,730,472,580,800 m. Write that in scientific notation rounded to two digits of accuracy.

1.4 The distance to Proxima Centauri, the nearest star to Earth, is 40,208,000,000,000 km. Write the distance (in kilometers) using scientific notation rounded to four digits of accuracy.

1.5 Electrical power is typically measured in watts (W). For everyday electronics, 1 pW would be an extremely small amount of power. If that were mistyped as 1 PW, how large of an error would have been made? (Note: The maximum output power of an electric power plant is typically a few gigawatts [GW]).

1.6 Explain why the equation 1 ft = 0.3048 m makes sense, but the equation that results when the units are omitted, 1 = 0.3048, does not make sense. How about the equation 3 kg = 3 s?

1.7 Many communities sponsor 5-k races, advertised as being 5 kilometers. At formal track meets, however, they have 5,000-m races. Describe the difference in the length of such races. (Hint: They are not simply "the same.")

1.8 In the classic 1939 MGM movie *The Wizard of Oz*, the following statement is made:

> *The sum of the square roots of any two sides of an isosceles triangle is equal to the square root of the remaining side.*

Write this statement as an equation, making sure to clearly define the symbols used with a drawing, and evaluate its mathematical correctness. Provide a rigorous mathematical proof supporting your conclusion.

1.9 The lowercase Greek letter π is commonly used to stand for the value of the ratio of the circumference to the diameter of a circle, $\pi = 3.14159\ldots$. Is that symbol used for anything else in physics? If so, what? How many physics uses can you find for the symbol T? How about G?

1.10 Figure 1.1 shows several examples of evidence of the presence of music in ancient cultures. Using a search engine, explore the Internet to find *additional* examples of similar archeological evidence (e.g., pictures of objects, etc.) showing the presence of ancient music from at least two different cultures.

2   Frequency and Rates

As was mentioned in Chapter 1, the Pythagoreans studied the musical scale scientifically around 2,500 years ago. The writings of another well-known Greek philosopher, Aristotle (384–322 BCE), also demonstrate a basic understanding of sound and how sound propagates through air. These works date back to long before the Industrial Revolution and long before any modern scientific apparatus would have been available to make measurements. One is left wondering what observations they might have been able to make, using materials at their disposal, which could lead them to this understanding.

That **the sounds we hear involve motion of the air** is evident by simply standing in the presence of a loud, low-pitched sound, such as that of a bass drum or thunder. The motion of the air can be felt like the wind and can, for example, cause motion in a candle flame. The extrapolation to sounds that are quieter and higher pitched may be a bit of a leap, but it would certainly seem reasonable even without additional investigation.

It is also relatively easy to observe that **sounds traveling through air take some travel time and that the travel time grows larger with a larger distance traveled**. There would be a noticeable delay in the sound of the hammer of a distant ironworker banging on an anvil, whereas the sound appears instantaneous when nearby. The delay is quite noticeable for distances of about 100 m or more. Anyone who has watched a professional baseball game from center field will have experienced the odd delay of about one-third of a second between when a batter hits a ball and when the sound reaches them. There is also a noticeable delay, often of a second or more, between the explosion of fireworks and the sound that is heard. The longest such delays that are regularly heard are those between lightning and thunder. Those delays can be long enough that even today, not everyone makes a direct connection between the two.

Frequency

It is perhaps less obvious how the ancients might have understood what is going on when a simple tone or musical note is present and how the different tones are

Figure 2.1 A stick and a wheel can produce clicks when rotated slowly and a tone when rotated quickly.

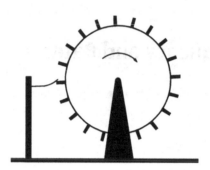

related to each other. Surely there were everyday events and materials that could have been used for such study. For example, the wheel was known to exist thousands of years before the ancient Greeks, and spoked wheels were common during the Iron Age. Hence, while no claim is made here that such wheels were used to study sound and music, they were available and *could* have been used.

Consider a simple wheel with spokes, pegs, or similar objects placed equally around the circumference (see Figure 2.1), and when the wheel is rotated, a small twig is held against the wheel. A modern version might use the spokes of a bicycle wheel and a small cardboard or plastic card.[1] When the wheel is slowly rotated, as each spoke or peg passes, the twig vibrates, and a distinct click is heard. If the wheel is spun at a slow, constant rate, a periodic clicking is heard. As the wheel is rotated faster and faster, the clicks become closer and closer together in time. While the basic mechanics of what is happening clearly do not change as the speed of the wheel increases, the perception gradually changes from that of distinct clicks separated in time to that of a tone. This illustrates a simple way to demonstrate that **tones are the result of a periodic repetition of some basic sound** – in this case, a "click." As the wheel goes even faster, the pitch of the sound is said to rise.

With modern instrumentation, it is relatively easy to show that the transition in perception from discrete clicks to a tone occurs when the clicks become less than about 50 ms apart. There is no exact value, and different people in different circumstances will hear the change at different rates, but 50 ms is a representative value. One click every 50 ms corresponds to 20 clicks per second.

The **period** is the time duration for each of the individual repeated events. The number of repeated events per unit time (e.g., seconds) is known as the **frequency**. Frequencies are usually expressed in units of **hertz** (Hz), named in honor of the German physicist Heinrich Hertz (1857–1894), where 1 Hz is "one event per second."[2] Thus, the gradual transition in perception between clicks and a tone occurs somewhere near a frequency of about 20 Hz.

---

[1] Not all bicycle wheels are equally well suited for this use.

[2] In some older literature, one might see cycles per second (cps) instead of hertz.

Example 2.1

If there is 25 ms per event for repeated events, the period is 25 ms, and the frequency is $1/25$ ms $= 40$ Hz.

If the frequency of repeated events is 250 Hz, then the period, the time per event, is $1/250$ Hz $= 0.0040$ s $= 4.0$ ms.

If the period is T, then the frequency is $f = 1/T$.

For a wheel with 90 spokes making a complete rotation 3 times each second, the frequency will be 270 clicks per second, or 270 Hz. If the tone produced is compared to the sounds of a standard piano, the closest match is for keys pressed near the middle of a piano keyboard. The sound from the piano is not the same, of course, but it can be surmised that what comes from pressing the middle keys of a piano also corresponds to some basic sound-producing event, although not likely a click, that is repeated a few hundred times per second. In the case of the piano, the source is, of course, a vibrating piano string rather than clicks from a twig or card.

Frequencies are a measure of the number of repeated events per unit time and are one example of a rate. Before going on, it is worth taking some time to step back and look more generally at the very important concept of rates.

Rates

A physicist will often describe the rules of nature based on changes. After all, given a starting point and knowing how things change, the end point can be determined. It is not that the physicist necessarily has to express the rules in that way, but it does seem to be a preference, and more importantly, it seems to work. As a consequence, the mathematics that is designed for dealing with changes, that is, the mathematics of calculus, is often the preferred tool for the physicist. Knowledge of the methods of calculus will not be presumed here; however, those interested in pursuing physics in more depth should certainly develop considerable skill in the use and practice of calculus.

In general, changes can often be specified using rates. A rate will describe how much something changes in relation to a change in something else. There are two important categories of rates that are common, simple **additive rates** and **multiplicative rates**. To compute overall changes, it is important to understand which category the rate belongs to.

Example 2.2

"Sixty kilometers per hour" is a speed. Speed is the rate distance is traveled as time progresses. "Twenty-five miles per gallon of gasoline" describes fuel consumption and is the rate of travel as the gasoline is used. "One hundred kilograms per person" is a rate expressing a quantity of some material used as the number of people is varied. "Two hundred and fifty clicks per second" is a rate used to characterize a sound that is heard. These are all examples of additive rates. That is, if the rate is 100 kg per person and there are 5 people, 500 kg is needed, and for twice as many people, twice as much is needed, or 1,000 kg. There are 100 cm per meter, or 100 cm/m, which is a rate. So, for example, to convert 5 m to centimeters, multiply by the conversion rate: 5 m × 100 cm/m = 500 cm. In each of these cases, the rate directly expresses how much is *added* (or removed) as the number of hours, gallons of gasoline, people, or meters – or whatever – is changed.

Example 2.3

If a financial investment earns 8 percent interest per year, an "interest rate," the calculation is not so simple. This is an example of a multiplicative rate. If the account sits for three years, the earnings are *not* 3 × 8% = 24%, and if the account sits twice as long, the earnings do not simply double. The calculation can be done year by year to see what happens. Suppose the account starts with $1,000. After the first year, it will have 8 percent more, or $1,080. The second year starts with $1,080, and so at the end of the second year, the account will have 8 percent more than that, or $1,166.40, and after the third year, it will have 8 percent more than that, or $1,259.71. The total increase, as a percentage, after three years comes to 25.97 percent. If the account is left in the bank for another three years, the ending balance will not be twice 25.97 percent but rather 58.69 percent, or about 2.25 times more.

Multiplicative rates show up in many types of physics problems, such as computations of signal losses in cables; in carbon dating; and as it turns out, in many problems involving human perception and many problems involving acoustics and music. Thus, a few more examples are in order.

Example 2.4

A 373-ft cable is to conduct a signal from one end to the other. The cable specifications say it will lose 5 percent of the signal for each 100 ft of cable. How much of the original signal is left after 373 ft?

Consider first the simpler problem where the cable is 400 ft. That is exactly four 100-ft lengths. After the first 100 ft, 95% = 0.95 of the signal remains. After the second, it is 95 percent of that, and so on. After four lengths of 100 ft, the fraction, f, of the original signal that *remains* is

$$f = 0.95 \times 0.95 \times 0.95 \times 0.95 = (0.95)^4 = 0.8145,$$

or just under 82 percent. The fraction of the signal *lost* is then about 18 percent. The "4" as a superscript is referred to as the *exponent*, and it represents the number of 100-ft units.[3] In order to compute the loss for 373 ft, you will, in essence, need to "multiply 0.95 by itself 3.73 times." That multiplication cannot be done one at a time but is easily computed using a handheld calculator with 3.73 instead of 4 as the exponent.[4] That is, there are 3.73 100-ft units, so the fraction remaining is

$$f = (0.95)^{3.73} = 0.826,$$

or about 83 percent. Remember that the quantity in parentheses is the amount *remaining*, not the amount lost. The second hundred feet of cable loses 5 percent of what it had to begin with, which is what remains after the first 100 ft, and cannot depend at all on how much was previously lost. After all, how can the second hundred feet of cable "know" how much was previously lost? Hence, the calculation will generally be based on how much remains, and from that, how much was lost can be computed, if needed.

A most common error while solving problems such as the previous example is to treat the loss as an additive rate – that is, to multiply the 5 percent loss per 100 ft by 3.73 units of 100 ft to get an 18.7 percent loss. For smaller changes, this erroneous method may actually give you an answer that is close. The reasons will be seen later (Chapter 12) in a different context. However, look at what happens for 2,100 ft of the same cable. The correct calculation gives

$$f = (0.95)^{21} = 0.34,$$

so about one-third of the original signal remains. However, if treated as an additive rate, this gives a loss of $21 \times 0.05 = 1.05$, or 105 percent, a loss larger than the starting signal! That simply cannot happen.

A quantity that increases based on a simple multiplicative rate, such as a financial investment, is said to **increase exponentially**. A quantity that decreases based on a simple multiplicative rate, such as in the cable loss problem, is said to **decrease exponentially**.

In more complicated situations, the rates themselves can change. For example, if a car is initially traveling at 60 km per hour (km/h) and then 10 seconds later it is traveling 70 km/h, the rate of travel has changed. In fact, the rate of travel has changed by 1 km/h per second, which is a rate of change of a rate. And of course, that rate of change of a rate can also change with time, and so on. Fortunately,

---

[3] If needed, more information about exponents is available in Appendix A.

[4] On calculators, the appropriate button to use might be labeled ^ or y^x.

although it is important to know they exist, such complications are not of particular concern here.

Musical Intervals

In music, an **interval** refers to two tones, or "notes," and is referred to as a *dyad*. The notes may be played simultaneously or in sequence, but in either case, tones with two different frequencies will be involved. One can create such tones with the spinning spoked wheel by changing the rotation rate or by changing the number of spokes (or pegs). The latter method allows for a simple, rapid comparison of two tones without critical timing – the Pythagoreans did not have stopwatches, to be sure.

Modern bicycle wheels with equally spaced spokes, where those spokes are alternately angled outward in opposite directions toward the axle, can be used to compare tones with frequencies a factor of 2 apart. A small piece of cardboard (or a twig) held near the rim of the wheel will contact all the spokes, but when the card is moved a bit closer to the axle, a click will occur only for every other spoke. Alternatively, of course, a special wheel can be designed with spokes or pegs that will serve the same purpose. With the wheel spinning rapidly so that a tone is heard, the sound from all the spokes (near the rim) and from every other spoke (away from the rim) can be compared. With the wheel spinning at a constant rate, the change in sound from one to the other compares well to the change heard for the first two notes of "Over the Rainbow," as heard in the well-known 1939 MGM movie *The Wizard of Oz*. Musically, those two notes are described as being one octave apart. The spinning rate of the wheel can be altered, and the change from every other spoke to all the spokes still "sounds like" those first two notes (although with an overall shift). The interval is still an octave, even though the starting tone has changed. It can be concluded that for music, what is called **an octave corresponds to a change in frequency by a factor of 2**. A change by an octave is multiplicative.

Using specially designed wheels or gears, it is straightforward to show that the frequencies for some other musical intervals are also related by a multiplicative factor. For example, a ratio of 3 to 2 will sound like a musical "fifth" no matter how fast the wheel is spinning (compare to the first two notes of "Twinkle, Twinkle Little Star" or to the introductory theme from the Lucasfilm/20th Century Fox movie *Star Wars*). Other musical intervals will be discussed in more detail in Chapter 3; the important point at the moment is that the relationship appears to be multiplicative. That is, if you listen to two musical notes, an "interval," the relative spacing of the frequency of the notes seems to be better characterized musically by a multiplicative constant, such as a factor of 2, or a factor of 3/2, and not a specific number of hertz.

The Notes on a Musical Keyboard

The frequency of a tone is an additive rate. If a spoked or pegged wheel produces 250 clicks per second – that is, clicks at a rate of 250 Hz – then after one second, there will be 250 clicks; after two seconds, 250 more, or 500 clicks; after three seconds, there will be another 250, for a total of 750; and so forth. In general, the number of clicks is the rate, 250 Hz, multiplied by the time in seconds. Using t for time and f for the frequency,[5] the number of events (in this case, clicks), N, is given by $N = f \times t$. The question to consider now is, How do those frequencies, when used for music – for example, from a keyboard – change from note to note?

The notes on a standard keyboard are played by pressing "keys." On a modern-day keyboard, some keys are traditionally white and some black.[6] The notes played with the white keys are labeled using letters of the alphabet, A through G, and then repeating. See Figure 2.2. The black keys are labeled as the sharp ($\sharp$) or flat ($\flat$) of the note just below or above them, respectively. A sharp means to move one key to the right, and a flat means move one key to the left. Thus, $F^{\sharp}$ is one key to the right of F. This convention can also be used for white keys that are not separated by a black key. For example, $E^{\sharp}$ would be one key above E, which is the same as F. Notes with different names but that are played using the same key are referred to as being **enharmonic**. Thus, $E^{\sharp}$ and F are enharmonic, as are $A^{\sharp}$ and $B^{\flat}$. Musically, there may be good reason to use one name rather than another, but that goes beyond the discussion here. **The interval between adjacent keys** (regardless of key color) **is called a** *half step*.

Each note produced by the keyboard can be characterized by a frequency. In principle, the frequency of the note could be measured by comparison to the spinning spoked or pegged wheel. In modern times, a microphone connected to an electronic device called, appropriately, a *frequency counter*, is probably a more convenient method. The first question to answer might be, What are the frequencies of the tones produced by a modern keyboard? The question of why those particular frequencies are used will be postponed for the moment.

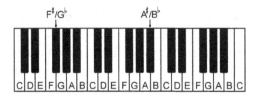

Figure 2.2 The alphabetic labels for the keyboard.

---

[5] Note that the variable f now refers to frequency, whereas previously it referred to a fraction.

[6] This is not universal but very common for modern instruments. Some harpsichords have the colors reversed.

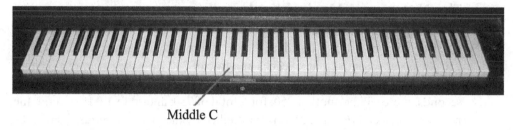

Middle C

Figure 2.3 A standard 88-key piano keyboard.

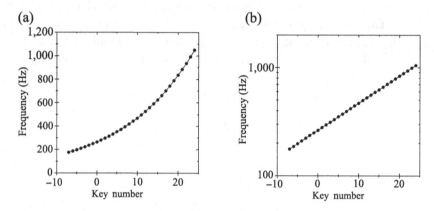

Figure 2.4 The frequencies of tones measured for a small electronic keyboard using (a) a linear vertical scale and (b) a logarithmic vertical scale.

A typical modern piano keyboard, shown in Figure 2.3, has a total of 88 keys. Near the middle is a note commonly referred to as *middle C* or sometimes as *C4*.[7] That note is often the first note that is taught to a new piano student, and it is common to use it as a reference point. To label the keys in a systematic *mathematical* way, label that as *key 0*. As you progress to the right (usually referred to as "up" the keyboard), number the successive keys, white or black, by adding 1 for each half step. As you go to the left ("down" on the keyboard), subtract 1 for each. Now press each key one at a time and measure the frequency. Figure 2.4a shows measured results, graphically, for a small electronic keyboard. Each dot on the graphic shows the results of a measurement, and the dots are connected with lines simply to help guide the eye from one to the next.

It is clear from this figure that as you go "up" the keyboard, the frequency increases (it "goes up"). Also, the change in frequency from one note to the next is not a constant amount but gradually increases. Going back to the intervals mentioned earlier, particularly the octaves, where a multiplicative rate is observed, it is tempting to assume that a similar process is occurring here as well. This

---

[7] For the standard 88-key keyboard, starting from the left, middle C is the fourth C. Note that the note just below C4 is B3, not B4.

assumption becomes a hypothesis that needs to be looked into further. In order to do that in a simple way, a slight digression is appropriate.

Logarithms and the Log Scale

Recall that if there are two values, label them x and y, and they are mathematically related by $y = 10^x$, then, by definition, x is the logarithm base 10 of y, which can also be written as $x = \log_{10}(y)$. So, for example, since $10^2 = 100$, then $2 = \log_{10}(100)$. Also, using the properties of the logarithm, each time y is multiplied by a constant amount, a constant amount is *added* to the logarithm of y. That is, the logarithm can be used to change a multiplication into an addition. If needed, more information about exponentials and logarithms can be found in Appendix A.

Example 2.5

$$\log_{10}(100) = 2,$$

$$\log_{10}(3 \times 100) = \log_{10}(300) = \log_{10}(100) + \log_{10}(3) = 2 + \log_{10}(3) = 2.47712,$$

$$\log_{10}(3 \times 300) = 2.47712 + \log_{10}(3) = 2.954243, \text{and so on.}$$

Early on in mathematics, students are often taught about the "number line" or something equivalent. Here, numbers are displayed graphically as being evenly spaced along a line. In order to add two numbers, simply move down the line the corresponding distance. That is, if the distance from 0 to 7 is 1.3 cm, then you can add 7 to any number by moving down the line by 1.3 cm, and you can subtract 7 by moving in the opposite direction (e.g., see Figure 2.5).

On the other hand, if the numbers are spaced so that their logarithms are evenly spaced, rather than the numbers themselves, you get something called a **log scale**. For such a scale, you can *multiply* by moving down the line a corresponding distance.[8] You can find that distance by starting at 1 and measuring to the value needed. For example, to multiply by 4, measure the distance from 1 to 4, then move up or down the line as needed to multiply by 4 (e.g., see Figure 2.6).

When using a linear scale, moving a fixed distance adds a fixed quantity, whereas for a log scale, a fixed distance corresponds to multiplication.

---

[8] This is the principle behind a computing device known as a *slide rule*, now generally considered obsolete.

Figure 2.5 Using
a number line to add 7.

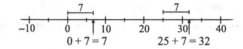

Figure 2.6 Using a log
scale to multiply by 4.

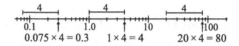

Now consider the measured frequencies from the keyboard, but where a log scale is used for the frequency (see Figure 2.4b). In this case, a constant distance in the vertical direction corresponds to a constant *multiplier* for the frequency. With this new type of plot (using the same data as in Figure 2.4a), the points lie along a straight line. That means that when moving a constant distance horizontally, which corresponds to a fixed number of keys along the keyboard, the frequency also increases by a fixed distance vertically, corresponding to a fixed multiplier. Since it is relatively easy to distinguish a straight line from one that is not straight, such a plot is very useful to test for multiplicative rates.

Further examination shows that there is an increase of a factor of 2, as precisely as can be measured, if you move 12 keys to the right. **A change in frequency by a factor of 2 is the octave**, and hence the modern keyboard "divides" the octave into 12 parts in such a way that when moving from one key to the next, the frequency is multiplied by a constant amount. That constant must be the value that, when multiplied by itself 12 times, gives the result 2. Mathematically, the constant is then, by definition, **the 12th root of 2**, written as

$$\sqrt[12]{2} = 2^{1/12} = 1.059463094\dots \tag{2.1}$$

In round numbers, the frequency changes by about 5.95 percent from one key to the next. That is, a half step corresponds to a change in frequency of just under 6 percent. Since the difference between adjacent notes is quite noticeable for music, a 6 percent change in frequency is quite significant musically. As stated previously, quantities with a rate of change that is a constant multiplier are said to change **exponentially**.

As you go up the keyboard, the frequency of the notes increases exponentially, and as you go down the keyboard, the frequencies decrease exponentially.

Knowing the (multiplicative) rate and any starting point, you can figure out the frequency of any other note.

Example 2.6

If a certain note labeled A has a frequency of 440 Hz, and the note C just to the right of that is three half steps above, its frequency can be computed by multiplying 440 Hz by 1.05946 three times to get 523.25 Hz. On the other hand, the note C below that A, which is nine half steps lower, can be computed by dividing 440 Hz by 1.05946 nine times – that is, it is $440 \text{ Hz}/(1.05946)^9 = 261.63$ Hz. Alternatively, use the fact that an octave is a factor of 2 in frequency, and then first go up to the nearest C, and then go down by an octave. That is, compute $440 \text{ Hz} \times (1.05946)^3/2 = 261.63$ Hz. Of course, that gives the same result.

With this scheme, one note must be defined, and then all the rest are determined. The one note that is often specified is the A above middle C ("A4"), and in recent times, a common choice for that note is 440 Hz, although sometimes other values (e.g., 442 Hz or 444 Hz) are used. In some other cases, the note C is defined (particularly "middle C"). There does not seem to be a sound scientific explanation for any particular choice for a reference frequency. It is a happenstance of history. A table listing notes determined in this way for several choices for A4 is included in Appendix C.

One very important question remains, however: Where did this strange value, $\sqrt[12]{2}$, come from? In addition, why is the octave divided into 12 rather than 15, or 9, or some other number of intervals? Is there a good reason to repeat the labels for each octave? For these questions, we will need to look at a little bit of music theory to see how these notes are used. Then, later, the connection between pitch, which is a perceived quantity, and frequency, which is certainly related and can be directly measured, will also need to be considered.

Summary

Frequency is the measure of events per unit time, and the period is the time per event. Hertz (Hz) is the unit used for events per second. The tones near the middle of a keyboard are repetitive sounds with a frequency of a few hundred hertz. Frequency is one example of a rate.

Much of physics is expressed in terms of rates of change. There are two important categories of rates: additive and multiplicative. Frequency is additive, meaning that if there are N events in a time t, there will be $2N$ in a time $2t$, $3N$ in a time $3t$, and so forth, so that N additional events are *added* for each additional time interval t.

Multiplicative rates are based on a unitless multiplier. If there is a first quantity, N, in a certain initial interval, there will be εN in the next, $\varepsilon^2 N$ in the one after that, then $\varepsilon^3 N$, and so forth, where ε is the multiplier. If $\varepsilon > 1$, the quantity increases exponentially. If $\varepsilon < 1$, the quantity decreases exponentially. For such a multiplicative process, the mathematics of logarithms is useful, including the log scale used for graphing. Examples of problems involving a multiplicative rate include computing compound interest for a bank account and determining signal losses in cables, among others.

Based on measurements, the frequencies of the sounds associated with musical intervals are based on a multiplicative rate. For example, each octave is a factor of 2 in frequency. Going up a keyboard, the frequency increases exponentially, and going down the keyboard, it decreases exponentially. The standard keyboard divides the octave into 12 half steps, so the frequency multiplier to go a half step up the keyboard is the 12th root of 2, $2^{1/12} = 1.0594631$, or a change just shy of 6 percent between adjacent keys on the keyboard.

ADDITIONAL READING

Harkleroad, L. *The Math Behind the Music*. Cambridge University Press, 2006.
Haynes, B. *A History of Performing Pitch*. Scarecrow Press, 2002.

PROBLEMS

2.1 It is said that J. S. Bach used pipe organs that had the note A above middle C tuned to 480 Hz. Approximately what note does that correspond to on a modern keyboard that uses 440 Hz for the A above middle C?

2.2 For the strings on a piano, if the note A above middle C has a frequency of 440 Hz, for that note, how many repeated string vibrations occur during 9.0 s? What is the time for each vibration?

2.3 If middle C has a frequency of 263 Hz, what is the frequency of the C three octaves lower? How about the C four octaves higher? These represent the lowest and highest Cs on a standard piano keyboard.

2.4 A common way to describe the diameter of wires and the thickness of plates is with a "gauge." The size is represented with an integer. For one wire gauge system in common use, the diameter of the wire changes in a multiplicative way, where a decrease in the gauge by 6 represents a doubling of the diameter. For example, gauge 20 wire is half the diameter of gauge 14 wire. For this gauge system, how much does the wire diameter change when the gauge decreases by 1?

2.5 Although it never caught on for performance, some have promoted a tuning scheme called "scientific pitch." That scheme fixes middle C to be exactly 256 Hz. What might be "scientific" about this choice? What is the A above middle C for this tuning scheme?

2.6 You have an open 2-L (2,000-ml) container of water. Some of the water evaporates each day. Compare the results for these two situations: (a) 100 ml of water evaporates each day for 20 days, and (b) 5 percent of the water evaporates each day for 20 days. Which description do you think most closely matches what really happens for an open container of water – a fixed amount per day or a fixed percentage per day?

2.7 For a certain guitar, the full length of the strings, from the bridge to the nut, is 651 mm, and fret positions, measured from the bridge, are 615, 581, 549, 518, 489, 462, 436, 412, 389, 368, and 347 mm (see Figure P2.7). Are the fret positions described by a multiplicative rate? If so, is it the same as what is seen for the keyboard (Figure 2.4b)? If not, how is it different?

Figure P2.7

3   The Notes We Use

A goal here is to try to understand why we use the musical notes we do. In Chapter 2, it was shown that there is a musical interval called an *octave* that corresponds to a factor of 2 in frequency and that a modern keyboard divides the octave into 12 parts using a multiplicative process. The result is that there is a change in frequency from one note to the next that is a factor of the 12th root of 2, a seemingly strange choice. To try to understand why this factor is used will require some reverse engineering. First, how notes are actually used in music will be discussed. Following that, some investigation into the mathematics behind the frequency intervals that arise will lead to a reasonable explanation as to why keyboards divide octaves into 12 parts, with note frequencies spaced equally in the logarithmic sense.

The discussion that follows is largely based on what is referred to as "Western music," the music that arose via the ancient Greeks and that has been practiced throughout Western Europe. That is where keyboards evolved. Many of the results also apply to music that has evolved from other starting points, even where keyboards are not used. For a discussion of some other musical scales, as well as a critique of Western scales, see Daniélou (1943).

Scales

One of the first observations that can be made about the notes used for songs, or even for elaborate works of music, is that in most cases, not all the available notes are used. The notes of the keyboard provide a palette, but the artists do not always use all the tones that are available. Aside from a genre of music known as *12 tone*, which by design uses all 12 notes from the octave, the music we hear uses only a subset of notes that are based on a **musical scale**. The most common scales in Western music are the major and minor scales. When a piece of music based on a particular scale uses a note from outside that scale, that note is often referred to as an "accidental."

Table 3.1 Examples of some musical scales. Refer to keyboard labels shown in Figure 2.2.

Scale	Half-Step Intervals	Keyboard Example
Major	2, 2, 1, 2, 2, 2, 1	White keys, starting on C
Natural minor	2, 1, 2, 2, 1, 2, 2	White keys, starting on A
Major pentatonic	2, 2, 3, 2, 3	Black keys, starting on F$^\sharp$
Minor pentatonic	3, 2, 2, 3, 2	Black keys, starting on D$^\sharp$
"Blues"	3, 2, 1, 1, 3, 2	Black keys, starting on D$^\sharp$, plus A
Phrygian dominant scale	1, 3, 1, 2, 1, 2, 2	White keys, starting on E, except use G$^\sharp$ instead of G
Whole tone	2, 2, 2, 2, 2, 2	Starting on C, three white keys, then three black keys

Different scales are listed in Table 3.1. This list is by no means complete and is intended only to illustrate some examples. With the exception of the whole-tone scale, all of these examples have a clear **root** or **tonal center** – a note that sounds like it is "home." That note is referred to as the **key** of the music. Simple songs will usually end on this note. More elaborate works may change their tonal center throughout the piece – often referred to as a *key change*. Hence, the word *key* in this sense is used to refer to a scale, not a physical key on a keyboard. Those interested in more details about these and other scales might wish to spend some time studying music theory.

These scales all rely heavily on the octave, a factor of 2 in frequency, and all are defined in terms of (multiplicative) intervals. Any note can be used as the root, going up or down by the appropriate number of intervals to produce the corresponding scale.

When studying music theory, it is not long before one comes across a diagram known as the **circle of fifths**. The circle of fifths shows up when looking at and analyzing music. It provides more information about how the notes are used – and thus additional clues as to why the notes are what they are.

In music, a *fifth* refers to an interval that is the same as that of the fifth note of the major or minor scale to the root (when played from lower to higher notes). Hence, a fifth is seven half steps. If you start with the C major scale, the fifth note is G. The fourth would be the same interval as that of the fourth note of the scale, and so on. When there is a difference, which type of scale is being used must also be specified. For example, the third note of the C major scale is E, whereas the third note of the C minor scale is E$^\flat$. Thus, the interval from C to E (four half steps) is referred to as a *major third*, and that from C to E$^\flat$ (three half steps) would be referred to as a *minor third*.

Figure 3.1 The circle of fifths.

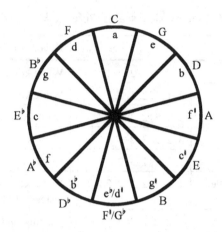

The circle of fifths arranges the notes found on the keyboard by fifths (see Figure 3.1). It is common to put C at the top of the circle, corresponding to the C major scale. Traveling around the circle clockwise, the interval increases by a fifth for each step. Starting with C at the top, the fifth note of the C scale is G, so that is the next entry clockwise. The fifth note of the G scale is D, and the fifth note of the D scale is A, and so on. Going counterclockwise from C by seven half steps, you get to F, a fifth below C. That is, C is the fifth note of the F scale. Continuing the process, F is the fifth note of the B♭ scale, and so on. Eventually, you end up at the same note (or at least an enharmonic[1] note name) whether you go clockwise or counterclockwise. Hence, it is a cyclic process, and thus it is natural to draw it in a circle.

On the inside of the circle, you will see lowercase letters. These refer to the (natural) minor scales, which use the same notes as the corresponding major scale but a different root. Thus, "a" refers to the a-minor scale, which uses the same notes as the C major scale but starts on A instead of C. The use of uppercase for major scales and lowercase for minor scales is relatively common.

The circle of fifths is useful for those learning their scales on a keyboard. The C major scale uses only the white keys. For each step away from C on the circle of fifths, you add one black key. As you go clockwise, you add the black key a half step up from two spots counterclockwise. That is, G major uses F♯; D major uses F♯ and C♯; A major uses F♯, C♯, and G♯; and so forth. As you go counterclockwise from C, you add the black key from one spot counterclockwise. That is, F major has B♭; B♭ major has B♭ and E♭; E♭ major has B♭, E♭, and A♭; and so forth.

---

[1] A half step up from C is C♯, and a half step down from D is D♭. On a keyboard, these are the same.

The circle of fifths is more than just a convenient way to organize the notes, however, because it also shows up when studying the chord structure of music.

Chords

A **chord** results when three or more different notes are played together. The simplest and most common chords are the major and minor **triads**. These consist of the root, the third, and the fifth notes of the scale played together.

Western music is highly chord oriented. Typically, a song uses a sequence of notes from a scale to provide a melody, along with an underlying set of chords. Sometimes the chords will be played one note at a time, known as a *broken chord*, or sometimes all at once. In some genres of music, composers may specify only the chords, and the players are free to play the corresponding notes in any manner they choose. In Baroque music, for example, accompaniments were often written using a "figured bass," which is a way of writing out chords. Jazz improvisation will also follow a predetermined chord pattern. Many popular and folk tunes can be performed using "lead sheets," a staple of the lounge pianist, which include just the melody and the associated chords. For these formats, the musician is entrusted to produce notes consistent with the chord that are also appropriate for the music at hand.

The understanding of how chords are used in music can be simplified if chord progressions are identified relative to the root. For music in the key of C major, the C major triad can be labeled with the Roman numeral I. The G major chord would then be based on the fifth of the C scale, and so it is labeled V. The F major chord would be IV, and so on. Minor chords are labeled using lowercase Roman numerals. Hence, since D is the second note of the C major scale, the d-minor chord would be labeled ii. Similarly, an a-minor chord would be vi, and so on. This labeling scheme works as long as the root of the chord is a note from the C major scale. Once the chords are all specified using Roman numerals, it becomes easy to change to another key (scale). For example, if you want to play in $B^\flat$ instead of C major, then I corresponds to the $B^\flat$ major triad, V is the F major triad, IV is the $E^\flat$ major triad, and so on. This scheme also allows an analysis of relative chord progressions in a way that is independent of the key (the root) of the piece.

The simplest tunes may have only the I chord. Other simple tunes will have the I and V or I and IV chords. Many popular, folk, and blues tunes can be played using only chords I, IV, and V – including many from the genre known as "three-chord rock." Slightly more complicated examples include I, IV, V, and vi, in various orders. Another "standard" chord progression that shows up in several mid-twentieth-century popular tunes is I, vi, ii, V. It is possible to find many more examples.

What is interesting about these chord progressions is that, going back to the circle of fifths, the chords that are actually used within a given song (or at least any significant section of a song) will usually be packed closely together on the circle of fifths. Most chord changes are no more than one spot away on the circle of fifths, sometimes two, but rarely do you see jumps larger than this.

The construction of some musical instruments, particularly those used for folk music, often uses a physical arrangement based on intervals of a fifth. For example, the left-hand buttons on many accordions are used to play chords. The buttons can be arranged according to the circle of fifths so that chord progressions generally involve only a small movement of the hand. The strings on a typical hammer dulcimer and the buttons on an autoharp are also arranged based on fifths.

The conclusion is that **there is something special about this interval called the fifth** and the role it plays in Western music. Major and minor thirds also have some importance. Along the way, a new question may have arisen regarding why some chords sound "nicer" than others. Still to be addressed is how this is related to the division of the octave into 12 parts. To address these issues, we turn again to the ancients, in particular, the Pythagoreans.

Pythagoras's Monochord

The Pythagoreans were keen to look at ratios. The record shows that to study musical intervals, a one-string device was constructed, often referred to as a **monochord** (Figure 3.2). In one simple implementation, the string has two sections that can be plucked or bowed either together or in sequence, and the resulting pitches can then be compared. A sliding bridge can be moved to change the relative lengths of the two sections. Although a monochord is easy to construct, the outer strings of a cello or guitar, along with an appropriately sized wedge, can also be used. The goal is to slide the bridge while listening to the sounds from both sections to find spots where it sounds "nice" ("consonant"). When those spots are found, the length of the string on either side of the bridge is measured, and the ratio is computed. Since, as will be shown in more detail later (Chapter 6), the frequency of the sound from such a vibrating string depends on $1/L$, where L is the length of the string between the supports, the ratio of the two lengths can directly yield the ratio of the frequencies of vibration.

Figure 3.2 A one-stringed "monochord."

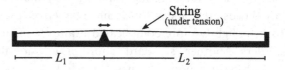

It is perhaps easiest to hear the effects if two bows, such as violin bows, are used and the two sections are played simultaneously. With this technique, as the bridge is moved near the center, the two lengths can easily be adjusted to match, at which point the tones produced no longer sound like two tones but just one. Bowing each section separately shows that, as one would expect, the pitches of the two sections are identical. That interval is referred to as **unison**.

A similar measurement can be made with the bridge about one-third of the distance from one end. After a slight adjustment, it can be made to sound not like two tones, but one. The tones are not the same when played separately but sound an octave apart, and the ratio of the lengths of the two sections will be 2 to 1. As the bridge is moved in toward the center from this point, some other "nice" spots can be detected. These are generally harder to find and hear until you get to two-fifths of the distance from one end, where there is another very nice spot. Playing the sections separately and comparing to a keyboard shows that the interval is a musical fifth. For that spot, the ratio of the frequencies (based on the ratio of the lengths) is 3 to 2. Other nice spots that can be detected have length ratios of 4 to 3, 5 to 4, 5 to 3, and so on, at least within the accuracy that the measurement of the lengths permits.

The "nice-sounding intervals" are found when two frequencies are related by simple rational numbers – that is, the ratio of two (relatively small) integers.

When listening for these nice intervals, the smaller the integers are, the easier they are to find. Almost everyone can easily find the unison (1 to 1), octave (2 to 1), and fifth (3 to 2), although fewer will hear the others clearly. A comparison of the intervals from a keyboard and the strings suggests that the fourth corresponds to 4 to 3, the major third to 5 to 4, the minor third to 6 to 5, and the major sixth to 5 to 3.

Before proceeding, it is appropriate to consider what is happening that causes these special spots to sound different – that is, what makes them sound "nice."

Beats

To understand what is going on, it is useful to look at mathematical functions originally derived for problems of plane geometry, the **sine** and **cosine**. It turns out that the sine and cosine can be used as a mathematically simple way to generate a periodic result. In particular,

$$A\sin(2\pi ft) \text{ and } B\cos(2\pi ft) \tag{3.1}$$

will be repetitive functions of the time, t, with a frequency, f, and amplitudes A and B, respectively. In general, these will be referred to as **sinusoidal functions** (whether a sine or cosine). Note that these functions are written assuming that "angles" are measured in radians, and the angles are allowed to take on any value. An angle of π radians corresponds to halfway through the repetition (i.e., 180°), 2π to the completion of one repetition, 6.5π to three complete repetitions – $(3 \times 2\pi)$ plus one-quarter of the way $(2\pi/4 = 0.5\pi)$ through the next repetition – and so on.

Since the mathematics of sine and cosine functions is well documented, that knowledge can be used to understand what might happen in other situations with periodicity, such as sounds. In particular, consider what happens if you have two repetitive signals added together, such as the sounds from the two sections of the monochord. Using the sine functions to act as stand-ins ("models") for the sounds, and using known identities,[2] you can predict what might happen. For example, the identity

$$\sin(2\pi f_1 t) + \sin(2\pi f_2 t) = 2 \sin\left(\pi(f_1 + f_2)t\right) \cos\left(\pi(f_1 - f_2)t\right) \qquad (3.2)$$

shows that **when you *add* two sinusoidal functions, the result is the same as the *product* of two other sinusoidal functions.** Of special interest here is the case when f_1 and f_2 are nearly the same, in which case $f_1 - f_2$ can be a very low frequency. If these functions represent sounds, and that low frequency is less than about 20 Hz or so, it will not be heard as a tone but as pulsing or throbbing. Mathematically, the low-frequency sinusoid on the right of Equation (3.2) acts as a time-dependent amplitude that multiplies a sinusoid that is at the average frequency. Some examples are shown in Figures 3.3 and 3.4. Note that the frequency of the "throbbing" or "beating" that is heard will be at the difference frequency, $|f_1 - f_2|$.

When two musicians are trying to play or sing the same pitch, it is relatively easy to hear if they are off a little bit due to this beating. If the **beat frequency** is 1 Hz, then you know the musician's tones differ by 1 Hz. In fact, this process of "beating one signal against another" can be used for any two signals, as long as you can hear or see the result, and it becomes easy to match signals to better than 1 Hz. So, for example, microwave radio signals, which have a typical frequency measured in gigahertz (GHz; 10^9 Hz), can also be matched to each other to better than 1 Hz if the sources themselves are stable enough.

While this explanation shows why the unison is easy to hear and "tune in," it is not enough to explain what happens for frequency ratios of 2 to 1, 3 to 2, and so on. The complete explanation for that will be left for Chapter 4, but at this point, suffice it to say that real musical sounds are repetitive, as are sinusoidal functions,

---

[2] See Appendix A for a list of some of these identities.

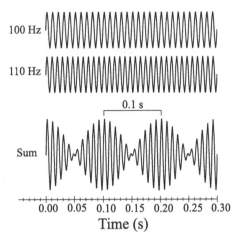

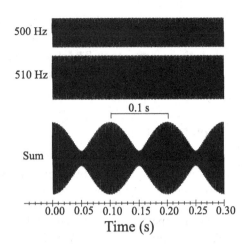

Figure 3.3 Adding signals at 100 Hz and 110 Hz results in a signal with a 10-Hz beat.

Figure 3.4 Adding signals at 500 Hz and 510 Hz results in a signal with a 10-Hz beat.

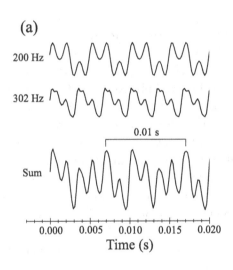

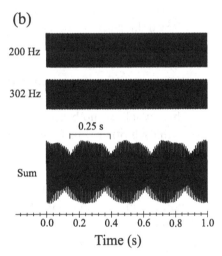

Figure 3.5 The sum of two musical signals shown for two different timescales. Slow beats are observed at $(2 \times 302 - 3 \times 200)$ Hz $= 4$ Hz.

but they are more complicated than sinusoidal functions. A sinusoid is too simple of a model for the sounds to explain all the ratios.

Figure 3.5 illustrates what happens if you have repetitive functions that are not simple sinusoidal functions but are more like what you expect for musical sounds. Looking closely, there seems to be a repetition at the difference frequency, as in Figure 3.5a. However, it is not, in fact, an exact repetition. Looking over a longer timescale, as in Figure 3.5b, a much slower change can be observed. For the sums of these signals, the slower repeats that occur are related not just to the difference in repetition frequencies but also to the differences in the integer multiples of those frequencies. **The slowest beats are those between the closest integer multiples of the separate frequencies.**

Example 3.1

If one trumpeter plays a note at 262 Hz and another plays a note at 394 Hz, the slowest beat expected is found by writing down the integer multiples of each of those frequencies,[3]

First trumpeter: 262, 524, 786, 1,048, 1,310, ... Hz,
Second trumpeter: 394, 788, 1,182, 1,576, ... Hz,

and then comparing those lists. Here, 788 and 786 are the closest, with a difference of 2 Hz. Hence, the slowest beating will be expected to be at 2 Hz, or twice a second.

Searching for "nice" spots on the monochord is listening for spots where there are no beats. That will happen whenever the two repetition frequencies are related by a ratio of integers, that is, by a rational number. In the previous example of the two trumpeters, if the second trumpeter had instead played at 393 Hz, which is exactly $262 \times 3/2$, the slowest beat frequency would have been at 0 Hz, meaning there would be no beating at all, and the trumpets would be considered in tune with each other. For typical musical signals, the smaller the integers, the larger will be the effect, and so the easier it will be to hear any beats that may result.

Table 3.2 summarizes some of the ratios between two frequencies, f_1 and f_2, that can be found using a well-trained ear and a monochord, as well as the corresponding musical intervals. These correspond to the spots where there are no slow beats.

Now there is a problem to resolve. The keyboard was found to have intervals using factors of $\sqrt[12]{2}$, which is an irrational number and hence, by definition, cannot be written as the ratio of integers. In fact, with the exception of the octaves (2, 4, 8, ...), all powers of this irrational value can *never* equal a rational number. It seems that two different standards are being used, one for the keyboard and one that arises from the monochord. The results from the monochord, at least, can be rationalized based on the phenomenon of beats, but the question of where this funny irrational value of $\sqrt[12]{2}$ comes from still remains unanswered.

---

[3] There is no need to go beyond the first half dozen or so multiples. In practice, beats from higher multiples become very difficult to hear.

Table 3.2 Frequency ratios and the name of the corresponding musical intervals.

When	Equals	Then a musician calls it:
$1\,f_1$	$1\,f_2$	Unison
$2\,f_1$	$1\,f_2$	Octave
$3\,f_1$	$2\,f_2$	Fifth
$4\,f_1$	$3\,f_2$	Fourth
$5\,f_1$	$3\,f_2$	Major sixth
$5\,f_1$	$4\,f_2$	Major third
$6\,f_1$	$5\,f_2$	Minor third

Temperaments

The choice of how the notes of a scale are tuned is known as a **temperament**. Based on the observations just described, a "nice" set of notes might be had by choosing frequency ratios that are based on the ratio of small integers. This is illustrated by the solid vertical lines in Figure 3.6 for integers 1 through 6. Because of the importance of the octave, these are repeated for each octave, and only one octave is shown. Note that a log base-2 plotting scale is being used, which is convenient when considering octaves.

Since the fifth, a ratio of 3/2, is also important, additional notes can be added, as needed, to ensure that additional fifths are present. A fifth above the minor third (6/5) is at 9/5, and a fifth above the major third (5/4) is at 15/8. A fifth above the fifth is at 9/4, which is in the next octave, since 9/4 > 2. Bringing that down an octave gives the note at 9/8. A note a fifth below the minor third is at 12/15, which is in the next octave lower. Bringing that up an octave yields the ratio 8/5. Likewise, the thirds (major and minor) are important, and additional notes can be added, such as the major third above the note at 9/8, giving 45/32. The ratios based on these intervals on top of intervals are shown in Figure 3.6 with the dashed vertical lines. This method of scale construction produces many notes that have a corresponding note a fifth and/or a third away, making it easy to produce those "nice" musical combinations.

If the process is continued to include ratios involving the (prime) number 7 (the dotted lines in Figure 3.6), there are now some closely spaced intervals that are not easily related to other notes via fifths and thirds. Closely spaced intervals will likely produce beats. Hence, those involving the number 7 (and 11, 13, and so on) are excluded in this scheme. To be practical, the process has to stop somewhere. The resulting temperament is called **just temperament** or **harmonic tuning**.

Figure 3.6 also illustrates the positions of the irrational frequency ratios deduced from measurements of a keyboard. On a log scale, they are evenly

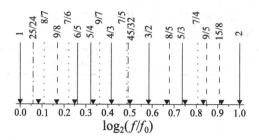

Figure 3.6 The frequency spacings obtained from rational numbers involving smaller integers are shown with the vertical lines. Note how the spacing between adjacent notes is unequal. The inverted triangles show the equally spaced values used for equal temperament.

spaced. This type of tuning, where such an even spacing (on a log scale) is used, is referred to as **equal temperament**. Equal temperament with 12 notes to the octave comes close to just temperament and, as a choice, can be used instead. Note in particular that the fifth, the frequency ratio of 3/2, which is so important music- ally and involves some of the smallest integers, is very closely matched due to the happenstance that

$$\left(\sqrt[12]{2}\right)^7 = 1.498307, \tag{3.3}$$

which is off from exactly 3/2 by only about 0.1 percent. The musical thirds are not in such good agreement, but they are reasonably close.

Yet another temperament can be constructed using only the prime num- bers up to 3 – that is, based only on octaves (factors of 2) and fifths (factors of 3/2). This is known as a **Pythagorean temperament**. Referring back to the diagram for the circle of fifths (Figure 3.1), if you have the note C, then the note a fifth higher, G, a factor of 3/2 higher in frequency, will sound nice along with C. Then if you have the note G, the note a fifth above that, D, will sound nice with that. That D will be in the next octave and can be brought back down an octave (by dividing the frequency by 2). This process can continue going clockwise around the circle. Going back to the C, but now going counterclockwise, F and C should sound nice together, as will B$^{\flat}$ and F, and so on. With this scheme, the frequencies of all the notes can be written as $3^n/2^m \times$ the frequency of C, for appropriately chosen values of the integers n and m.

There is a bit of a problem with this scale because the clockwise and counter- clockwise steps do not match up. If traveling all the way around the circle back to the top, that "C" is not the same frequency as the starting value. That mismatch ultimately occurs due to the behavior of prime numbers – there is no way to get an exact octave (a factor of 2) that uses a nonzero power of 3. Prime factorizations are unique.

Reference to the "Pythagorean comma" is a reference to this mismatch. Suggestions to address this issue include having a single unequal interval, say, roughly halfway around the circle, or spreading out the mismatch by using equal intervals that are close to 3/2 but not exactly 3/2. Alternatively, the mismatch can be left alone, turning the circle of fifths into a spiral, resulting in a "noncyclic temperament." It is interesting to notice at this point that this method of generating musical notes, assuming the mismatch has been addressed, naturally results in 12 different notes.

So why does a keyboard use equal temperament when, for example, it seems that temperaments based on rational numbers will sound "nicer"? Because just temperament uses a scheme with unequal half steps, it will be in tune for just one scale. For example, if you tune your keyboard for the C major scale, then that scale will be nicely in tune. However, if you now want to play using B or D scales, the tuning will be off. A similar problem occurs for Pythagorean temperament, which also has unequal half steps. On the other hand, if you use equal temperament, the tuning will be just as good (or bad) for the C scale as it is for the B or D scales. **Equal temperament is a compromise useful for keyboard instruments.**

There are many other temperaments that have been used. Many are defined not by specific mathematical frequency ratios but by the tuning procedure used to tune a specific keyboard instrument, such as a piano or harpsichord. Through much of the development of classical music, up to the twentieth century, a compromise called *meantone temperament* was often used. There are actually many different subtypes of meantone tuning. Meantone tuning is somewhat in between just and equal temperaments and allows one to change scales around the circle of fifths while maintaining reasonable tuning, as long as one does not venture too far from the top. Although known hundreds of years earlier, equal temperament did not become the de facto "standard" until the early part of the twentieth century, and its adoption was not without controversy. It was, by the standards of the time, simply out of tune, and so why would anyone want to use it? Now its use is routine. We are used to it.

Specific tunings are often described numerically by further subdividing the equal-tempered half step into 100 parts, called **cents**. Thus, for example, 50 cents would be considered a "quarter tone." Cents are multiplicative, as are the half steps, so one cent is the number that, when multiplied by itself 100 times, gives a half step. That is,

$$1 \text{ cent } = \sqrt[1200]{2} = 1.0005777895. \tag{3.4}$$

Even well-trained musicians rarely hear a difference smaller than 1 cent, so it is usually quite safe to round off values to the nearest cent. Table 3.3 compares equal and just temperaments using cents.

Table 3.3 A comparison of equal and just temperaments.

Interval	Just Scale	Equal Temperament	Just Scale (cents)	Equal Temperament (cents)
Unison	1/1 = 1	1.00000	0	0
Minor second	25/24 = 1.0417	1.05946	71	100
Major second	9/8 = 1.1250	1.12246	204	200
Minor third	6/5 = 1.2000	1.18921	316	300
Major third	5/4 = 1.2500	1.25992	386	400
Fourth	4/3 = 1.3333	1.33483	498	500
Diminished fifth	45/32 = 1.4063	1.41421	590	600
Fifth	3/2 = 1.5000	1.49831	702	700
Minor sixth	8/5 = 1.6000	1.58740	814	800
Major sixth	5/3=1.6667	1.68179	884	900
Minor seventh	9/5 = 1.8000	1.78180	1,018	1,000
Major seventh	15/8 = 1.8750	1.88775	1,088	1,100
Octave	2/1 = 2	2.00000	1,200	1,200

Why 12 intervals within the octave rather than some other number? The 12-note division does seem to arise somewhat naturally from the nice-sounding intervals that led to the just and Pythagorean temperaments, and so if equal temperament is to be a good approximation, it is not surprising that it might also have 12 notes. More generally, to divide the octave into N equal parts using equal temperament would result in a half-step frequency ratio of $\sqrt[N]{2}$, the Nth root of 2. Our ears would very much like to have a note with a frequency ratio of 3/2 present as well.[4] That is, some note m half steps above the root should have a frequency ratio to the root that is close to 1.5000. Simply running through all the possibilities shows that a 0.3 percent or better match is present for $N = 12$, 17, 24, 29, 31, and so forth. That is, if, in addition to the octave, the desirable ratio of 3/2 is taken into account, 12 is the smallest value that works well, and 17 is the next smallest candidate. Practicality then limits the equal-tempered keyboard to a more manageable 12 keys. The use of extra keys has been tried, to be sure, but never has it caught on. Hence, we use 12 notes for each octave on the standard keyboard.

---

[4] Experiments have confirmed this preference in infants, who have had as limited exposure to music as is humanly possible, which suggests humans are born with this preference.

Nice Chords

As mentioned previously, chords will have three or more notes played together. Looking at just temperament, chords can be described using frequency ratios involving integers. The **major triad** (or simply *major chord*) will have notes with frequencies in the ratios of 4:5:6. The **minor triad** will have the ratios of 10:12:15, which, in terms of pairs of notes, can be reduced to 4:5, 2:3, and 5:6. Chords can also be played out of sequence, which is known as an *inversion*. For example, the C major triad has the notes CEG, normally in that order going up the keyboard. The chord GCE would be an inversion of that chord – the same note names in a different order. The GCE inversion would have frequency ratios of 3:4:5. The major and minor chords, and their inversions, are all generally considered "nice" chords.

On the other hand, chords with notes placed a full step (a "second") or a half step apart are considered dissonant and in need of resolution to a nice chord. The second gives a ratio of 8:9, and the half step involves even higher-valued integers.

These observations led Wild (2002) to propose a "rule of eight" as a way to divide chords into consonant (nice) and dissonant (not so nice) based on frequency ratios. The idea is that when using just temperament, the frequency ratios between all *pairs* of notes in the chord are reduced to the smallest integer values possible – that is, all common factors are removed – then if all integers that remain are 8 or less, the chord is "nice." Of course, the result is somewhat subjective and will depend on circumstances, especially if equal temperament is being used. Equal temperament will never produce nice integer ratios. However, this "rule" seems to be reasonable if just temperament intervals are substituted. It is interesting that what seems to matter most is the ratio of the frequencies of isolated pairs of notes, even though more than two notes are being played.

For musicians who are not tied to a keyboard, some adjustment of the tuning is always possible and frequently occurs. This is not surprising because the musician will tend to adjust to make the music sound "nice." One place where this occurs regularly is in barbershop quartet singing.

One common chord that uses four notes is called the *seventh chord*. That is, a major (or minor) triad is played, and in addition, the seventh note from the (major or minor) scale is added (e.g., CEGB or CEGB$^\flat$). Such a four-note chord, with a minor seventh added to a major triad, occurs frequently in barbershop quartet harmony. Analyzing that chord using just temperament yields frequencies in the ratios of 20:25:30:36, the smallest values that are all integers. According to the rule of eight, this should be dissonant (i.e., "not nice"). The first three of these integers share a common multiple of 5, and the reduced frequency ratios are 4:5:6. The fourth note would then be 36/5, which is not an integer. However, that highest note, the seventh of the minor scale, is very close to 35/5 = 7, which would give rise to ratios of 4:5:6:7. Since that would include only integers smaller than 8 for all possible pairs of notes, it can be expected to sound "nicer" than if the original seventh were used. Since a singer can easily change pitch to make the music sound nicer, that is often what happens. The seventh of the chord is sung a bit low in frequency (a bit "flat"), and this gives rise to what is called the **barbershop seventh** or **harmonic seventh**. Since the equal-tempered seventh is already a bit flat compared to the just seventh, these seventh chords are common and are at least "somewhat nice" when played with a keyboard.

To bring this discussion to a conclusion, it should be stressed that while the notes used for music can be discussed and rationalized using mathematics and the important physical phenomena of beats, the notes used actually evolved through some ill-defined socio-evolutionary process. There was no congressional decree that established or defined the notes, and there is no standard that is universally recognized. Musical notes can be analyzed, and possibly understood, using mathematics, but the notes are not defined by the mathematics – they are defined by the musicians and their audience.

Summary

Music tends to use a subset of the notes available. Western music is highly chord oriented. Pythagoras's monochord can be used to show that "nice" musical intervals are those where the frequencies are related by the ratio of relatively small integers. A value that is equal to a ratio of integers is called a *rational number*. Tones that are close together produce slow beats. Using those rational values removes undesirable beats from the sound.

A musical tuning, a temperament, based on rational-frequency multipliers, called *just* or *harmonic tuning*, will avoid beats but is not practical for keyboard instruments. A compromise tuning is used instead. A modern keyboard will use equal temperament, where the half step is found using a frequency multiplier that is the 12th root of 2, an irrational number, and so beats will result. Historically, the adoption of the equal-tempered scale is relatively recent. Singers and players of many non-keyboard instruments can, and often do, adjust their tuning from note to note to avoid these beats.

The perception of nice-sounding (consonant) chords appears to be based on individual pairs of notes used within the chord. If the frequencies for each pair are related by the ratio of two integers, each less than about 8, the chord will usually be considered consonant.

ADDITIONAL READING

Barbour, J. M. *Tuning and Temperament: A Historical Survey*. Dover, 2004.
Daniélou, A. *Introduction to the Study of Musical Scales*. Indian Press, 1943.
Duffin, R. W. *How Equal Temperament Ruined Harmony*. W. W. Norton & Co., 2007.

Durfee, D. S., and J. S. Colton. "The Physics of Musical Scales: Theory and Experiment."
American Journal of Physics 83, no. 10 (2015): 835–842.

Jorgensen, O. H. *Tuning: Containing the Perfection of Eighteenth-Century Temperament,
the Lost Art of Nineteenth-Century Temperament, and the Science of Equal
Temperament, Complete with Instructions for Aural and Electronic Tuning.* Michigan
State University Press, 1991.

Wild, J. "The Computation Behind Consonance and Dissonance." *Interdisciplinary
Science Reviews* 27, no. 4 (2002): 299–302.

PROBLEMS

3.1 Two clarinets play notes with frequencies 1.2 Hz apart. What beat frequency
will be heard? What interval are they likely trying to play?

3.2 If an oboe plays middle C at 263 Hz and a clarinet plays the E above
middle C using equal temperament, what slow-beat frequency would
you expect?

3.3 If an oboe plays a note at 440.0 Hz and a tenor sings a note at 349.2 Hz, what
slow- beat frequency (<20 Hz) would be expected?

3.4 Use the rule of eight and the intervals for just temperament to determine all
the "nice" three-note chords where the lowest note is middle C and no note is
an octave or more above middle C. How about four-note chords? Do they
sound "nice" when you play them on a keyboard?

3.5 For a Pythagorean tuning that is based on frequency ratios of 3/2, and if no
correction for the mismatch has been included, then a complete clockwise
trip around the circle of fifths will result in a total change in frequency by a
factor of $3^{12}/2^m$, where the integer m is adjusted so that the result is as close to
1.00 as is possible. What is the value of m? If you start at middle C (262 Hz),
what is the frequency of the note obtained after one full trip around the
circle?

3.6 Two musicians tune so that the A above middle C is at 440 Hz, and then they
play a C major scale together. If one plays with the just-tempered scale and
the other uses the equal-tempered scale, how far apart are the frequencies for
their middle Cs?

3.7 If the root of a chord is middle C, at a frequency of 261.0 Hz, what are the
frequencies of the corresponding just- and equal-tempered minor sevenths
and the barbershop seventh?

3.8 Violin strings are normally tuned to be a fifth apart and are often tuned to
each other by ear, by listening for the beats. Typically, a violin player tunes
their A string to a standard such as 440 Hz, then tunes the next-lower string
(the D string) to be exactly two-thirds of that, and then the lowest string (the
G string) to be two-thirds of that. Compare the result of that tuning procedure to what one gets by tuning the lower strings using an electronic tuner that
uses the equal-tempered scale.

3.9 Can you find two (or more) songs that are very different but use the same basic chord sequence?

3.10 Use sound generators or sound-editing software to produce sets of two tones (dyads) that correspond to an interval of a just-tempered minor third and an equal-tempered minor third. How do they compare?

4   The Frequency Domain and Pitch

Life is experienced second by second, minute by minute, hour by hour, and day by day. Hence, it is only natural to describe what goes on around us in those terms. You might have a doctor's appointment at 9:00, a meeting at 10:00, and so on. A television guide will list the available shows according to when they are expected to appear. A physicist might say, "An object released from rest at $t = 0$ will have fallen a distance of $4.9t^2$ meters after t seconds." Such descriptions tell what happens at successive times and are referred to as descriptions "in the **time domain**." The time is specified, and then what has happened, or is going to happen, at that time can be computed, looked up, or otherwise ascertained.

When, for example, a sound is recorded, such as the sound of a musical instrument, the recording consists of values of the received signal saved at successive points in time – that is, values are recorded in the time domain. That set of data can be represented symbolically, for example, as $S(t)$, where the letter t stands for *time*, and S stands for *signal*. The "function" $S(t)$ accepts values of time, t, and returns the value of the signal at that time. That is, "S is a function of t," and "t is the argument of the function S." It is noted that the choice of which letters and symbols to use is entirely up to the writer, although in every case, the symbols must be well defined for the reader. Using t to stand for time is, understandably, a very common practice.

The function $S(t)$ will contain the entire set of data values and, in addition, how the data are organized according to time, t. A plot or graph of data values, such as of $S(t)$ versus t, is often used to visualize and understand how the values are organized relative to one another in the time domain. The sequence of data values in time – how they change and at what rates – is at least as important as the individual data values.

There is another very important way data can be organized, which is "in the **frequency domain**." If various musical instruments are playing different notes of a chord, it is natural to describe that sound by describing the different pitches, that is, frequencies,[1] being played. That description can be made visual using musical notation and a conductor's score. There is at least one problem, however, in that two different instruments playing the same pitch do not sound the same. Some better definition of what is meant is necessary before frequency can be used as the argument for a function that is to be expressed in the frequency domain. It is necessary to know the answer to "the frequency of what?"

The Fourier Series

Near the beginning of the nineteenth century, French mathematician Joseph Fourier (1768–1830) developed many important theorems, including several related to vibrations. Fourier showed that for any periodic signal that can be expressed in the time domain, such as $S(t)$, there is a choice for constants A_0, A_1, A_2, and so forth called *amplitudes*, and φ_1, φ_2, and so forth (phi-one, phi-two, etc.), called *phases*, such that

$$S(t) = A_0 + A_1 \cos(\omega_0 t + \varphi_1) + A_2 \cos(2\omega_0 t + \varphi_2) + A_1 \cos(3\omega_0 t + \varphi_3) + \dots,$$

$$(4.1)$$

where the periodic function has a repeat frequency f_0, and to shorten the notation, the **angular frequency**, $\omega_0 = 2\pi f_0$, with units of radians per second, is being used. The series of terms on the right would be referred to as the **Fourier series representation** of $S(t)$. With minor adjustments, sine functions or a combination of sine and cosine functions can be used instead of cosine functions. For our purposes here, it will be sufficient to use the cosine form.

In general, the Fourier series representation for a periodic signal that has a frequency f_0 may include any or all of the cosine terms that repeat after a time $T_0 = 1/f_0$. Most will repeat more often than that – for example, twice, three times, or more – in the same time, but they all repeat in unison after a time T_0. Therefore, any sum of those cosines also must repeat after a time T_0. It is also true that any cosine term that does not repeat after a time T_0 must be excluded if the sum is to repeat after any time interval T_0.

In some cases, mathematical equality between the series and $S(t)$ is only achieved when the number of constants and sinusoidal (e.g., cosine) functions becomes infinite. It is a remarkable mathematical result that, in many cases, an ever-increasing number of values can be added together and still result in a finite sum; however, that mathematical gem will not be pursued here. In practice, a finite number of values will suffice to achieve any desired level of accuracy for any signals measured in the real world.

---

[1] Frequency and pitch are not the same but are closely related. For the moment, that distinction is set aside.

If enough of the constants are known, $S(t)$ can be computed for any time, t, using Equation (4.1). All that is required is a calculator with a "Cos" function and the skill to use it. An important part of Fourier's theorem is that if $S(t)$ is known for at least one repeat cycle, it is relatively straightforward to determine each of the constants for the series. In practice, if N different values of $S(t)$ are known, equally spaced in time over one repeat cycle, that is sufficient data to compute N constants. The method used to compute the N constants from N values measured in the time domain is known as a **Fourier transform**.

One consequence of Fourier's theorem is that there is no difference between a periodic function produced directly using $S(t)$, wherever it came from, and using the result obtained by adding a large number of cosine functions with appropriately chosen amplitudes and phase constants. Hence, if it is convenient to do so (and it often will be, indeed), any periodic signal can be considered to be made up of, or composed from, a summation of a number of sinusoidal signals, no matter how that signal was actually generated in the first place.

Consider three simple mathematical examples: the sinusoidal, square, and triangle functions, each with a repeat frequency of f_0, as shown graphically in Figure 4.1. Their Fourier series representations, using cosine functions and using $\omega_0 = 2\pi f_0$ as before, can be written as follows:

Cosine function with amplitude 1:

$$S(t) = \cos(\omega_0 t). \tag{4.2}$$

Square function with amplitude $\pi/4$:

$$S(t) = \cos(\omega_0 t) - \frac{1}{3}\cos(3\omega_0 t) + \frac{1}{5}\cos(5\omega_0 t) - \frac{1}{7}\cos(7\omega_0 t) + \dots \tag{4.3}$$

Triangle function with amplitude $\pi^2/8$:

$$S(t) = \cos(\omega_0 t) + \frac{1}{3^2}\cos(3\omega_0 t) + \frac{1}{5^2}\cos(5\omega_0 t) + \frac{1}{7^2}\cos(7\omega_0 t) + \dots \tag{4.4}$$

The amplitudes for these particular functions were chosen so that the form of the series can be easily seen. Other amplitudes can be obtained by simply multiplying

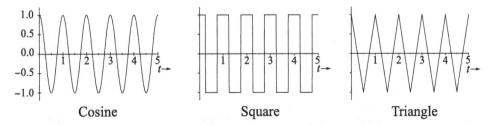

Figure 4.1 The amplitude of the cosine, square, and triangle functions shown as functions of time, t, with t in units of $1/f_0$.

each term by a constant value. For example, the series representation for a square wave with amplitude 7.52 will be as in Equation (4.3), but with each term multiplied by $7.52 \times 4/\pi$.

Laboratory function generators are designed to produce reasonably accurate sinusoidal signals, and many can also produce square and triangle signals. If such a function generator, or an equivalent source, is connected to a speaker, a difference in the sound quality (i.e., **timbre** or **tone**) for these three signals is easy to hear. Since all three series representations begin with the same term, a cosine at frequency f_0 – the so-called **fundamental frequency** for these signals – any differences heard must be due to the additional terms (or lack thereof). The sound from a single cosine is referred to as a **pure tone**. Sounds that require more than one cosine in their Fourier series representation are referred to as **complex sounds** or **composite sounds**. The sounds associated with the additional cosine terms are called **overtones**, and each cosine term considered separately is sometimes referred to as a **partial**.

A frequency that is an integer multiple of another is, by definition, a **harmonic** of that other frequency. The fundamental, which is 1 times itself, is normally referred to as the *first harmonic*; the second harmonic would be at twice the fundamental frequency, and so on, so that if the frequency is N times that of the fundamental, it is the Nth harmonic. The terms *overtone*, *partial*, and *harmonic* are often used interchangeably, although that may not be accurate, and the meaning must be discerned from the context.

The cosine function by itself has only one partial, which is at the fundamental frequency. The square and triangular signals have many overtones, or partials, which are at frequencies that are odd harmonics ($N = 3, 5, 7, \ldots$) of the fundamental frequency. Overtones at even harmonics ($N = 2, 4, 6, \ldots$) are absent from these particular signals. Figure 4.2 illustrates the construction of the square and triangle functions using a sum of cosine functions from the Fourier representation shown previously.

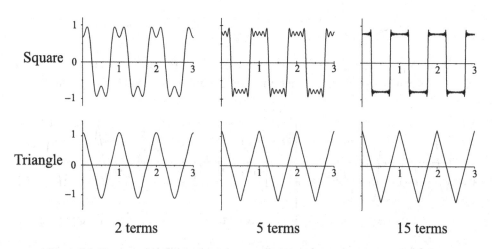

Figure 4.2 Truncated series approximations for the square and triangle functions.

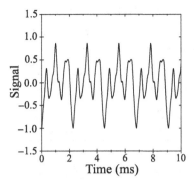

Figure 4.3 Recorded oboe sound.

The process of adding sinusoidal functions together to create another function is sometimes referred to as **Fourier synthesis**, and it is one of several methods used by musical synthesizers to create tones. Here, the triangle function is well approximated with only a few cosine functions. The square function requires many more to achieve an approximation of similar accuracy. It is not a coincidence that most people would describe the sound, or tone, of a triangle signal as being, in some sense, between that of the cosine and the square signals.

Figure 4.3 shows a portion of a signal recorded (in the time domain) for an oboe playing the note A with $f_0 = 440$ Hz. The repetition with a period of $(1/440)$ s = 2.2727 ms is clearly visible, although there is much more going on than a simple sinusoidal signal. Using a computer and a Fourier transform routine to determine the constants, this oboe signal, $S(t)$, can be very closely approximated with the series

$$
\begin{aligned}
S(t) = {} & 0.1506\cos(\omega_0 t + 2.832) + 0.1291\cos(2\omega_0 t + 2.344) \\
& + 0.2212\cos(3\omega_0 t - 2.458) + 0.0625\cos(4\omega_0 t + 2.712) \\
& + 0.0270\cos(5\omega_0 t + 3.084) + 0.0041\cos(6\omega_0 t + 2.806) \\
& + 0.0538\cos(7\omega_0 t - 0.579) + 0.0234\cos(8\omega_0 t + 2.968) \\
& + 0.0125\cos(9\omega_0 t - 1.284) + 0.0252\cos(10\omega_0 t + 2.882) \\
& + 0.0059\cos(11\omega_0 t + 1.643) + 0.0013(12\omega_0 t + 2.612)
\end{aligned}
\tag{4.5}
$$

and for a more precise match, many more terms can be included, and the numerical values can use even more decimal places. However, this description is sufficient at the moment and is already complicated enough. A similar description for two oboes, a flute, a French horn, and a string section, all playing different pitches at the same time, would certainly make matters even more complicated.

Spectra

The Fourier series representation for a periodic signal is completely determined by the constants, and hence a table of those values is sufficient to describe the signal. If the fundamental frequency is known, then only the amplitudes and

Table 4.1 Coefficients for the cosine Fourier series representation of a recorded signal of an oboe playing the note A at 440 Hz.

Harmonic, n	Frequency (Hz)	Amplitude, A_n	Phase, φ_n
1	440	0.1506	2.832
2	880	0.1291	2.344
3	1,320	0.2212	−2.458
4	1,760	0.0625	2.712
5	2,200	0.0270	3.084
6	2,640	0.0041	2.806
7	3,080	0.0538	−0.579
8	3,520	0.0234	2.968
9	3,960	0.0125	−1.284
10	4,400	0.0252	2.882
11	4,840	0.0059	1.643
12	5,280	0.0013	2.612

phases for each of the partials by harmonic number will need to be specified. More generally, the frequency of each partial can be specified. Table 4.1 is thus an equivalent description of the oboe sound and would be considered a description of the oboe sound in the **frequency domain**.

Although there may be some loss of precision, the amplitudes and phases can also be presented graphically. Such a graphic provides a way to easily compare the relative contributions from each of the partials and shows what the signal looks like when sorted by the frequency of the cosine terms. Such a plot is called the **spectrum** of the signal, and a device that produces spectra[2] is called a **spectrometer**. Figure 4.4 shows such a graph for the first 26 amplitude coefficients of the oboe sound. A logarithmic scale is used for the vertical axis to accommodate the large range of values encountered.

A similar analysis could be done for a flute, a violin, or any other instrument that can produce a continuous tone. For each of those instruments, a table (or graph) could be created showing the partials for that instrument. It is then straightforward to describe the sound of multiple instruments playing different notes at the same time by simply merging their tables of constants. This is easier to do if the partials are ordered by frequency rather than harmonic number. Since every instrument has a fifth harmonic, for example, it would be necessary to also keep track of which instrument it came from, whereas the frequency of the partial is unambiguous.

---

[2] *Spectra* is the plural of *spectrum*. *Spectrums* is also considered an acceptable plural.

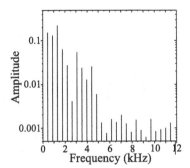

Figure 4.4 Frequency spectrum of recorded oboe sound.

Note also that while each instrument is assumed, at this point, to produce a periodic signal, the sound from a combination of instruments might not be periodic, and that is just fine.

Now, getting back to the issue at hand, when speaking of the frequency domain and asking "the frequency of what?," it seems natural to use the frequencies of each of the cosine terms in the Fourier representation – the partials. Each cosine by itself is just that, a cosine, and by itself would sound the same no matter which instrument it came from. The sound for each instrument can be regarded as being built up from these basic sinusoidal building blocks. There are other functions, aside from cosines and sines, that could be used to make such a series representation, but there are good reasons to stick to the trigonometric ("sinusoidal") functions. First, they are relatively simple mathematical functions that also arise in many simple physical situations (e.g., the simple pendulum and mass on a spring, seen in Chapter 5). Second, human hearing seems to respond to sinusoids as independent building blocks of sound, at least in some cases.

To see how, or if, the phase factors affect the sound, they can be changed and the signal resynthesized. For example, the phase constants for the recorded oboe sound can all be changed to zero, leaving the amplitudes alone, as shown in Figure 4.5. Now combine the terms back together and play that signal as a sound and compare it to the original. Alternatively, a minus sign can be added in front of any term, which is equivalent to changing the phase constant by π (=180°), or the phase constants can even all be replaced using a random-number generator. When compared, the resulting signals will all sound identical.[3] That result is evidence that the partials – that is, the individual cosine (or sine) terms – are contributing independently to what is heard, and it is not so important how they add to each other. A change in a phase constant will have the same effect as a shift

---

[3] This will be generally true as long as the signal levels are not too large and a quality sound-reproduction system is used. If the signal is particularly large or the sound produced is particularly loud, nonlinear effects may create an audible difference when phases are changed.

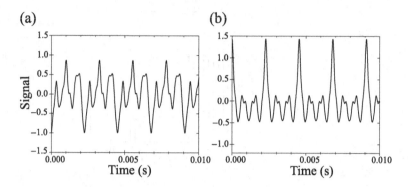

Figure 4.5 Oboe signal synthesized from (a) the amplitudes and phases from Table 4.1 and (b) the amplitudes from Table 4.1 with all phases set to zero. When played back through a speaker, these signals sound nearly identical.

of the cosine term in time relative to the others. If the individual cosine terms are what is important, then how they are shifted in time relative to each other is not important. Thus, an amplitude spectrum, without concern for phases, is usually sufficient in such cases.

An effective way to analyze a sound, or any periodic signal, is to divide that signal up into partials, the building blocks, using a Fourier series representation, and then consider what happens to each of the partials individually. This is referred to as **analysis in the frequency domain**. For example, consider the absorption of sound during its reflection off a wall. The reflective properties of the wall can be measured and characterized using a single sinusoidal signal having a frequency that can be varied. The results of measurements made using many different frequency settings could then be used to describe the "frequency-dependent properties of the wall." The frequency being referred to is that of the single sinusoid used for testing, not of the wall itself. Once the results are known for the single sinusoid over the relevant frequency range, those results can be applied to complex sounds, such as the sound from the oboe, term by term, using the Fourier series representation.

There are several cases where Fourier series representations and analysis in the frequency domain must be done with care. Included are cases where there are nonlinearities and cases where two series representations are combined that contain one or more individual terms that are at exactly the same frequency. Also, for a signal that is changing in time – for example, it is turned on and/or off – the addition of nonperiodic time dependence can also make matters more complicated. These and related issues will all be considered later, especially in Chapter 9.

Frequency-domain analysis illustrates one reason why the slow beats for two simultaneous composite signals might be determined using the closest *multiples* of the signals' frequencies. Periodic composite signals will have partials at integer multiples of the fundamental frequency, and if any two partials are nearby in frequency, slow beats can be heard. For example, Figure 4.6 illustrates the

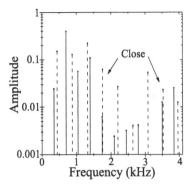

Figure 4.6 Combined spectrum of oboe playing A and tenor singing F showing closely spaced partials.

Fourier series amplitudes for the oboe and a tenor at the same time. The oboe is playing the note A ($f_0 = 440$ Hz) as described previously. The tenor is singing the note F just below that, tuned according to equal temperament ($f_0 = 349.23$ Hz). These notes are nominally a major third apart (the A is the third note of the F major scale). The fifth partial of the tenor and the fourth partial of the oboe are very close in frequency, giving rise to a slow beat at $4 \times 440.00 - 5 \times 349.23 = 13.85$ Hz, which may make them sound somewhat out of tune.

Power Spectra – Decibels

The frequency-domain plots just presented are simple plots of the amplitude of the terms in the Fourier series. Often, such plots will be presented in terms of relative power (or energy or intensity). In general, the power (or energy or intensity) of a signal is proportional to the amplitude squared. It is often convenient to use **decibels** for such a description. The decibel is a way of expressing ratios of two values using a logarithmic scale. A decibel (dB, pronounced "deebee") is 1/10th of a bel (B), although the bel is rarely used by itself. The decibel was named for Alexander Graham Bell (1847–1922) and arose through work to describe exponential losses in cables and related problems during the early development of the telephone industry.

Since power depends on amplitude squared, the power for a composite signal can be found by squaring the Fourier series representation for that signal. The result of that operation will be a series expression for the instantaneous power from the signal. Such an expression will usually be quite complicated. If, however, only the **time-averaged power** is of interest, averaged over a cycle, then the power is simply obtained from the sum of the squares of the individual amplitudes. All other terms that are present in the expression for the instantaneous power will contribute zero to the average power. The zero average is due to trigonometric identities such as

$$\cos(\omega_1 t)\cos(\omega_2 t) = \frac{1}{2}\Big(\cos\big((\omega_1 + \omega_2)t\big) + \cos\big((\omega_1 - \omega_2)t\big)\Big), \qquad (4.6)$$

where the terms on the right clearly average to zero over time (since a cosine spends as much time positive as negative) unless the two frequencies are equal. When the two frequencies are equal, the average over time is simply one-half since $\cos(0) = 1$, independent of time. Thus, the total time-averaged power of the signal is the sum of the time-averaged power from each of the partials – the contribution to the time-averaged power from each partial can be considered independently of the other partials. Note that it is usually the case that time averaging over a cycle is implied, and so the words "time averaged" are often omitted.

If P_1 is the power of signal 1, and P_2 is the power of signal 2, then the difference between signal 1 and signal 2 expressed in dB is given by

$$(\textbf{Difference in dB}) = \textbf{10log}_{10}\left(\frac{P_1}{P_2}\right), \qquad (4.7)$$

and if A_1 is the amplitude of signal 1, and A_2 is the amplitude of signal 2, then the difference is

$$(\textbf{Difference in dB}) = \textbf{20log}_{10}\left(\frac{A_1}{A_2}\right). \qquad (4.8)$$

Note that the difference between two signals expressed in decibels will be the same whether amplitude or power (or energy or intensity) is used. Also note that here, the word *difference* is being used in its more general sense, not to mean the result of a subtraction. Since decibels are always based on a ratio, all the units cancel; however, the stand-in "unit" dB is kept, as needed, to remind the reader where the value came from and what it represents.

Example 4.1

For the oboe signal represented in Table 4.1, the difference between the third partial and the first, the fundamental, is

$$20\log_{10}\left(\frac{0.2212}{0.1506}\right) = 3.34 \text{ dB.}$$

That is, the third partial is 3.34 dB larger than the first. It is straightforward to show that a factor-of-2 power difference is 3.0103 dB, so for the oboe signal, the power from the third partial is a bit more than a factor of 2 larger than the power from the first.

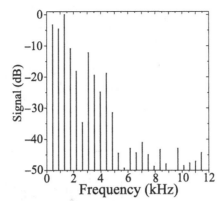

Figure 4.7 Oboe spectrum with decibels used for vertical axis.

Many examples using decibels will appear later. It is worth pointing out now that when talking of a signal in dB it is necessary to remember that the signal is what is inside the logarithm. For example, consider a signal that is initially 3 dB larger than a reference. That signal is then increased by a factor of 6. When expressed in decibels, the value *inside the logarithm* is what is multiplied by 6, not the number of decibels. The signal with six times more power will be $10 \log_{10}(6) = 7.78$ dB larger. Thus, the signal that starts at 3 dB (larger than some reference) and is increased by a factor of 6 results in a signal that is 10.78 dB. In particular, it is *not* 6×3 dB $= 18$ dB (larger than the reference). Also, when considering the power from a composite sound, it is common to refer to the individual contributions from each partial using decibels; however, the total power is *not* found by adding those decibel values together – remember that adding decibels (i.e., logarithms) is equivalent to a multiplication for what is inside the logarithm.

When plotted using decibels, the oboe signal shown earlier would appear as in Figure 4.7. For this particular plot, the amplitude of the largest partial was used as the common reference. Hence, by definition, the third partial, which is the largest, will be at 0 dB, and all others will be less (i.e., negative when expressed in dB).

Spectrum of Noise

Noise might be considered to be, in some sense, the opposite of a "nice" periodic signal. In this context, *noise* refers to a signal that is random in character and should not be confused with interference. **Interference** is any undesired signal that is present, random or not. Noise will be random and may or may not be desirable, depending on the circumstances.

Randomness implies the inability to predict a later value based on previous values. For a simple sine, cosine, or even the oboe signal shown previously, such a prediction is straightforward. Of course, any real signal will have some continuity from one point to the next. The value a very, very short time later will be

Figure 4.8 Random noise with (a) a longer and (b) a shorter correlation time.

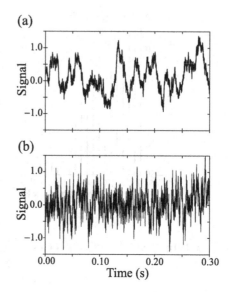

almost unchanged because a real signal will not change instantaneously, and hence any signal is at least somewhat predictable. One characteristic of a signal is the **correlation time**, which describes the timescale over which a reasonable prediction of a future value can be made based on past values. For a perfectly periodic signal, that correlation time is infinite. For noise, that time becomes very short. Figure 4.8 illustrates two random signals with differing correlation times.

When the noise spectrum includes all frequency components with equal amplitude, at least over a relevant range of frequency, it is referred to as **white noise**. The name is in analogy to the colors of light, where white light is an equal mixture of all frequencies of visible light, which, when viewed separately, correspond to a rainbow of visible colors. When white noise is measured with a spectrometer for a sufficiently long period of time, the spectrum will approach a simple horizontal line, at least over a specified frequency range, representing an equal contribution at all frequencies. For shorter times, random variation from a straight line will be observed – after all, it is a random signal.

If the frequency components of the noise are not of equal magnitude, it is referred to as **colored noise**, again in analogy to visible light. When the power level *decreases* by a factor of 2 (−3 dB) for each octave *increase* in frequency, it is referred to as **pink noise**.[4] That is, the pink-noise spectrum corresponds to −3 dB/octave, which is the same as saying the noise power (measured in the frequency domain) is proportional to $1/f$. When the amplitude decreases by a factor of 2, which means the power level decreases by a factor of 4 (−6 dB) per octave, this is referred to as **red noise**. On the other hand, if the power *increases* by a factor of

---

[4] A factor of 2 is not exactly 3 dB, but it is common practice to treat the two interchangeably. In this context, the rounded value of 3 dB is being used to refer to a change that is exactly a factor of 2, −3 dB, to a factor of 1/2, and so forth.

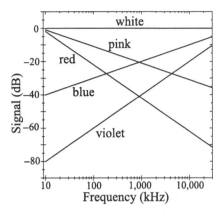

Figure 4.9 Examples of noise spectra for different types of colored noise. Note that a logarithmic scale is being used for both axes.

2 per octave, it is called **blue noise**, and if the amplitude increases by a factor of 2 per octave, it is **violet noise**. When a logarithmic scale is used for the horizontal axis, the spectra for all of these are straight lines, as shown in Figure 4.9.

Noise can be useful for some measurements since all of the frequency components of interest are present at once. Hence, instead of scanning a sine-wave generator through all possible frequencies and recording the results one value at a time, all frequencies can be present at the same time, and the components are only separated after the measurement using a spectrometer. Pink noise, which has equal power per octave, as opposed to equal power per hertz, is often convenient to use for such purposes.

How a Fourier Transform Works (Optional)

There is considerable mathematical theory behind the Fourier transform that goes beyond this discussion. However, the principle that makes the transform work can be understood by considering the trigonometric identity mentioned earlier in a different context:

$$\cos(\omega_1 t)\cos(\omega_2 t) = \frac{1}{2}\cos\big((\omega_1 + \omega_2)t\big) + \cos\big((\omega_1 - \omega_2)t\big). \qquad (4.9)$$

When such terms are averaged over time, the result will be zero unless $\omega_1 = \omega_2$. This is shown by the example in Figure 4.10.

Now start with a signal $S(t)$ and multiply it by a cosine function that is at any fixed frequency, ω. Mathematically, that is equivalent to multiplying the Fourier series representation of $S(t)$ by that same cosine. That is:

$$\cos(\omega t)S(t) = A_0 \cos(\omega t) + A_1 \cos(\omega t)\cos(\omega_0 t) + A_2 \cos(\omega t)\cos(2\omega_0 t) + \ldots . \qquad (4.10)$$

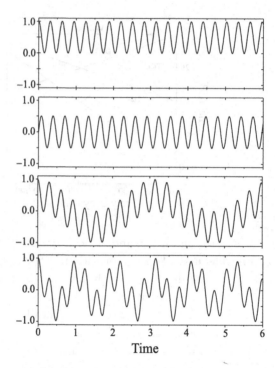

Figure 4.10 From top down: graphs showing $\cos(10t) \times \cos(10t)$, $\cos(10t) \times \sin(10t)$, $\cos(9t) \times \cos(11t)$, and $\cos(7t) \times \cos(13t)$. Only the first case, where the frequencies are equal, will have a nonzero time average.

The phase factors have been left off for clarity. More generally they are included, in one way or another, for both the signal and the additional cosine term. Now average the result (Equation [4.10]) over time. On the right, it can be seen that everything will average to zero unless ω is equal to one of the frequencies $0, \omega_0, 2\omega_0, 3\omega_0, \ldots$, and so on. Now sweep ω starting from zero, taking a new time average for each new value of ω. When ω is zero, the result is proportional to A_0. As ω is increased, the average is zero until $\omega = \omega_0$, at which point the result is proportional to A_1. Increase ω more, and the result is zero until $\omega = 2\omega_0$, at which point the result is proportional to A_2, and so on. The time-averaged response as a function of the variable ω is the Fourier transform (aside from an overall constant).

With digital sampling of a signal and the use of a computer, the transform can be obtained relatively quickly. If the total number of data values is chosen appropriately, a common choice being 2^N, where N is a positive integer, an algorithm known as a **fast Fourier transform** (FFT) can be used. A FFT is a particularly efficient way to do the numerical calculation.

Spectral Resolution and Range

Before finishing this discussion about the frequency domain and spectra, it is important to point out that, in general, any real spectrometer will not provide infinitely fine frequency divisions and will not cover an infinite range of frequencies. The frequency resolution will be limited by the total measurement time, and the frequency range will be limited by the measurement resolution in the time domain.

Frequency resolution is related to beats, discussed in Chapter 3, and to the "uncertainty principle," discussed in Chapter 9. The basic cause of the limitation in resolution is due to the important trigonometric identity discussed previously – that the product of two sinusoidal signals will average to zero over time unless they have identical frequencies. If two sinusoidal signals have frequencies that are very close to one another, it will take a long time to average to zero. If a measurement is made over a shorter time period, those two signals may look like they have the same frequency since they do not average to zero for that shorter time. The simplistic result is that you should not expect a frequency resolution finer than about $1/(T_m)$, where T_m is the time to make the measurement. Thus, if you sample a signal for one-tenth of a second, the frequency resolution will be no better than about 10 Hz, in round numbers.

The range of frequencies that can be accurately represented is limited by the resolution of the measured data in the time domain. The limitation on the resolution in time may be due to the finite speed of response of the electronics used for recording, the choice of sampling time, or possibly other causes. In any case, there is a smallest resolution in the time domain, say, Δt, and given that, the highest frequency you can measure will be no better than about $1/(2\Delta t)$. The factor of 2 arises from the fact that to see an oscillation, you need to see the signal go positive and then negative, and thus at least two measurement intervals are required to see one oscillation.

Pitch

It is tempting to equate frequency with **pitch**. Clearly, the two are somehow related. However, what is heard as the "pitch" of a composite signal is a **perceived quantity**. The ear and brain work together as an amazing sensor, and that sensor is not completely understood. Two things about hearing that are well known are (1) all the simple physical models are too simple, and (2) it is normal for people to hear things somewhat differently.

Pitch can be quantified using a matching experiment. That is, an individual is asked to match the frequency of a sinusoidal signal (or a periodic signal comprising short clicks or some other reference signal) with the pitch of a composite sound. The "pitch number" assigned to the composite sound is the frequency of the reference signal that is perceived to match. Under many circumstances, the pitch number is just the repetition frequency for the composite signal – that is, what one would expect from the signal's spectrum. This is not surprising. What is surprising is that there are circumstances where the pitch number and the frequencies of the Fourier components of a composite sound seem to have little relationship. Those who study hearing are especially keen to look at these auditory puzzles as a way to test various models for the hearing process. A few examples are presented here.

It is often surprisingly difficult to discriminate octaves. A male can sing harmony "above" a female singer, and yet at the same time, a spectrum will show that his fundamental frequency is actually lower than that of the female singer. Two musical instruments can play a note at the same fundamental frequency, and yet if the note is at the low end for one of the instruments and the high end for the other, they may not be perceived as being in the same octave. This difficulty in perceiving octaves is exemplified by an auditory illusion known as the **Shepard scale**, named for Roger Shepard, who presented it in 1964. The illusion is often compared to the optical illusion shown in M. C. Escher's lithograph "Ascending and Descending," wherein it appears there is a set of stairs that one can climb (or descend) indefinitely, and yet one never really gets very far. The pitch of the notes of the Shepard scale will seem to continually rise (or fall) indefinitely, and yet the pitch is never particularly high or low.

The tones of the Shepard scale are constructed with Fourier synthesis using partials that are related by octaves. That is, the components include a partial with a frequency, f_1, and then also partials at the octaves above, $2f_1$, $4f_1$, $8f_1$, $16f_1$, and so forth, as well as the octaves below, $f_1/2$, $f_1/4$, $f_1/8$, and so forth. The relative amplitudes of the overtones are set by a fixed, bell-shaped function that emphasizes the overtones with midrange frequencies. The perceived octave associated with the tone will tend to be that of the strongest partial. As f_1 is raised, at some point, two strong partials will become equal in amplitude, leading to some ambiguity in the perception. As f_1 is increased further, the lower partial will take over. This effect is illustrated in Figure 4.11. The Shepard scale can be created with discrete frequency steps or in a continuous manner (a "glissando").[5]

Another auditory illusion important for music is that of the **missing fundamental**. In this case, the spectrum of the sound includes signals at the harmonic frequencies of a particular fundamental frequency, but the amplitude of the fundamental itself is very small or even zero. The perceived pitch number is often (but not always) that of the fundamental. For example, Figure 4.12 shows the spectrum of a tenor who has matched their pitch to the note F above middle C. The fundamental does indeed match the expected frequency associated with F above middle C (350 Hz); however, its intensity is quite low. Using Fourier synthesis, this tone can be recreated with the fundamental completely removed, and few (if any) listeners will even notice the difference. It is known that some listeners will hear missing

---

[5] The continuous glissando is often referred to as the *Risset* or *Shepard–Risset glissando*.

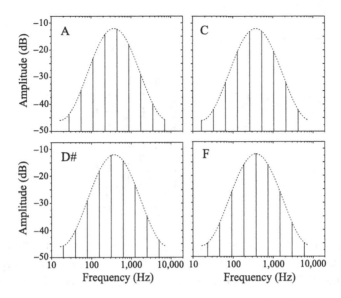

Figure 4.11 Representative spectra for composite sounds used for Shepard's tones. The overall envelope of the amplitudes remains constant while the individual spectral lines gradually move up in frequency.

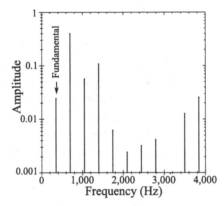

Figure 4.12 Spectrum of a tenor matching the note F above middle C.

fundamentals more often than others. In fact, composite tones can be synthesized such that roughly half the population hears a missing fundamental and half does not. This is clear evidence that not everyone perceives sounds in the same way.

Another very interesting experiment in pitch perception is to take a fundamental and the corresponding sequence of harmonics, then offset each of them by a fixed amount and compare. For example, a composite tone constructed from a few harmonics of 200 Hz of equal amplitude, such as

$$S(t) = \cos(3\omega_0 t) + \cos(4\omega_0 t) + \cos(5\omega_0 t) + \cos(6\omega_0 t) + \cos(7\omega_0 t) + \cos(8\omega_0 t),$$

$$(4.11)$$

Figure 4.13 Spectra of a signal with harmonics of 200 Hz (*solid*) and one where the partials have been offset by 40 Hz (*dashed*).

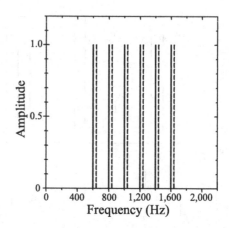

where $\omega_0 = 2\pi \times 200$ Hz, will have a pitch number of 200 Hz, which is the frequency of the missing fundamental. Now, offset the frequency of each partial by the same *additive* constant, say, $\Delta\omega = 2\pi \times 40$ Hz, so that the tone is now described by

$$S(t) = \cos\left((3\omega_0 + \Delta\omega)t\right) + \cos\left((4\omega_0 + \Delta\omega)t\right) + \cos\left((5\omega_0 + \Delta\omega)t\right)$$
$$+ \cos\left((6\omega_0 + \Delta\omega)t\right) + \cos\left((7\omega_0 + \Delta\omega)t\right) + \cos\left((8\omega_0 + \Delta\omega)t\right), \quad (4.12)$$

and many listeners will say that the pitch rises. The spectra for these signals are compared in Figure 4.13. Note that while the partials continue to have the same frequency spacing, they may not be harmonics of a common fundamental. For this particular example, the shifted frequencies are actually harmonics of 40 Hz – the 16th, 21st, 26th, 31st, 36th, and 41st harmonics – although the 40-Hz fundamental is not heard in this case. What is most interesting is what happens when the pitch number is determined for the shifted signal by comparison to a reference signal. The result is usually near, but slightly above, 200 Hz, in the vicinity of 210 Hz, and varies somewhat from person to person.

In addition to the Fourier transform and spectrum, another way to look for repetitive signals using mathematics is to use something called the **autocorrelation function**. This is a way of comparing a signal with a time-delayed version of itself. It should not be surprising that the autocorrelation function and the spectrum are closely related; however, the full details of the autocorrelation function are a bit beyond this discussion. It will be sufficient to know that if a periodic signal is written as a Fourier series using (any) sinusoidal functions, then the corresponding autocorrelation function is formed by squaring the coefficients, changing any sine functions that may appear to cosines, and setting all phase coefficients to zero. The final result can be "normalized" by dividing by the sum of the squares of the coefficients so that at a delay time $t = 0$, the function is 1. In terms of the

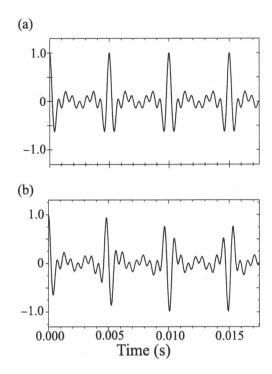

Figure 4.14 Autocorrelation for the 200-Hz missing fundamental example (a) and when each partial is shifted up by 40 Hz (b).

general cosine series given previously (Equation [4.1]), the corresponding (normalized) autocorrelation function is

$$R(t) = \frac{A_0^2 + A_1^2 \cos(\omega_1 t) + A_2^2 \cos(\omega_2 t) + A_3^2 \cos(\omega_3 t) + \dots}{A_0^2 + A_1^2 + A_2^2 + A_3^2 + \dots}. \tag{4.13}$$

When $R(t) = 1$, the signal and its time-delayed counterpart match perfectly. When $R(t) = 0$, it is said, possibly somewhat inaccurately, that there is "no correlation," and when $R(t) = -1$, the time-delayed version is different only by a minus sign.

The similarity between some features of the autocorrelation function and the results of pitch-perception experiments suggests hearing perception may be, at least in part, related to this autocorrelation. For example, using the 200-Hz missing fundamental signal described previously, the autocorrelation shows a match when $t = 1/200$ Hz $= 5$ ms. This is shown in Figure 4.14. When the partials are shifted by 40 Hz, the autocorrelation shows a first peak, close to 1, a short time sooner than 5 ms, which would be suggestive of a repetition frequency of a little over 200 Hz. Note that in this case, the signal does not actually repeat this quickly, although it is perceived to repeat.

In the previous example, where the frequencies of the partials are shifted by 40 Hz, a pitch is still discernable; however, the idea that there is a definite pitch

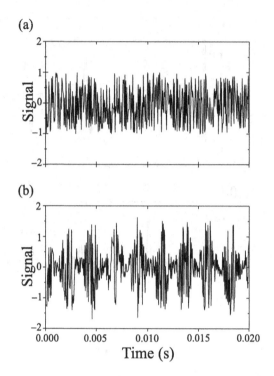

Figure 4.15 Noise (a) and noise modulated in amplitude by a 440-Hz sinusoidal signal (b). The spectrum and autocorrelation function for these signals are identical, but they do not sound the same.

number starts to become questionable. A more extreme case occurs when white noise is amplitude-modulated by a sinusoidal signal. That is, the amplitude of the noise is made to change in time sinusoidally, as illustrated in Figure 4.15. The spectrum and the autocorrelation function for such a signal are the same as they are for white noise; however, these modulated signals can be used to produce identifiable melodies, at least if the frequency of the sinusoid is not too high. Identifying a pitch number for such signals is even more difficult. In some cases, one is left with pitch as simply being "a property of a sound that allows melodies to be perceived" but that is otherwise not particularly quantifiable. For such circumstances, pitch may not be very well suited for study by practitioners of physics.

Mathematically, the Fourier transform and the autocorrelation are defined based on averages over all time. Pitch perception, however, clearly occurs moment to moment, at least on a human timescale. One suggestion is that the spectra and/or correlation functions should be computed using a **window function**, which includes only data within a specified time interval. Data outside that window are treated as zero. Then the spectra and correlation become time dependent as well. The pitches heard for the modulated noise are due to the time-changing amplitude of the result. Unfortunately, this explanation does not work well for some other composite sounds. The discussion here will proceed using frequencies (e.g., of vibration, etc.), and further details related to pitch perception will be left for the cognitive scientists to consider.

Physical oscillators are considered in Chapter 5. In particular, it will be most important to look at the fundamental prototype for the vibrational behavior of many musical instruments, the vibrating mass on a spring or, more generally, the "harmonic oscillator." Following that, more complicated oscillators, such as strings and chimes, will be presented.

Summary

Any periodic signal can be described using the sum of time-dependent sinusoidal functions with appropriate amplitudes and phases. Such a description is called the *Fourier series representation* of the signal. The frequency of the original signal is the fundamental frequency, and the lowest-frequency sinusoid that can appear in the sum is at that fundamental frequency. The rest of the terms will be at frequencies that are harmonics – that is, integer multiples – of that fundamental frequency. For sounds, the terms are referred to as *partials* and the higher-frequency terms as *overtones*.

A periodic signal can be described by specifying the amplitudes and phases of the sinusoids in the Fourier series representation. A spectrum is a description of the amplitudes, often presented graphically. For a sustained, periodic sound that is not too loud, the perceived sound does not depend on the phases, only on the amplitudes. Spectra are often displayed using decibels (dB), a logarithmic scale based on ratios between two values.

Spectral resolution depends on the timescale and time resolution used for the measurements of the signal.

Pitch is a perceived quantity closely related to the frequency; however, there are a number of notable exceptions. Those exceptions show that hearing is a complicated process and that not everyone hears the same.

ADDITIONAL READING

Deutsch, D. "A Musical Paradox." *Music Perception* 3, no. 3 (1986): 275–280.

Heller, E. J. *Why You Hear What You Hear*. Princeton University Press, 2013.

Shepard, R. N. "Circularity in Judgements of Relative Pitch." *Journal of the Acoustical Society of America* 36, no. 12 (1964): 2346–2353.

PROBLEMS

4.1 The intensity of a signal is increased by a factor of 5. What is the increase in decibels?

4.2 The amplitude of a signal is increased by a factor of 5. What is the increase in decibels?

4.3 Use a spreadsheet or a similar plotting application to show graphically that

$$S(t) = \sin(\omega_0 t) - \frac{1}{3^2}\sin(3\omega_0 t) + \frac{1}{5^2}\sin(5\omega_0 t) - \frac{1}{7^2}\sin(7\omega_0 t) + \ldots$$

is a triangular wave. How does this wave differ from the wave described in Equation (4.4)? (For simplicity, use $\omega_0 = 1$ unit.)

4.4 For a linear amplifier, the output is simply proportional to the input. If nonlinearities are present, the situation will involve higher powers of the input – for example, it may contain terms that are proportional to the square and/or cube of the input signal, and the output signal becomes distorted.

a. Using trigonometric identities (see Appendix A), compare the spectra (in decibels) for the input and output signals for a simple two-tone input

$$S(t) = 2\cos(628t) + \cos(942t) \quad \omega_0 = 1 \text{ unit}$$

sent through a nonlinear amplifier where the output is related to the input by

$$S_{out} = 3S_{in} + 0.2S_{in}^2.$$

b. It was stated that relative phase shifts for terms in a Fourier series representation of a continuous periodic tone will not make a difference in how it sounds. The statement was qualified to say that this would only be true if the sound was not too loud, so nonlinearities do not show up. Why would nonlinearities change the sound, for example, for the oboe signals shown in Figure 4.5?

4.5 The sound of a human voice involves the vibration of the vocal folds ("chords") in the larynx combined with the frequency-dependent response of the sound throughout the vocal tract, nasal passages, and mouth. The signal from the larynx is often described as having harmonics that decrease at a rate of about 6 dB per octave. If you were going to make a simulated voice sound using a signal played through a speaker placed within cavities approximating the vocal tract, nasal passages, and mouth, should the signal be more like a pure tone (a sinusoidal signal), a triangular signal, or a square signal? That is, which of these decreases closest to 6 dB per octave?

4.6 There are a number of resources available (audio editors, Java scripts, etc.) that can be used for Fourier synthesis. Using one of these resources, construct a complex sound using a sum of pure tones (sinusoids), including overtones at several harmonic frequencies. Now change the phase of one or more of the overtones (e.g., change sines to cosines, add a time delay, or similar) and compare the sounds produced. Under what circumstances can you hear a difference?

5   Harmonic Oscillators and Resonance

The **harmonic oscillator** is perhaps the simplest vibrating system and is used as the starting point for the discussion of virtually all types of vibration, including the vibrations of and within musical instruments. The prototype harmonic oscillator is a mass connected to a so-called **ideal spring**. The resulting motion is derived from **Newton's laws of motion**. While most musical applications are more complicated than the mass on the spring, there are exceptions, including tuning forks, the ocarina, and some speaker enclosures.

A harmonic oscillator has a natural oscillation frequency. When driven by a periodic force at, or very near, that frequency, the system can produce a large "resonant" response. The driven harmonic oscillator provides some additional fundamental background for the behavior of some musical instruments. The sympathetic vibration of strings in a piano is one example based on the driven oscillator.

To start, then, consider Newton's laws and how they apply to a mass on a spring.

Newton's Laws and Gravity

Sir Isaac Newton (1642–1727) is one of the most well-known physicists of all time. He made many extremely important contributions to science, especially to mathematics, optics, and mechanics. One of his most well-known results is sometimes referred to as the "discovery of gravity." The legend has him reach a great epiphany after observing (or in some retellings, being struck on the head by) a falling apple. Of course, everyone knew that objects fell to the ground. It did not take a genius to figure that out. What Newton discovered was a way to develop a single, highly accurate mathematical model for the motion of objects that was the same for a falling apple and for many other motions, including the motion of the moon about the earth. He developed his "laws of motion" based on the concept of **forces**, and he found a "universal law of gravity" that describes the forces for both the apple and the moon. The falling apple and the orbiting moon are two specific cases of the same basic physics problem – he unified the two seemingly different problems into one.

The motion of an object is described using its **position**; its **velocity**, which is how rapidly the position is changing and in what direction; its **acceleration**, which is how rapidly the velocity changes and in what direction; and so on. Adding a quantity called **mass**, which is a property of an object related to the quantity of matter in the object, and the concept of forces, Newton's laws of motion are often expressed as follows:

1. An object in motion stays in motion, and an object at rest stays at rest, unless a force acts on it.
2. The total force, F, on an object equals its mass, m, times its acceleration, a.
3. For every action, there is an equal and opposite reaction.

The first law does not only mean that an object that is moving stays moving; it also means that the object will maintain a constant motion. That is, the object will travel with constant speed in a constant direction unless a force acts on it. The second law defines the relationship between forces and any *changes* in velocity that occur – how the motion differs from constant motion. The third law might be more clearly expressed as "if object 1 pushes on object 2 with a force F in one direction, then it must also be the case that object 2 is simultaneously pushing on object 1 with a force F in the exact opposite direction."

The first law is regarded by some as a special case of the second. However, the first law by itself is a very bold statement. Everyday experience would seem to show that the natural state for any object is to be stationary. If you slide a book on a desktop, it will come to a stop. If you roll a ball, it will eventually come to a stop. If you drop a rock over the side of a cliff, it will eventually come to a stop. This first law says that the natural state of an object is to maintain whatever motion it has. Something must be done to change that motion, and that something is called a *force*. Hence, if you slide a book on a desktop and it comes to a stop, there must have been a force involved.

While the first law gives a starting point for the case where there is no force, the second law deals with the rate of change from that starting point and how to compute it. Written as an equation, the second law is $F = ma$. Hence, the units used for forces will be the units of mass multiplied by the units of acceleration. For SI units, those units are kilogram-meters per second squared (kg·m/s^2), called newtons (N), whereas in English (or "imperial") units, forces are in pounds (lb). The conversion is 1 lb = 4.4482 N or 1 N = 0.2248 lb. The second law says that when there is more mass, a larger force is required to get the same changes in motion. That is, mass acts to resist changes in motion. More generally, the tendency to resist *changes* in motion is referred to as **inertia**.

Notice that a velocity includes both the distance traveled with time (a speed) and a direction. Hence, a change in *either* results in a change in the velocity. When we speak of acceleration in everyday language, we tend to think only of the speed and how that changes. When applying Newton's laws, a change in direction is just as important. A force must be applied to change the direction of travel. That force will also have a direction. Notice that the direction of motion is not necessarily along the direction of the force, and vice versa. However, the acceleration, the rate of *change* in the velocity, will always be in the direction of the force.

Rather than getting into more technical details, some simple examples are presented. In each case, the average force or average acceleration is considered. The average gives the net result after some specified time or event and avoids consideration of the details along the way.

Example 5.1

A 1,000-kg car is traveling south at 30 m/s when it becomes necessary to slow down to 24 m/s for a construction zone. It takes 12 seconds to slow down. What is the (average) acceleration?

In this case, the initial velocity is 30 m/s to the south, and the final velocity is 24 m/s to the south. The change in velocity to the south is −6 m/s, which can also be expressed as a change of +6 m/s to the north. The acceleration is the rate of change with time, or $(6 \text{ m/s})/(12 \text{ s}) = 0.5 \text{ m/s}^2$ to the north. Hence, the average force on the car is 500 N to the north. Notice that in everyday language, an object that slows down is said to "decelerate"; however, in physics, all changes in velocity with time are referred to as *accelerations* – some may cause a speedup, some a slowdown, and some simply a change in the direction of travel.

Example 5.2

A 1,500-kg car is traveling south at 30 m/s when it makes a turn to the west. The car maintains a speed of 30 m/s throughout and takes 15 s to make the turn. What is the acceleration?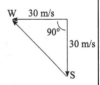

In this case, the speed has remained constant, but the direction has changed. Drawing a picture will help to determine the total change in velocity. Draw an arrow pointed south to represent the initial velocity. Draw another arrow *from the same starting point* going to the west to represent the final velocity. The length of the arrows represents the respective speeds. Here, the speeds are the same, so the arrows should be the same length. Now draw another arrow from the end of the first to the end of the second. That third arrow shows the change, in this case, pointed to the northwest. The length of that arrow shows the magnitude of the change.

In this case, these arrows form a right triangle, so the Pythagorean theorem can be used to get the magnitude of the change in the velocity:

$$\Delta v = \sqrt{(30\text{m/s})^2 + (30\text{m/s})^2} = 42.4 \text{ m/s}.$$

Remember that here, the uppercase Greek delta, Δ, is being used to signify that a change in a quantity is being considered. Hence, "Δv," read as "delta-vee," is a single variable that stands for "the change in v." The acceleration is the change in velocity divided by how long it took to change, or $(42.4 \text{ m/s})/(15 \text{ s}) = 2.83 \text{ m/s}^2$. The average force on the car is then $1{,}500 \times 2.83 \text{ kg·m/s}^2 = 4{,}250$ N, to the northwest. Both the average acceleration and the average force are toward the northwest. Note that during the turn, the car was never headed toward the northwest, and the speed of the car was actually always less than the magnitude of the change, 42.4 m/s. This may seem counterintuitive; however, this result is correct and thus demonstrates the importance of including direction in these calculations.

Gravity Near the Surface of the Earth

Observations, dating back to at least Galileo's time (~1600), show that when the effects of the air are negligible, **objects fall to Earth with a constant acceleration**, regardless of their mass and initial velocity. Careful measurements show that the acceleration, accurate to much better than 1 percent around the world, is 9.8 m/s^2, and by definition, it is in the direction "down." That acceleration is commonly referred to using a lowercase g.

Newton's second law then says that the force of gravity on these falling objects, F_g, must be in the downward direction, with a magnitude $F_g = mg$. This result can be used as a model for the force of gravitation near the surface of the earth. The same model, with appropriate values substituted for g, can be used for the surface of other planets and their moons, where in each case, "down" is toward the center of the planet or moon.

To digress just a bit, when large distances are considered, such as the distance from the earth to the moon, this model is no longer valid, and a better model, such as Newton's universal law of gravitation, is necessary. With that better model, the moon's acceleration toward the earth is found to be 0.0027 m/s^2, far less than 9.8 m/s^2. Note also that even though the moon's acceleration is always directly toward the earth, the moon "falls around" the earth and never travels directly toward it. The moon remains safely in orbit even though it is constantly accelerating toward the earth.

Figure 5.1 If an object does not fall, there must be a force equal to its weight pulling up on the object.

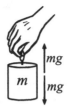

Near the surface of the earth, the simple gravitational model, $F_g = mg$ downward, is fine for falling objects – or "projectiles" more generally – but what if you hold an object stationary with your hand? In that case, the object's acceleration is clearly zero, and hence the force on the object must also be zero. If you let go of the object, it falls with an acceleration g. Does gravity suddenly turn on when you let go? Surely, holding a heavier or lighter object does not feel the same, even though neither may be accelerating, and intuitively, it does not seem appropriate to have gravity turn on and off in such a manner. Instead, the force of gravity is taken to be unchanged, and some other contribution is found in the opposite direction that makes the **total force** on the object zero. Thus, to keep an object of mass m from accelerating downward as a result of gravity, a force mg in the upward direction must be supplied (e.g., see Figure 5.1). The magnitude of this upward force necessary to keep the acceleration at zero is what we call the *weight* of the object. When the object is released, the upward force on the object "turns off," and the total force becomes mg in the downward direction.

The idea that the "total" or "net" force on an object can be divided into separate, noninteracting contributions ascribed to different sources is a great convenience and seems to work very well. This idea also allows forces to be determined even in the absence of motion – such as the force needed to hold an object stationary – by balancing an unknown force with a known force. Note the very important facts that when adding the forces together, the directions of each of the forces must be taken into account, and the force referred to in Newton's laws is always the total force from all contributions acting on an object – an object can have but one acceleration. In fact, since only the total force will ultimately be important, all forces can be divided into as many pieces as is convenient. As long as the sum is unaltered, there will be no change in what happens, so if it is more convenient, why not?

Hooke's Law for Springs

Among many important experimental results, Robert Hooke (1635–1703), an English scientist and contemporary of Isaac Newton, examined the behavior of many types of springs, at least in part in an effort to make better clocks to aid in maritime navigation. When navigating by the stars, it is relatively easy to determine one's latitude, but in order to accurately determine longitude, it is necessary to know what time it is. Each one-minute error in time can correspond to about a 10- to 15-mile error in position. In Hooke's time, ocean voyages across the Atlantic could last a month or more, so an error of even a few seconds per day could become quite significant. In the process of his investigation, Hooke came up with what is now called **Hooke's law for springs**.

Hooke's law is not really a law of nature but rather a mathematical model for the behavior of springs that seems to work reasonably well. In simple terms, the model says that the additional force needed in order to stretch (or compress) a spring from a given starting point is proportional to the distance of the stretch (or compression). The result is usually expressed in terms of the force provided by the spring. From Newton's third law, that force has the same magnitude, and is in the direction opposite to, the force needed to stretch (or compress) the spring.

Hooke's law can be expressed mathematically as follows. If you change the length of the spring by an amount Δx, where a positive value means an increase in length (a stretch) and a negative value means a decrease (a compression), then the change in the force provided by the spring, ΔF_s, the **return force**, is given by

$$\Delta F_s = -k\,\Delta x, \tag{5.1}$$

where k is a positive constant and its value depends on which spring is being used.

The constant k in this context is referred to as the **spring constant**, and it will have SI units of N/m = kg/s^2. A larger spring constant corresponds to a stiffer spring since more force is needed to make the same change in length. The minus sign in the equation symbolizes the fact that the change in the force from the spring is in a direction to return the spring to where it was. If a particular spring's length is increased by pulling to the right – that is, Δx is positive as you pull to the right – then the additional force from the spring will pull back to the left.

A spring that follows Hooke's law is sometimes referred to as an **ideal spring**. Most real springs follow the law reasonably well as long as they are not stretched or compressed too far. That is, real springs can usually be modeled as ideal springs, and Hooke's law can then be applied if the spring constant is known.

The spring constant for a particular spring can be determined experimentally using known forces and the measurement of distance. For example, consider a spring with (unknown) spring constant k that first supports a known mass M_1 and then, in turn, M_2 against the pull of gravity (see Figure 5.2). In each case, the mass is stationary – any oscillations are allowed to die out. When the mass is stationary, its acceleration is zero, and so the total force must be zero. The vertical position of the end of the spring is measured for both cases, resulting in values x_1 and x_2. When computing a change, the first value is subtracted from the second, so the change in position is $\Delta x = x_2 - x_1$. The force of the spring on the masses in each case must be equal and opposite to the force of gravity. Hence, $\Delta F = (M_2 - M_1)g$. Then the spring constant is given by

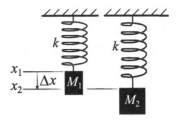

Figure 5.2 A spring constant can be determined by applying two different forces and measuring the change in length of the spring.

$$k = \frac{\Delta F}{\Delta x} = \frac{(M_2 - M_1)g}{x_2 - x_1}. \tag{5.2}$$

Note that the resulting spring constant should always be positive.

Example 5.3

When a 20-g mass is hung on a spring, the spring length is 30 cm. When 40 g is hung, the spring length is 40 cm. What is the spring constant?

$$k = \frac{(0.04 \text{ kg} - 0.02 \text{ kg}) \times 9.8 \text{ m/s}^2}{(0.40 \text{ m} - 0.30 \text{ m})} = 2.0 \frac{\text{kg}}{\text{s}^2}.$$

The measurement could be done with $M_1 = 0$ – that is, with no mass at all. If the spring remains within the range where the ideal spring model is valid, the position of the end of the spring when no mass is attached can then be defined to be the position $x = 0$, and increasing x (more stretch) would be in the downward direction. With these additional conditions, Hooke's law can be written simply as $F = -kx$, a form commonly seen. This form can also be used for a mass hanging on a spring if $x = 0$ is taken to be the equilibrium position. At equilibrium, the pull downward due to gravity is canceled by the pull upward due to the spring, so in that case, $F = -kx$ will represent the total combined force of the spring and gravity – gravity has seemingly disappeared from the problem when this specific choice for $x = 0$ is used.

The Harmonic Oscillator

The simplest oscillator is the one-dimensional **harmonic oscillator**. The harmonic oscillator serves as a useful prototype for virtually all oscillating systems. By definition, for a harmonic oscillator, something changes in time with a sinusoidal time dependence. What changes depends on the particular type of oscillator. It is common to start by considering an ideal mass and spring system where the changes are of the position (and velocity) of the mass.

Mass on a Spring

Consider a mass hanging on a spring. If the system is started by releasing the mass from a position away from equilibrium, there will be a return force pulling the mass toward the equilibrium position (see Figure 5.3). If there is a force, there is acceleration in the same direction, so the mass will build up speed. However, when the mass reaches the equilibrium position, the force is now zero, and hence there is nothing to stop the mass at that point. In the absence of a force, the mass continues to move with constant velocity right through the equilibrium position. Once the mass is past the equilibrium, the force now turns around the other way and slows the mass first to zero, then in the reverse direction back toward the equilibrium. The mass shoots past the equilibrium as before, and the process repeats. In the absence of any friction or air resistance, this periodic motion will continue indefinitely.

Rather than attempt a detailed derivation of the frequency of such an oscillation, consider the following analysis based only on the dimensions (the units) of the quantities involved. There are three potentially important contributing factors: the acceleration due to gravity, g; the spring constant, k; and the mass M. These are combined to get a period, T, or frequency, $f = 1/T$. When the quantities are combined, their units must combine in the same way. If the two sides of the equation are to be equal, the units on both sides of the equal sign must match. So, make a guess that the formula is

$$T = g^p \, k^q \, M^r, \tag{5.3}$$

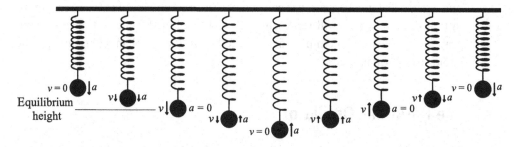

Figure 5.3 An illustration of the motion of a mass hanging on a spring when released from rest at a point just above the equilibrium height.

where the exponents, p, q, and r are values to be determined. Looking only at the units, this formula can be correct only if

$$s = \left(\frac{m}{s^2}\right)^p \left(\frac{kg}{s^2}\right)^q (kg)^r = m^p \ kg^{q+r} \ s^{-2(p+q)}, \tag{5.4}$$

so it must be that $p = 0$, $q + r = 0$, and $p + q = -1/2$. Remember that an exponent of 1/2 is the same as the square root. There is only one such combination that can work. It must be the case that

$$T = const.\sqrt{\frac{M}{k}} \quad \text{or} \ f = \frac{1}{const.}\sqrt{\frac{k}{M}}, \tag{5.5}$$

where "const." is, at this point, an unknown numerical value that has no units and is the same for all such problems. A more detailed analysis[1] shows that in this case, that constant should be equal to 2π.

Example 5.4

If a mass $M_1 = 0.10$ kg is hung on a spring, and the frequency of oscillation is $f_1 = 2.5$ Hz, what new mass, M_2, should be used to get a new frequency of $f_2 = 3.0$ Hz?

This is readily solved by creating a ratio to get rid of the constant, then putting all the known quantities on one side of the equal sign and the unknowns on the other:

$$f_1 = const.\sqrt{\frac{k}{M_1}}, \ f_2 = const.\sqrt{\frac{k}{M_2}} \rightarrow \frac{f_1}{f_2} = \sqrt{\frac{k}{M_1}\frac{M_2}{k}} = \sqrt{\frac{M_2}{M_1}} \rightarrow \sqrt{M_2} = \frac{f_1}{f_2}\sqrt{M_1}.$$

Square both sides to get rid of the square root, and then put in the known values to get

$$M_2 = \left(\frac{f_1}{f_2}\right)^2 M_1 = \left(\frac{2.5 \text{ Hz}}{3.0 \text{ Hz}}\right)^2 0.10 \text{ kg} = 0.069 \text{ kg}.$$

For a mass and ideal spring, with motion in only one dimension and ignoring friction, the position of the mass at different times follows a sinusoidal time dependence. For example, if the mass starts from rest (velocity = 0) at position $x = A$, where $x = 0$ is the equilibrium position, the position, x, of the mass at a later time, t, written as " $x(t)$," is given by[2]

---

[1] The constant can depend on the units used. Here, SI units are assumed for all quantities.
[2] In this context, the t in the parentheses of $x(t)$ is there to simply remind the reader that the position x depends on time.

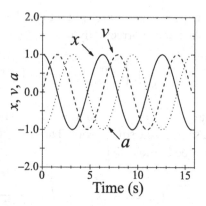

Figure 5.4 The relative time dependencies for the position, x; velocity, v; and acceleration, a, of the oscillating mass on a spring. Here, the graph units are chosen so that the maximum for each is 1 unit.

$$x(t) = A \cos(2\pi ft) = A \cos(2\pi t/T) = A \cos(\omega t), \qquad (5.6)$$

where the "angular frequency" $\omega = 2\pi f$ is being used. If f is in hertz, then ω is in radians per second, or simply 1/s or s^{-1}. From this expression, the velocity, v, and acceleration, a, of the mass can be determined to be

$$v(t) = -\omega A \sin(\omega t); \quad a(t) = -\omega^2 A \cos(\omega t) = -\frac{k}{M}x(t). \qquad (5.7)$$

It is straightforward to show that these solutions satisfy $F = Ma$ for this problem. These time dependencies are illustrated in Figure 5.4. Note, for example, that when at the equilibrium position, the acceleration is zero, but the speed is a maximum.

Near the surface of Earth, the spring constant and/or the mass can be determined using a stationary system, where the downward pull of gravity is balanced by the upward pull of the spring. In the absence of gravity, the oscillation frequency can be used. If the oscillation frequency is measured for a known mass, the spring constant can be determined, and if the spring constant is known, the frequency can be used to determine the mass.

While the "mass on a spring" serves as a basic prototype harmonic oscillator for understanding many oscillating systems, other systems may not have a spring or an easily identified mass, and they may not even be mechanical in nature. However, those systems will have components that play these roles. The more general form to remember for the frequency of oscillation is

$$f = \text{const.} \sqrt{\frac{\text{return force per unit displacement}}{\text{inertial constant}}}, \qquad (5.8)$$

where the inertial constant is what tries to keep the system moving when in motion, and the return force acts similarly to the spring, in that it tries to move the system back to an equilibrium position and exhibits a larger force the farther the system is from that equilibrium. The constant out front may change, depending on the circumstances and units used.

The Pendulum

The simple pendulum, a mass on a string that is free to swing, is a simple oscillating system. The result for the mass on a spring (Equation [5.8]) can be used to determine a pendulum's oscillating frequency. If the mass of the string is negligible, the inertial term must just be the mass, M. The position of the mass can be described by an angle, θ, as shown in Figure 5.5. The displacement from the equilibrium position is the arc length, $s = L\theta$, when θ is expressed in radians.

The return force is that part of the total force that pulls the pendulum toward $\theta = 0$. The forces here are the downward pull due to gravity, Mg, and the upward pull at an angle by the string, F_s. A convenient way to find the return force is to divide the downward force due to gravity into two pieces, which, when added together, give Mg in the downward direction. The pieces chosen include one that is along the direction of the string that simply pulls against the string and will not change the displacement and one at a right angle to the first that is along the direction of motion. This latter piece is the return force.

Trigonometry and plane geometry can be used to show that the return force is $Mg \sin\theta$. For harmonic motion, the return force should be proportional to the displacement. This is achieved approximately by the pendulum if the angle is kept small enough. An approximation that can be verified with a hand calculator is that if θ is in radians, and θ is significantly smaller than 1 radian (about 57°), then $\sin\theta$ is approximately equal to θ, or $\sin\theta \approx \theta$. The smaller the value of θ, the better the approximation.[3] Hence, as long as the angle is not too large, the return force can be written as $Mg\theta$, making the problem look just like the mass on a spring. The displacement is $L\theta$, so the predicted frequency is

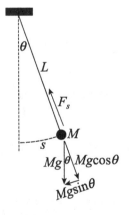

Figure 5.5 Geometry for a simple pendulum.

---

[3] See Appendix A. Additional approximation techniques are discussed in Chapter 11.

$$f = \frac{1}{2\pi} \sqrt{\frac{Mg\theta/L\theta}{M}} = \frac{1}{2\pi} \sqrt{\frac{g}{L}}. \tag{5.9}$$

It is interesting to note that the frequency does not depend on the mass since both the return force and the inertia depend on M. If the motion is small, there is also no dependence on the amplitude of the motion.

Despite the approximation, the result for the pendulum is accurate and reproducible enough that pendulums have been used extensively for timekeeping. The measurement of the frequency (or period) of pendulums[4] is also one method that has been extensively used to measure the small variations in g that occur around the world.

The Tuning Fork

The **tuning fork** is a stable and highly portable device used to provide a reference at a fixed frequency. The invention of the tuning fork is attributed to the British musician John Shore in about 1711. It consists of two tines, or prongs, connected in a U-shape (see Figure 5.6). When struck, the two tines move in opposite directions with a (nearly) sinusoidal time dependence.

A derivation of the frequency of oscillation for a tuning fork is beyond the level of the presentation here, but the basic behavior is similar to that of the harmonic oscillator. The return force will be due to the properties of the material used and the thickness and shape of the tines. A thicker rod is clearly harder to bend than a thinner rod, and it will have a larger return force for the same displacement. The inertial part will depend on the mass of the tines, with more emphasis on the mass at the ends because at that point the motion is greatest. Since the mass and return force both depend on the thickness, shape, and material used, it is more difficult to separate the two effects. Nevertheless, the frequency can be lowered by using longer or thinner tines or by adding mass at the ends. The frequency can be increased by shortening the tines or by making them thicker.

Figure 5.6 A tuning fork is a simple mechanical oscillator that can be used as a frequency standard.

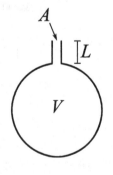

Figure 5.7 Geometry of a simple Helmholtz oscillator.

---

[4] *Pendulums* is usually considered the correct plural of *pendulum*, although sometimes *pendula* is used.

The Helmholtz Resonator

A **Helmholtz resonator** consists of an enclosed volume of air, V, with a smaller neck or pipe to the outside. Named for Hermann Helmholtz (1821–1894), these resonators are acoustic examples of the mass-on-a-spring problem.

The elements needed for a more detailed derivation of the return force will appear in Chapter 11, after the appropriate gas laws have been discussed. To understand the basic idea, however, consider a balloon of volume V. When it is compressed a bit, it will provide a return force back toward equilibrium. If the compression is not too large, then the return force will be roughly proportional to the amount of compression. A similar return force is present in the resonator as air is added or removed through the opening. Adding a small amount of air causes a compression of the air already inside. During that small compression, the air within the volume V does not need to move very much, so the inertia associated with that motion is not so important. However, as the air streams in and out through the narrow neck, that air moves almost like a piston and provides the inertia.

Referring to Figure 5.7, a simplified result for the natural oscillation frequency of the resonator is given by

$$f = \frac{c}{2\pi} \sqrt{\frac{A}{VL}}, \qquad (5.10)$$

where V is the internal volume, A is the cross-sectional area of the neck, L is the "effective length" of the neck, and c is the speed of sound. The speed of sound includes the relevant properties of the gas (e.g., air). It is difficult to predict accurate results from this equation since the "effective length" of the neck is not always well known – and it is somewhat of a fudge factor to begin with. The effective length takes into account the inertia of the air not just inside the neck but also a bit beyond both ends of the neck, where the air must also be moving somewhat rapidly. Thus, the effective length will be longer than the actual length.

An empty soda or wine bottle is a simple everyday example of such a resonator, although it is crude compared to those designed and used by Helmholtz. By blowing appropriately across the top, you can estimate the natural oscillation frequency. If the bottle is then partially filled with water, the volume, V, is decreased, and so the frequency goes up.

The ocarina, a simple, small instrument in the flute family, is also based on a Helmholtz resonator. The resonator, which is most important for determining the pitch, consists of a volume to hold gas but with many holes of varying sizes to the outside. As the holes are covered and uncovered by the player's fingers, the

Figure 5.8
Example of a simple ported speaker enclosure.

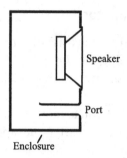

total area, A, to the outside is changed. Using combinations of different-sized holes, the different notes of the musical scale can be produced, at least approximately. Unlike almost all other flutes, the locations of the finger holes for an ocarina are largely irrelevant for tuning purposes.

Some speaker enclosures include a pipe, or "port," to the outside (see Figure 5.8). Such an enclosure can act as a Helmholtz resonator. A longer length of pipe (larger "L") will decrease the resonant frequency of the enclosure, which can improve, or at least change, the bass response.

In architecture, Helmholtz resonators can be included as part of the construction, for example, by having small openings in a wall, with a larger enclosed volume within the wall, to influence the acoustics of a room. The resonator will preferentially absorb acoustic energy for frequencies near its resonant frequency. Even an appropriately sized and shaped vase, or other similar decorative object, can be used for this purpose.

The resonant behavior of a violin, acoustic guitar, and many other stringed instruments will include the response from the internal volume of air that is connected to the outside through small openings. The violin has two openings, which are traditionally a stylized S-shape, whereas guitars and many other similar instruments usually have a single circular opening. However, the full resonant behavior of these instruments also includes a significant motion of the walls of the resonator, making it a much more complex problem than that of a simple Helmholtz resonator.

Electronics

In electronics, there are devices called *inductors* and devices called *capacitors*. A simple inductor can be made from a wire wound into a coil called a *solenoid*, a spring-like shape. A capacitor consists of two electrically conducting surfaces ("plates") that are near each other but are not connected. There will be more discussion of these devices in Chapter 18. In electronics, one measures "voltages," which are somewhat analogous to mechanical forces, and "currents," which are somewhat analogous to mechanical motion in response to a force. If an inductor and capacitor are simply connected together, as illustrated in Figure 5.9, the

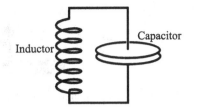

Figure 5.9 Schematic of an electronic version of a simple harmonic oscillator.

combination becomes an electronic version of a simple harmonic oscillator, where the measured voltage and current will oscillate sinusoidally in time. The inductor plays the role of the mass, and the capacitor plays the role of the spring. Since inductors are commonly referred to using a variable L and capacitors using C, this circuit is referred to as an *LC circuit*.

This simple electronic circuit is the basis for Italian inventor Guglielmo Marconi's (1874–1937) earliest radio transmitters and receivers. The basic physics had been previously studied and developed in some detail, as a laboratory demonstration, by the German scientist Heinrich Hertz (1857–1894). Hertz designed his device to provide a test of the then-new electromagnetic theory of the Scottish scientist James Clerk Maxwell (1831–1879); however, Hertz thought the device had no practical use. In 1909 Marconi was honored with a Nobel Prize in Physics in recognition of his work on long-distance radio communication. Radio electronics is also used to create eerie science fiction sounds, such as with a theremin (see Chapter 20).

The Driven Harmonic Oscillator – Resonance

In use, some oscillators are simply disturbed from equilibrium and then allowed to oscillate freely. In other cases, an oscillator may be subjected to the continuous application of a time-dependent force, and the response to that force is of consequence. Since the basic underlying physics of the oscillator is the same for both, there are strong connections between the two different behaviors.

An oscillator driven periodically at, or near, its natural frequency will be of particular interest. That will give rise to a phenomenon known as **resonance**. A common picture to have in mind, which will be used here with some later qualification, is to imagine a person swinging on a swing set. Pushing at the natural frequency of the swing will give rise to a large response. In that case, the swing is being driven at its resonant frequency. Pushing much more slowly or much more rapidly will give only a small response – it is not nearly as effective.

When considering a driven harmonic oscillator in detail, it becomes imperative to add some form of friction, or "damping," to the model. This will help avoid impossible solutions (e.g., infinities) and, at the same time, will make the model more realistic. Friction is a velocity-dependent

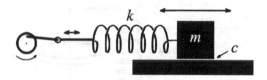

Figure 5.10 Driven mass on a spring oscillator, with friction.

force that opposes motion and that tends to make the velocity go to zero. In general, friction can be very complicated. For the purposes here, the actual model for friction used is not all that important. It is only important that some energy be lost to counteract the energy supplied by an applied force.

To model the situation, consider once again the mass on a spring but with a source of friction added (see Figure 5.10). That might be from air resistance, from mechanical sources, or a combination. For the ideal spring, which tries to return the position to zero, the force is $-kx$. Hence, for friction, a simple model might be $-cv$, where c here is a friction constant, and this will tend to return the velocity to zero.[5] A larger value of c corresponds to more friction.

Since the basic motion of the oscillator is sinusoidal in time, and in light of the results from Chapter 4 regarding the frequency domain, consider the case where the applied force is sinusoidal in time with frequency f. The so-called "natural frequency" of the harmonic oscillator is labeled f_0.

Putting the various forces together and using Newton's second law gives the following:

$$F_s = \text{Force due to spring} = -kx,$$
$$F_f = \text{Model for force due to friction} = -cv,$$
$$F_d = \text{Driving force} = F_0 \cos(2\pi ft),$$
$$F_s + F_f + F_d = ma,$$

where equilibrium at the position $x = 0$ is assumed, v is velocity, and a is the acceleration. Here, the amplitude of the sinusoidal driving force is F_0. It is common to rearrange Newton's second law so that all the terms that explicitly involve the motion of the mass are on one side; all the rest are on the other. Hence, the **equation of motion** is given by

---

[5] This model is appropriate for air resistance at slow speeds and has the advantage that, when used, the driven oscillator problem is readily solvable. Another model for friction will be discussed later, in Chapter 10, in the context of nonlinear physics. The constant c here is being reused for a new purpose – it is not the speed of sound.

$$ma + cv + kx = F_0 \cos(2\pi ft) . \tag{5.11}$$

Methods to find solutions to this equation of motion are well known. There are two pieces to the solution, a **transient solution** that solves

$$ma + cv + kx = 0 \tag{5.12}$$

and a **steady-state** solution that solves the original equation and is what is present after the system has been running for a while. The transient solution is relevant for some time after the force is *changed*, for example, when the force is turned on and off. The total solution is simply the sum of the transient and steady-state solutions.

There are three different types of transient solutions, depending on how much friction is present. The solutions are categorized as follows:

Overdamped
Friction is large – left on its own, the system will not even oscillate.
Underdamped
Friction is small – left on its own, the system will oscillate for a while when disturbed from equilibrium.
Critically Damped
Friction has a certain critical value, which is the dividing line between the other cases.

If a new variable, $\gamma = c/2m$, is defined, critical damping occurs when $\gamma = 2\pi f_0 = \omega_0$.[6] Larger values give rise to overdamping, and smaller values result in underdamping. The motion for several different values of damping are shown in Figure 5.11. To have a system near critical damping is important in many applications, although perhaps not any that are musically related. When a system with damping near the critical value is disturbed from equilibrium, it has the fastest return to equilibrium. So, for example, the combination of the shock absorbers and the springs on a car would be designed to respond quickly, without significant oscillation, and thus would be designed to be near critical damping.

Musically, the underdamped case, far from critical damping, is of particular interest. For that case, the transient solution will have a significant number of oscillations before losing most of its energy. The solution for the transient term looks like Figure 5.12a and is described mathematically as

$$x(t) = A_0 e^{-\gamma t} \cos(2\pi f_0 t - \varphi_0) , \tag{5.13}$$

where the two constants, A_0 and φ_0, depend on the starting conditions at $t = 0$. The main addition due to friction is the exponential piece, $e^{-\gamma t}$, which causes the

---

[6] The variable c is specific to the mass-on-a-spring problem. The variable γ, which expresses the ratio of the loss to inertial terms, can be used more generally. As before, ω_0 is the natural frequency in radians per second.

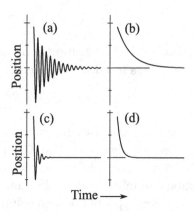

Figure 5.11 Position as a function of time for a harmonic oscillator for different damping conditions: (a) and (c) are underdamped; (b) and (d) are overdamped; (c) and (d) are nearer to critical damping than are (a) and (b).

amplitude to decrease exponentially with time as the friction "burns up" the oscillator's energy. These underdamped transient solutions will be discussed in more detail.

The steady-state solutions for the *driven* underdamped case look very similar and are given by

$$x(t) = A \cos(2\pi f t - \varphi), \tag{5.14}$$

where f is the driving frequency, and

$$\left.\begin{array}{l} A = \dfrac{F_0/m}{\sqrt{\left(\omega_0^2 - \omega^2\right)^2 + 4\gamma^2\omega^2}}, \\[2ex] \varphi = \tan^{-1}\left(\dfrac{2\gamma\omega}{\omega_0^2 - \omega^2}\right), \end{array}\right\} \tag{5.15}$$

where once again, the angular frequencies are used to shorten these already-complicated expressions. Here, A gives the overall amplitude, and the **phase angle**, φ, is a measure of the time delay between the sinusoid of the driving force and the sinusoid of the response. It is usually sufficient to know that these results (Equations [5.14] and [5.15]) exist and refer to them as needed. However, the general behavior that is represented by these equations is important. That behavior is shown graphically in Figure 5.12. In these plots, A is in units of F_0/k, and γ is in units of ω_0.

The amplitude of the steady-state response becomes large when the driving frequency equals the natural frequency (i.e., where $f/f_0 = 1$). This behavior, when a large response occurs as a result of the presence of a driving force at a natural frequency, is what is generally referred to as *resonance*. The phase angle, φ, makes

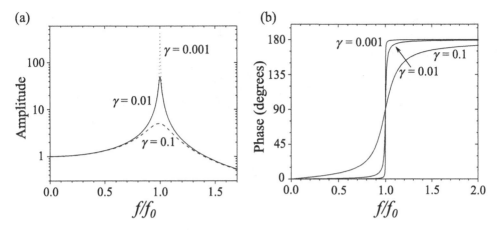

(a)

(b)

Figure 5.12 A graphical representation of the relative amplitude and phase of the response of a harmonic oscillator to a sinusoidal driving force. Here, f_0 is the natural frequency of the oscillator. Note that a logarithmic scale is used for the amplitude, A.

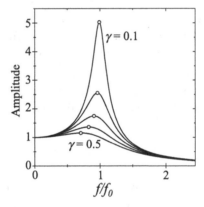

Figure 5.13 Resonant response when the damping approaches critical damping.

a transition from $0°$ at low frequencies to $180°$ at high frequencies, with the halfway point, $90°$, at the natural, or resonant, frequency.

With even more damping, the steady-state solutions become a bit more complicated, but the solution is known (see Figure 5.13). The main difference is that the peak position moves to lower frequencies. This effect is important for the design of some speaker enclosures, for example, where sound-absorbing material is included in the enclosure.

Quality Factor

Returning to the case with smaller damping, the width or "sharpness" of the frequency response is an important characteristic. Since the curves are smooth, some definition of where the width is to be measured is necessary. A common

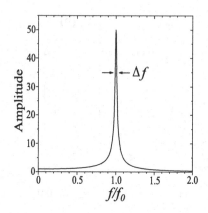

Figure 5.14 The half-power points can be used to define a width of a resonance.

choice is to measure at the **half-power points**. Since the energy of a harmonic oscillator is proportional to its amplitude squared, and power is energy per second, the half-power points will be where the amplitude squared is one-half the maximum amplitude squared. That will occur when the amplitude is $1/\sqrt{2}$, or about 71 percent, of the maximum (see Figure 5.14).

The ratio of the width of the response to the frequency of maximum response, $\Delta f/f_0$, will be a unitless value that describes the relative width. The smaller this value, the "sharper" is the resonance, and the narrower is the range of frequencies where the response is large. The sharpness of a resonance is often expressed in terms of a **quality factor**. The quality factor, Q, for a resonance is defined as[7]

$$\frac{1}{Q} = \frac{\Delta f}{f_0}, \tag{5.16}$$

so a narrower resonance is considered to have a higher quality factor, Q. The quality factor is a measure of the relative size of the damping (γ) compared to the natural frequency (ω_0) and generalizes to all such resonances. Everything else being equal, a higher quality factor means less friction.

Since the quality factor involves characteristics of the oscillator only, and not of the driving force, it can also be useful for characterizing the transient solution. For example, as an approximation, **Q will be the number of oscillations made by a system during the time it takes for the amplitude of the transient to decay most of the way**. "Most of the way" is a bit vague, but somewhere around 90 percent of the initial amplitude is appropriate. Consider the example shown in Figure 5.15, where the parameters used correspond to $Q = 10$. Simply counting the oscillations until the signal is about 10 percent of its starting value provides an estimate of Q to be about 10 ± 2, which is pretty close for such a simple method.

If the number of oscillations to die down (by about 90 percent) is roughly Q, and the time per oscillation is $T = 1/f_0$, then the time to die down is roughly Q/f_0.

---

[7] This definition is strictly valid for smaller damping (less friction) only.

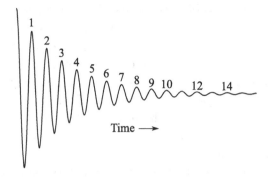

Figure 5.15 The amplitude of oscillation as a function of time for an oscillator with $Q = 10$. Note that roughly 10 oscillations occur before most of the energy is gone.

Example 5.5

If a 440-Hz tuning fork takes five seconds to die down most of the way, what is its quality factor and half-power frequency width?

The number of oscillations during five seconds is 5 s × 440 Hz = 2,200 so $Q \approx 2,200$. If this fork were to be driven by a sinusoidal driving force, the width of maximum response, Δf, would be about $f_0/2,200 = 0.2$ Hz.

Note that the determination of the behavior of the time-dependent transient solution also determines the behavior of the driven oscillator as the driving frequency changes. The two solutions are connected.

Aside from the constants determined by the starting values, the transient solution is the same whether or not there is a driving force. Hence, once a driving force is turned on, the time needed before the system reaches steady state will be about Q/f_0, the time for the transient solution to die down most of the way.

It is worth pointing out another relationship that will be considered more generally in Chapter 9. The timescale for the presence of the transient solution is $\Delta t = Q/f_0$. The half-power frequency width for the steady-state solution is $\Delta f = f_0/Q$. Multiplying the left and right sides of these two relations, it is then always the case that $\Delta f \, \Delta t = 1$. A system with a wider frequency response will always respond faster in time, and vice versa.

Sympathetic Resonances

While it is important to understand the results for the harmonic oscillator driven by a single sinusoidal source in order to make sense of at least some of what follows, there are only a few musical examples where this actually occurs. One such case is called **sympathetic resonance**. Sympathetic resonance occurs when there are two (or more) oscillators with natural frequencies that are the same within their respective frequency ranges (Δf). The motion from one oscillator can act as a driving force for the other. For example, if you put two identical pendulums next to each other on a common support, and if that support is not completely rigid, then the motion of one pendulum can drive the other pendulum, and vice versa. The piano, and its cousins, as well as some older traditional string instruments, such as the sitar, use sympathetic resonances to modify their sound.

Sympathetic resonances can be easily demonstrated using a piano. Normally, when no keys are depressed, dampers press on the piano strings to silence them. If the right-most foot pedal (the "sustain pedal") is depressed, the dampers are removed from all the strings. If, with that pedal depressed, a singer or instrumentalist next to the piano sings or plays a note for a few seconds and then stops, the tone will continue to arise from the piano. Releasing the damper pedal stops the sound. What happens is that the corresponding string or strings that match the note played, and its overtones, will be driven into oscillation by the sound. Once the driving force is stopped, it takes a while for the string motion to stop. The time it will take for the string motion to build up and then die down can be determined experimentally by plucking the string with the damper removed and listening to how long it takes for the sound to die down "most of the way."

A second example with the piano requires access to the strings. If the sustain pedal is pressed and a key is struck, and then the specific string or strings that were struck are damped, say, with a finger, the strings that oscillate at harmonic frequencies will continue to ring. The overtones of the struck string provide a periodic driving force at harmonic frequencies. More details of string vibrations will be considered in Chapter 6.

Many of the notes in a piano use multiple strings tuned as a group. For some pianos, the middle pedal causes the hammer mechanism to move sideways, so the hammer does not hit all of the strings in the group. This causes a change in the tone quality. A string that is not struck will, in time, gradually build up in amplitude as a result of sympathetic resonance with its neighbors that were struck. The gradual buildup at the beginning makes the notes sound like they were hit with a softer hammer.

Sympathetic resonances also occur between the bars used for xylophones, marimbas, vibraphones, and the like, as well as in the air column in the tuned resonating pipes near the bars. These resonances are discussed more in Chapter 7.

Most musically related examples of oscillators and resonance will be more complicated than the simple mass on a spring with or without a sinusoidal driving force. Those complications will be addressed in the next couple of chapters. However, most of those more complicated situations can be successfully explained, starting from the simple mass on a spring.

Fictitious Forces and Physics Demonstrations (Optional)

Some of the simpler physics demonstrations used to show the oscillatory behavior discussed previously may appear to be somewhat of a cheat. For example, the driven oscillator can be demonstrated using a pendulum, held at the top by hand. As the top support of the pendulum is moved horizontally with a periodic time dependence, the resonant behavior (amplitude and phase) of the mass can be observed. A similar demonstration will use a mass on a spring and a vertical hand motion. The cheat is, of course, that the top is supposed to be stationary – how does moving the top of the pendulum (or spring) equate to a periodic force on the mass at the bottom? A common argument, which should be greeted with skepticism, is that the motion is "kept small," so it is negligible. However, there is a better explanation.

Newton's laws are designed for use in **inertial reference frames**.[8] For a typical simple pendulum measured on the surface of the earth, the reference frame (i.e., coordinates fixed in the room in which the experiment is conducted) is close enough to being inertial that Newton's laws can be expected to work well. A reference frame that is accelerating (relative to an inertial frame) will not be an inertial frame. If Newton's laws are used in a non-inertial frame, **fictitious forces** arise – that is, extra forces need to be added to the problem to compensate. These are not real forces.

Example 5.6

If you are driving in a car at a constant speed while you make a 90° left turn, the seat of the car will apply a force on you that causes you to turn left with the car. That real force of the seat on your body is called the **centripetal force**. However, if you use the reference frame of the car, an accelerating reference frame, it will feel like there is a force on your body to the right – everyone in the car will tend to lean right when you turn left. That behavior can be "explained" using a fictitious force. Such a force does not exist in the inertial reference frame. There is no apparent source, or cause, for a fictitious force, and hence these forces do not obey Newton's third law. For this type of circular motion, the fictitious force is often referred to as the **centrifugal force**. Another well-known fictitious force associated with rotation is the **Coriolis force**, important for Earth's weather systems.

---

[8] It is not obvious that any such reference frames actually exist. Turning the argument around, an inertial frame is one in which Newton's laws work. In many cases, there are reference frames where those laws work extremely well, so the corrections necessary due to noninertial behavior are small enough to neglect.

Example 5.7

Astronauts viewed in the reference frame of the International Space Station (ISS) appear to be weightless. However, the force on the astronauts due to Earth's gravity is almost the same as it is on the ground. In the orbiting reference frame, the fictitious force matches that of the gravitational force, but it is in the opposite direction. Hence, most of Earth's gravitational force appears to be canceled out. The remaining gravitational forces are referred to as *microgravity*. Those forces include attraction to other masses on the ISS (walls, other astronauts, etc.), as well that due to the gradient of Earth's gravity. That is, the pull from Earth is stronger by a very small amount on the side closer to Earth, and it is weaker on the far side. That gradient can result in a rotational force, a torsion, on the astronauts.

In a reference frame that is stationary on the surface of the earth, such fictitious forces are small (compared to other everyday forces) but can be observed for pendulums if the motion is monitored over a longer time period ($\gtrsim 20$ min). A common apparatus for such an observation is the so-called **Foucault pendulum**.[9] A Foucault pendulum is a pendulum where precautions have been taken in the construction to eliminate other small effects that can mask those of the small fictitious forces that arise as a result of Earth's rotation. One way to help accomplish this is to use a very long pendulum, which is why Foucault pendulums are (usually) quite long. In fact, the adoption of the heliocentric model of the solar system – the claim that the sun is the center of our solar system – is really saying that if you use the sun as the center, with the planets orbiting around it, the coordinate system is closer to being inertial than if another coordinate system is chosen, say, one that is geocentric, or Earth centered.

Returning to the "cheat," there is a theorem from advanced mechanics that shows that motion in any accelerating frame, when described mathematically within that frame, can be treated using Newton's laws, but with the addition of appropriate fictitious forces. If you imagine yourself in that accelerating frame, the fictitious force is the extra force that exists only because of the acceleration of the frame.

For the problem where a pendulum is driven using a horizontal motion at the top, the problem can be viewed in a reference frame that also moves horizontally in such a manner that the top of the pendulum is stationary – imagine watching images from a video camera attached so that it moves along with the motion at the top. To describe the motion in that reference frame, the mathematics is equivalent to a pendulum fixed at the top but with the addition of a (fictitious) horizontal driving force on the mass. That is, viewed in that non-inertial reference

---

[9] Named for the French scientist Jean-Bernard-Léon Foucault (1819–1868), often referred to as Léon Foucault. He is included among the 72 French scientists and engineers whose names are inscribed on the Eiffel Tower.

frame, the pendulum will act the same as if there were a (real) driving force on the mass in an inertial frame. Hence, such a demonstration is not as much of a cheat as it might first appear to be.

Summary

Newton's laws describe the motion of objects in terms of the net, or total, force on the object. When using Newton's laws, it is often useful to divide the net force into separate contributions, such as the force from the gravity of Earth, the forces from a stretched spring, or the directional components of forces in more than one dimension. Mass describes the quantity of matter, and more mass requires more force to change the motion.

Hooke's law is a model for a spring where the force supplied by the spring is proportional to the amount of stretch or compression of the spring. The proportionality constant is called the *spring constant*. A larger spring constant corresponds to a spring that is stiffer – that is, it requires more force to stretch and compress than does a spring with a smaller spring constant. The spring constant describes force per unit displacement.

The natural oscillations of a mass on a spring will be sinusoidal (harmonic), with a natural frequency that depends on the square root of the spring constant divided by the mass (Equation [5.5]). The general result that can be applied to other oscillators is that the oscillation frequency, f, is given by

$$f = \text{const.} \sqrt{\frac{\text{return force per unit displacement}}{\text{inertial constant}}}. \tag{5.8}$$

Other simple systems that act as harmonic oscillators include pendulums, tuning forks, Helmholtz (acoustic) resonators, and the simple electronic LC circuit important for radio communications.

A harmonic oscillator driven by a periodic force will exhibit resonant behavior – the response to the force becomes very large when the frequency of the driving force is very near to the natural oscillation frequency. The maximum size of the response is determined by the relative size of the damping (i.e., friction-like) forces, which can be quantified with a quality factor, Q. An undriven oscillator set into motion will make roughly Q oscillations before it has lost most of its energy.

If their natural frequencies are close, one harmonic oscillator can act as a driving force for others, resulting in a sympathetic resonance.

Newton's laws describe motion in inertial reference frames. If used in other frames, particularly those that accelerate or rotate, fictitious forces – a result that acts somewhat like a force but that has no cause – may be present.

ADDITIONAL READING

Many introductory physics texts cover Newton's laws, harmonic oscillators, and pendulums. Consult any such text. Examples include the following:

Knight, R. D. *Physics for Scientists and Engineers*, 3rd ed. Pearson, 2013. (See Chapters 5 and 14.)

Urone, P. P., and R. Hinrichs. *College Physics*. OpenStax, Rice University, 2020. (See Chapters 4 and 16.)

PROBLEMS

5.1 A car weighing 1 ton crashes head-on into a truck weighing 20 tons (see Figure P5.1). Consider three different cases: (a) the car is traveling at high speed and runs into a stationary truck, (b) the truck is traveling at high speed and runs into a stationary car, and (c) both vehicles are traveling at the same high speed when they collide. In each case, which vehicle experiences the larger force during the collision?

Figure P5.1

5.2 An astronaut in a spacecraft orbiting Earth appears to be "weightless." A typical spacecraft with astronauts aboard will be in a (nearly) circular orbit that is about 250 km above Earth's surface (6,600 km from Earth's center). What is the direction of the acceleration of an astronaut aboard such a spacecraft? What force is causing that acceleration? Why does the astronaut appear to be weightless even though there must be a force on the astronaut?

5.3 A ball is tossed straight up. What is the acceleration of the ball (a) when the ball is on the way up and is halfway to its highest point, (b) when the ball is at its highest point, and (c) when the ball is halfway back down?

5.4 On Earth, a simple pendulum 24.8 cm long has a period of 1.0 s. What is the period of that pendulum on the surface of the moon? How about Mars? The acceleration due to gravity on the surface of the moon is 1.63 m/s^2, and on Mars, it is 3.72 m/s^2.

5.5 Suppose you want to double the oscillation frequency of a mass and spring system. If you change only the spring constant, by how much should it change? If you change only the mass, by how much should it change?

5.6 A simple pendulum 24.8 cm long, just shy of 1 ft, has a period of 1.0 s. In 1901, members of the faculty of the Michigan College of Mining created a pendulum 4,440 ft long in a very deep vertical mine shaft. Roughly how long would it take for such a pendulum to make one full swing? According to a newspaper report of the event, the pendulum "came almost to a standstill" in 20 minutes. Estimate the quality factor for this pendulum.

5.7 The middle C (261-Hz) key of a piano is pressed and held to produce a sound that decreases approximately exponentially with time. (a) If it takes 10 s for the sound to decrease most of the way, approximately what is the quality factor, Q, for the motion? (b) If the string for the note two octaves higher has the same Q, how long will it take for its sound to decrease most of the way?

5.8 You hold both ends of a spring, one end in each hand, as you stretch the spring by 1 cm. The spring constant is 100 N/m. What is the total force you are applying to the spring? How does that compare with the forces applied separately by your left and right hands?

5.9 What is the phase difference, in degrees, between the driving frequency and the steady-state response of a harmonic oscillator if the driving force is at a frequency $\pm f_0/(2Q)$ away from the natural frequency, f_0? How does the amplitude of the oscillation at those points compare to the amplitude at f_0?

5.10 If a kilogram (kg) is a unit of mass but weights are forces, what does it really mean when someone says, "I weigh 66 kg"?

6   String Theory

When a physicist refers to *string theory*, it is most likely a reference to a theoretical framework wherein the fundamental objects in our universe can be thought of as (very small) one-dimensional objects, like strings, that can vibrate. That theoretical framework is also sometimes referred to as an attempt to find "the theory of everything." A primary need driving this theory is the desire to find a connection between, and ideally to unify, quantum mechanics, a very successful framework for objects on a small scale, with general relativity, which describes gravitational forces on a much larger scale. In the process of developing such theories, the mathematics becomes quite abstract and apparently may include many more dimensions than we normally experience. Those extra dimensions are somehow hidden from us. While interesting to contemplate, such a theory goes well beyond what will be presented here. What will be presented here is the classical description of the vibrations of a real string, such as that found on a violin or guitar. While such a description is obviously important for music, let us not forget that it may also form the fundamental basis for "the theory of everything."

The mass on a spring, the harmonic oscillator described in Chapter 5, provides the basic understanding for simple vibrating systems. The vibrating string under tension serves as the prototype for somewhat more complicated systems. For music, such systems include the string, woodwind, and brass instruments.

In what follows, the basic behavior of the vibrating string will be derived by first treating it as a simple harmonic oscillator. Then it will be shown that a single string can be considered, in a sense, to act as a collection of simple harmonic oscillators. As will be seen, especially in Chapters 7 and 14, the result for the vibrating string can also be applied, after some further generalization, to just about any one-dimensional system that vibrates.

The Frequency of Vibration for a String Under Tension

It is well known that if you pluck a string under tension, such as a violin or guitar string (or rope or cord), it will oscillate. To adjust the tuning, that is, to adjust the frequency of the oscillation, the **tension** in the string is adjusted, for example, with

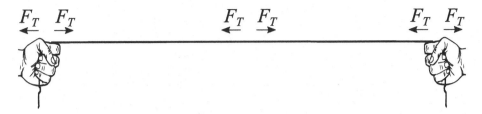

Figure 6.1 The force in a string is referred to as *tension*. At any point along the string, the force pulls equally in opposite directions.

a tuning knob. The length of the vibrating portion of the string is changed to play different notes, for example, by the appropriate placement of a finger. Tension and length are clearly important factors for determining the frequency of oscillation.

The magnitude of the tension in a string, F_T, is the magnitude of the force necessary to hold the end of the string (Figure 6.1). From Newton's third law, the force exerted by the string will be equal to, and in the opposite direction of, the force necessary to hold it. At the opposite end of the string, the force will have the same magnitude but will be in the opposite direction.[1] The string essentially transfers the force from one end to the other. Of course, a single string can always be thought of as being two, three, or more strings attached end to end, where the tension of each string pulls against the tension from the next. Hence, the tension can be regarded as being present throughout the entire length of the string.

The mass on a spring, the prototype harmonic oscillator, was presented in Chapter 5, and it was found that the oscillating frequency, f, could be determined using

$$f = \text{const.} \sqrt{\frac{\text{return force per unit displacement}}{\text{inertial constant}}}. \tag{6.1}$$

For the string, the inertial term is clearly related to the mass of the string, M. There may be some constant factors as well, which would be the same for different strings. Since it will also be desirable to consider different lengths of the same string, the mass is usually described using a linear mass density. The **linear mass density** is simply the total mass divided by the length, L, of the string. It is common to use a lowercase Greek rho, ρ (or, alternatively, mu, μ), for this density. Hence, the mass of the string is written as mass per unit length multiplied by length, $M = (M/L) \times L = \rho L$. An assumption here is that the string is uniform along its length.

The return force for sideways (**transverse**) motion of the string clearly involves the tension. The larger the tension, the harder it will be to displace the string. Note

[1] This result, and later results, assume the force of gravity on the string or rope is negligible compared to the tension.

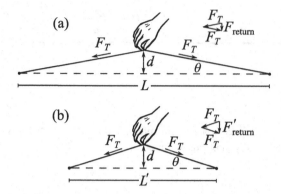

(a)

(b)

Figure 6.2 The return force will depend on the tension and the length of the string. Note that the displacements shown are exaggerated for clarity.

Figure 6.3 A closeup of one of the force diagrams from Figure 6.2.

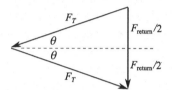

that it is assumed here that the string is extremely flexible, so there is no return force for the string in the absence of tension. Perhaps it is less clear that the return force per displacement also depends on the length of the string.

Consider a simple displacement for a string under tension pulled sideways at its center. If not pulled too far, the tension will not change significantly, and as an approximation, the tension can be taken to be unchanged. The restoring force will be due to the imbalance in the *direction* of the tension because the string is no longer straight. The net force required to move the string will be counterbalanced by the sum of the tensions on either side.

As shown in Figure 6.2, when the same displacement distance, d, is used, **the return force for a shorter string will be larger than that for a longer string**. To get the length dependence, refer to Figure 6.3, which shows a closeup of one of the force diagrams. A comparison of the similar triangles in Figures 6.2 and 6.3, and using $\tan\theta \approx \sin\theta \approx \theta$, valid for small angles (θ in radians, as was seen for the pendulum in Chapter 5), yields

$$\tan\theta = \frac{d}{(L/2)} \approx \theta, \quad \sin\theta = \frac{F_{\text{return}}/2}{F_T} \approx \theta,$$

$$\frac{F_{\text{return}}}{d} = \frac{F_T}{L}. \tag{6.2}$$

That is, if the displacement is not too large, the return force per unit displacement will scale as F_T/L. So, for example, a string with half the length will have twice the return force for a given displacement and tension, and so on. While this simple analysis shows that the return force for a unit displacement depends on F_T/L, there may be other numerical factors that contribute to the return force, the same for all strings, that do not depend on tension or length and that cannot be determined from this simple derivation.

Combining the results for the force and inertia, the predicted relationship for the frequency of vibration, f, of a string under tension is

$$f = \text{const.} \times \sqrt{\frac{F_T/L}{\rho L}} = \text{const.} \times \frac{1}{L}\sqrt{\frac{F_T}{\rho}}, \tag{6.3}$$

where the numerical value of the constant ("const.") term must be determined by other means. This result is enough to show the basic relationship between length and frequency, discussed previously, and is important for understanding Pythagoras's monochord (Chapter 3). A more careful derivation, which is usually done using calculus, or simply through experimentation, will show that the constant term is one-half for a uniform string fixed at both ends.

The situation is not quite so simple, however. The motion of a string can be more complicated than that of a simple oscillator. Consider a string of length L, but pulled equally in opposite directions at $L/4$ and $3L/4$ and then released simultaneously (see Figure 6.4). The net motion of the string at the halfway point, $L/2$, will be zero. Due to the symmetry of the problem, where the left side of the string is always opposite to the right side, it must be the case that the string at the very center will stay at zero. If that point stays at zero, then the left half of the string looks just like the original string pulled at the center, except with half the length. A similar argument applies to the right side of the string. The center of the string could be fastened down, and there would be no change in the motion.

Using the same analysis as previously used (Equations [6.2] and [6.3]), but for a string started in this fashion, thus predicts double the frequency since the vibrating length is effectively cut in half. By dividing the string into three, four,

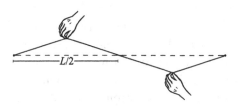

Figure 6.4 Plucking *symmetrically* will effectively cause the two halves to act as two strings, each with half the length of the original. A string with half the length vibrates at twice the frequency.

(a)   (b)

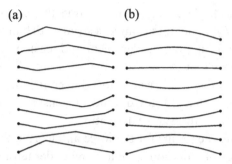

Figure 6.5 Snapshots in time showing a comparison of the motion of (a) a plucked string and (b) a sinusoidal wave for a string. Time is increasing from top to bottom. ★

or five sections, and pulling symmetrically in opposite directions, frequencies of three, or four, or five times the original frequency are expected. Clearly, the string does not have a single frequency of vibration associated with its motion; it can have many. If the lowest vibrational frequency is labeled f_1, then vibrations can also occur at the harmonics of f_1, that is, at the frequencies f_1, $f_2 = 2f_1$, $f_3 = 3f_1, \ldots, f_n = nf_1, \ldots$, and so on. In a musical context, this set of frequencies is sometimes referred to as the **harmonic series**.[2]

While the plucked string is a useful starting point to connect the vibrational frequencies of the string with those of a mass on a spring, there are good reasons to consider a sinusoidal starting displacement instead. Figure 6.5 illustrates one of those reasons. The subsequent motion after the string is released is much more complicated for the plucked string than for a string that is, somehow, started in the shape of a sine wave. That fact, combined with Fourier's theorem, discussed in Chapter 4 for a time-dependent signal, suggests that perhaps these sinusoidal vibrations can serve as the appropriate building blocks for more complicated motions. An explicit example will be shown later in this discussion. A second reason is that the sinusoidal motion connects more directly to the mass on a spring. That can be seen by considering resonances in response to a sinusoidal driving force.

String Resonances

If a string vibrates like a mass on a spring, then it should also have a resonance when driven by a periodic source. Since the string has many natural frequencies, it should not be surprising to find many resonances, each of which, considered separately, behaves like a mass on a spring. One of the simplest ways to search for these resonances is with an electronically controlled vibrator at one end of a string under tension. In light of the results for the mass on a spring and the discussion of

[2] In mathematics, the term *harmonic series* is similar in a general way, but it does not have the same meaning.

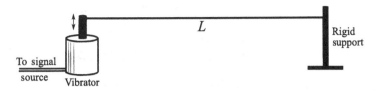

Figure 6.6 An example setup to view string resonances.

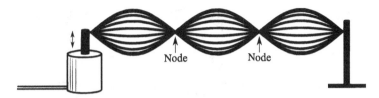

Figure 6.7 For a typical string, the vibrations are rapid and appear something like those illustrated here. Positions with little motion are called *nodes*. The amplitude of the string's motion is greatly exaggerated.

the frequency domain, it makes sense to use a sinusoidal signal to drive the vibrator. The setup might look like that shown in Figure 6.6.

As the frequency is increased from a very small value, a small and rather random-looking sort of string motion might be observed most of the time. However, as the frequency is gradually increased, a larger response will be observed near certain frequencies. One such response, as it will typically appear to the eye, is illustrated in Figure 6.7. There are locations on the string that appear to vibrate very little, if at all, that are called **nodes** of the oscillation. Halfway in between the nodes are locations where the motion is a maximum. These positions are called **anti-nodes**.

A snapshot with a camera or a freeze-frame video can be used to view what is going on at any instant. For the driving frequencies where a maximum in the response is observed, that is, for the resonances, those freeze-frame images will look something like those in Figure 6.8. The corresponding set of resonant frequencies is the same as those found above by plucking the string. The shapes of these frozen images are not the sharp-cornered triangular shapes used earlier to derive the natural frequencies but are smooth and sinusoidal. The lowest frequency corresponds to a snapshot with one-half of a sine wave between the supports. **An additional half sine wave is added for each successive resonance.** The snapshots shown are all drawn such that the first half wave is upward, the second down, and so on. Of course, at a later time, this will be reversed, so the first half will be down, the second up, and so on.

A plot showing the vibrational amplitude for the string as a function of frequency will look something like what is shown Figure 6.9. Near each peak is a sketch of the corresponding string motion, frozen in time, that is observed. At frequencies between the peak responses, the motion may appear to be quite complicated. That motion is actually just the combination of the (simpler) motions associated with several different resonances.

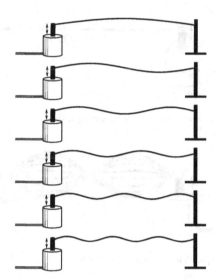

Figure 6.8 The resonant string motion frozen in time. The lowest-frequency resonance is illustrated at the top. Successively higher-frequency resonances are shown as one goes down in the figure.

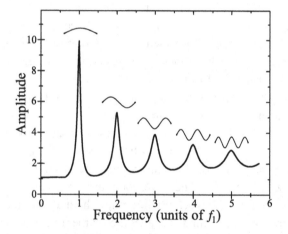

Figure 6.9 The maximum amplitude of the string as the vibrator frequency is swept. The type of resonant motion observed is shown above each maximum.

In summary, each of the separate resonances has associated with it a particular sinusoidally shaped oscillation. These are referred to as **modes of oscillation**. Each resonance peak is associated with a different mode. Each mode, considered by itself, behaves like a harmonic oscillator. For an ideal string[3] under tension, there will be an infinite number of modes.

---

[3] *Ideal* here means that no factors other than those already assumed are, or become, important.

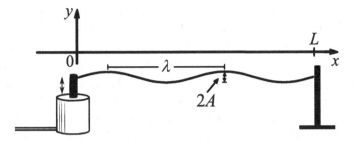

Figure 6.10 A snapshot of a standing wave.

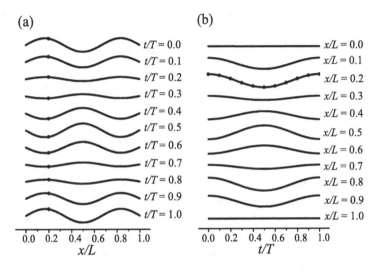

Figure 6.11 Showing the motion (a) over the length of the string for different times and (b) over one period in time at different positions. Note that the extra dots shown in (a) from top to bottom correspond to the extra dots shown in (b) from left to right.

Standing Waves

The solutions in the presence of a sinusoidal driving force consist of a sinusoidal displacement along the length of the string. At the same time, each point along the string undergoes a sinusoidal oscillation with time. These waves on the driven string do not appear to be moving *along* the string and hence are referred to as *standing waves*. In contrast, waves coming into the shore at a beach are examples of traveling waves – they seem to be going somewhere, and there is a sense of direction to them. Traveling waves will be discussed in Chapter 8.

Figure 6.10 illustrates one of the resonance "snapshots" from Figure 6.7 for use as an example. The distance along the string is quantified using the coordinate x, and the transverse (sideways) motion of the string is quantified using the coordinate y. The string is of length L. Figure 6.11 illustrates the motion of the string as a function of time and position along the string.

The transverse motion is periodic with distance when moving away from $x = 0$ at any fixed time. That is, the up-and-down (or side-to-side) motion repeats. With the exception of the special case that occurs during the instant when the string is perfectly straight, the repeat distance does not depend on time. The repeat distance is called the **wavelength**, often symbolized using a lowercase Greek lambda, λ. The wavelength can be measured from peak to peak (as shown in Figure 6.10), from trough to trough, by using the zero-crossings, and so forth. The amplitude of the wave, A, is the maximum distance from zero displacement in the y-direction. The **peak-to-peak amplitude** is the distance from the maximum (e.g., highest) to minimum (e.g., lowest) displacement. The peak-to-peak (pp) amplitude is often the easiest to measure for many types of waves. The amplitude, A, is one-half of the peak-to-peak amplitude.

The mathematical description of such a standing wave must include the sinusoidal variations along the x-direction at any fixed time, as well as the sinusoidal variations for any particular position, x, as time is allowed to proceed. Such a description is

$$y(x,t) = A \sin\left(\frac{2\pi}{\lambda}x\right) \times \cos\left(\frac{2\pi}{T}t\right) = A \sin(kx)\cos(\omega t), \qquad (6.4)$$

where the displacement, y, depends on both x and t. The particular sinusoidal functions (sine or cosine) may be changed, although the particular combination shown will be useful for the string if the coordinates shown in Figure 6.9 are used.

It is common in physics to use **angular frequency** for the time dependence, $2\pi/T = 2\pi f = \omega$, as has already been seen, and similarly, a **wave number**, k, for the spatial dependence, $k = 2\pi/\lambda$. Note that despite the close resemblance, the kx here has nothing to do with the kx previously seen when discussing Hooke's law for springs and harmonic oscillators (Chapter 5). The symbols are the same, but what they refer to here is quite different. Once again, context is important. For historical reasons, it is common to use those same symbols in these two different contexts.

Recall that the sine and cosine functions repeat each time the angle goes through 360°, or 2π radians. That is, if x equals one wavelength, then $kx = 2\pi x/\lambda = 2\pi\lambda/\lambda = 2\pi$ (radians), and the sine function repeats. Likewise, if the time, t, is one period, T, the cosine repeats, so it must be that $\omega t = 2\pi t/T = 2\pi T/T = 2\pi$.

Example 6.1

Consider the standing wave described by

$$y(x,t) = 0.04 \sin(17x)\cos(11t),$$

assuming all units are Standard International (SI). In that case, the amplitude of the wave is 0.04 m = 4 cm, and the frequency and wavelength are computed as follows:

$$11T = 2\pi \rightarrow T = 2\pi/11, \quad f = 1/T = 11/(2\pi) = 1.75\,\text{Hz},$$
$$17\lambda = 2\pi \rightarrow \lambda = 2\pi/17 = 0.37\,\text{m}.$$

Note that the wavelength is a measure of the spatial variations in the wave, and it is found using the value that multiplies x, the spatial coordinate. The period and frequency are related to the time dependence of the wave and are found using the value that multiplies t, the time.

The mathematical description for the standing waves on a specific string must include one more fact. Since the string is fixed at both ends, it must be true that $y(0, t) = y(L, t) = 0$.[4] That is, at all times, the *ends* have not moved from zero. Since $\sin(0) = 0$, Equation (6.4) is already satisfied at $x = 0$. That is why the sine function was chosen. However, the sine function must also *always* give zero at the other end of the string where $x = L$. That in turn requires that between $x = 0$ and $x = L$, there must be an integer number of half wavelengths. Hence, a valid description for this string must include the requirement that

$$L = n\frac{\lambda}{2} \;\rightarrow\; \lambda = \frac{2L}{n} \;\rightarrow\; k = \frac{2\pi}{\lambda} = \frac{n\pi}{L}, \tag{6.5}$$

where n is any positive integer.

The frequencies for the possible standing waves, the **modes**, will be an integer times the lowest frequency; that is, they are harmonics of the lowest frequency, as was derived earlier in a less mathematical way for the plucked string. Thus, for this string fixed at both ends, there are standing waves, mathematically described in Equation (6.4), with

$$\lambda_n = \frac{2L}{n} \quad \text{and} \quad f_n = \frac{n}{2L}\sqrt{\frac{F_T}{\rho}}, \tag{6.6}$$

where n is a positive integer that is being used in two different ways – both as a label and as a numerical value. For example, the wavelength of the third harmonic is labeled λ_3, and it has the value $2L/3$. The latter says that for the third harmonic, there are 3 half wavelengths over the length L.

At this point, it is worth pointing out an interesting relationship between the wavelength and frequency for these standing waves. If the wavelength and frequency are multiplied, the result is the same constant value, call it c, for *all* the allowed standing waves on that string,

$$\lambda_n f_n = \frac{2L}{n}\frac{n}{2L}\sqrt{\frac{F_T}{\rho}} = \sqrt{\frac{F_T}{\rho}} = c, \tag{6.7}$$

---

[4] The slight motion of the mechanical oscillator that provides the driving force can be considered negligible or can be treated using fictitious forces, as described at the end of Chapter 5.

and that constant, c, has units of speed (i.e., m/s). Although the waves are "standing," there is a characteristic speed associated with the motion that is the same for all the modes. Note that this cannot be the speed of the sideways motion since that clearly must depend on the amplitude and frequency of the vibration. This speed does not even depend on the length of the string. This characteristic speed will be considered more in Chapter 8.

These sinusoidal standing-wave solutions for the driven string will be very useful for describing the motion of the string in other cases as well. Since the standing waves are sinusoidal, Fourier's theorem can be used to describe any motion of the string as a sum of standing waves. Thus, the resonant standing-wave solutions become the building blocks for all solutions. As a musically relevant example, the plucked string is reconsidered.

The Plucked String

A simplified analysis of the motion of a lossless plucked string starts with the string being pulled at a single point along the string, then released. The initial state of the string looks like what is shown in Figure 6.12, where the string is of length L, and it is plucked a at distance L/b from one end ($b > 1$). The initial triangular shape is used to determine a Fourier series of sinusoidal functions for the initial displacement along the string, and then the frequency of oscillation for each of those sine waves will be a simple harmonic progression. That is, once the string is released, each of the standing-wave solutions will oscillate at its corresponding natural frequency. The resulting sum of standing waves corresponding to the plucked string is:

$$y(x,t) = \frac{2b^2}{(b-1)\pi^2} \left(\sin\frac{\pi}{b} \sin\frac{\pi x}{L} \cos\omega t + \frac{1}{4}\sin\frac{2\pi}{b}\sin\frac{2\pi x}{L}\cos 2\omega t \right.$$

$$\left. + \frac{1}{9}\sin\frac{3\pi}{b}\sin\frac{3\pi x}{L}\cos 3\omega t + \dots \right)$$

$$= \frac{2b^2}{(b-1)\pi^2}\sum_{n=1}^{\infty}\frac{1}{n^2}\sin\left(\frac{n\pi}{b}\right)\sin\left(\frac{n\pi x}{L}\right)\cos(n\omega t), \tag{6.8}$$

so the amplitude of the standing wave corresponding to the nth harmonic is

$$A_n = \frac{2b^2}{(b-1)(n\pi)^2}\sin\left(\frac{n\pi}{b}\right), \tag{6.9}$$

Figure 6.12 A string plucked at a distance L/b from one end. The motion of the string is greatly exaggerated.

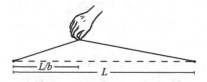

which may look a bit complicated; however, it is a simple matter to put in values for any particular term.[5]

Example 6.2

What is the motion for a string plucked at the center?

Plucking at the center corresponds to $b = 2$, so

$$y(x,t) = \frac{8}{\pi^2} \left(\sin\left(\frac{\pi x}{L}\right) \cos(\omega t) - \frac{1}{9} \sin\left(\frac{3\pi x}{L}\right) \cos(3\omega t) + \frac{1}{25} \sin\left(\frac{5\pi x}{L}\right) \cos(5\omega t) - \ldots \right).$$

Compare this to the expression for the simple triangle wave (Equation [4.4]).

More generally, if b is an integer, then the corresponding harmonic, $n = b$, and all multiples of that harmonic are missing. For example, if you pluck at one-seventh of the way from one end of the string ($b = 7$), the 7th (and 14th, etc.) harmonic will be very small. For an instrument where the string is struck, rather than plucked, such as a piano or hammer dulcimer, the size of the hammer and the finite time of contact of the hammer will complicate the analysis, although a similar reduction in these harmonics can still occur.

Figure 6.13 shows the theoretical spectrum of amplitudes for the different standing-wave modes for three selected plucking positions. These show that for plucked stringed instruments, it should be expected that the higher overtones will

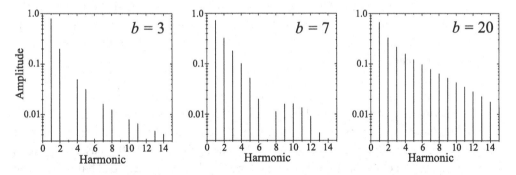

Figure 6.13 Spectra for the plucked string showing the effect of the pluck position on the amplitudes of the harmonics. Note that the larger the value of b, the closer the pluck is to the end of the string.

---

[5] The use of the uppercase Greek sigma to indicate a sum, such as in Equation (6.7), is common. More information about this use is included in Appendix A.

contribute more to the sound when the string is plucked near the end (larger b) than when the string is plucked closer to the center.[6]

The opposite effect can also be of use for stringed instruments. A finger lightly placed at any spot along the string will dampen the motion at that point. If the position is chosen to be at a place where some of the standing waves have nodes (i.e., no motion), then those waves will be undamped, whereas the standing waves that have motion at that spot will rapidly disappear. For example, if the center of the string is used, all the odd harmonics, including the fundamental, will be damped, whereas the even harmonics will continue to sound. If a finger is lightly placed one-third of the distance along the string, only overtones that are at multiples of 3 times the fundamental will survive. This musical effect is referred to as **artificial harmonics**, although there is really nothing "artificial" about the harmonics.

End Conditions

The frequencies and wavelengths for the standing waves on a string derived earlier were for a string that is tied down, or "fixed," on both ends. For real strings, that is generally the only condition that is practical. However, the vibrating string can be used as a starting point for a discussion of many other types of waves where this restriction does not exist. Examples seen later include sound waves in pipes, such as organ pipes, where the end can be open or closed. Hence, it is useful to consider other end conditions for strings, even if such conditions cannot be easily reproduced in the lab.

The opposite of "an end that is tied down," a **fixed end**, would be one that is totally free to move. The string must still be under tension, however, so the motion should be free only in the direction of wave motion, in this case the transverse direction. When discussing wave motion, such an end is referred to as a **free end**.

To picture such a free end for a string, imagine a very small loop at the end of the string that is free to slide, without friction, along a rod (Figure 6.14). Since the loop at the end has negligible mass, any net force applied to it would cause a very large and rapid motion. The result is that the loop will rapidly slide to the point where there will be no force along the direction of the rod. In this case, that means that the end of the string will always be perpendicular to the rod. That is, were the string not perpendicular, tension would cause the loop to rapidly move along the frictionless rod until the string becomes perpendicular.

---

[6] The full analysis of the sound produced must also include how these vibrations are turned into sound, which is an additional complication and is certainly different for acoustic instruments compared to instruments with electronic pickups. At this point, only the vibration of the string is being considered.

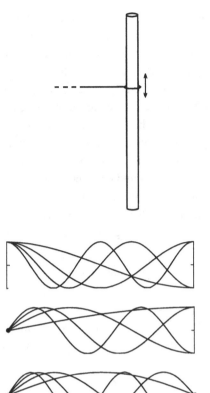

Figure 6.14 A free end for waves on a string under tension is imagined to result from a loop that slides without any friction along a rod. In such a case, the string near the loop will always be perpendicular to the rod.

Figure 6.15 Free–free (*top*), fixed–free (*middle*), and fixed–fixed (*bottom*) end conditions for a vibrating string determine which standing waves can be present. These figures show the longest-wavelength mode and the next three successively shorter-wavelength modes.

Now there are three possibilities to consider for the standing waves on the string – a string with both ends fixed, a string with both ends free, and a string with one end fixed and the other free. The case where both ends are fixed was already considered. In that case, only solutions where both ends are always zero were allowed. When both ends are free, the solutions should correspond to those sinusoidal functions that are at a maximum or minimum at both ends. If one end of the string is fixed and the other free, sinusoidal functions are always required to be zero at the fixed end and a maximum or minimum at the free end. In each case, there are an infinite number of possibilities that are harmonically related.

Figure 6.15 illustrates the first few standing-wave solutions for different combinations of end conditions. For both the fixed–fixed and free–free strings, the lowest-frequency solution has a half wavelength along the length of the string, L. That is, $\lambda/2 = L$, or $\lambda = 2L$. **Each successive higher-frequency solution includes one**

more half wavelength. Thus, all the solutions will have an integer number of half wavelengths. The fixed–free situation is more complicated. The lowest-frequency solution has a quarter wavelength along the length of the string, so $\lambda/4 = L$, or $\lambda = 4L$. Each successive solution has one additional *half* wavelength. Or to say it another way, the fixed–free solutions correspond to a quarter wavelength, three quarter wavelengths, five quarter wavelengths, and so on. Hence, the fixed–free string can be expected to have only the odd harmonics of the lowest frequency.

This is a first example illustrating that **the series of overtones that occurs depends strongly on the conditions at the ends**. The conditions at the ends are more generally referred to as **boundary conditions**.

A mathematical description of the wavelengths and frequencies for the fixed–fixed string, and hence also for the free–free string, was given in Equation (6.6). The frequencies and wavelengths for the fixed–free string will be

$$\lambda_n = \frac{4L}{n} \quad \text{and} \quad f_n = \frac{n}{4L}c, \ n = 1, 3, 5, 7, \cdots, \tag{6.10}$$

where n is any positive *odd* integer, and c is the constant from Equation (6.7). Using the fact that for *any* integer, $2n$ will be an even value and hence $2n - 1$ will be odd, the result for the fixed–free string is sometimes written

$$\lambda_n = \frac{4L}{2n - 1} \quad \text{and} \quad f_n = \frac{2n - 1}{4L}c, \ n = 1, 2, 3, 4, \cdots, \tag{6.11}$$

where n is *any* positive integer. The set of solutions is unchanged; however, the meaning of n has changed. In the latter case f_n and λ_n correspond to values for the $(2n - 1)$th harmonic and not the nth harmonic.

For strings having the same tension, density, and length, the fixed–free string will have a fundamental frequency that is a factor of 2 lower than the fixed–fixed and free–free string. That is, the fundamental is an octave lower. Although it is difficult to demonstrate with strings, this octave change can be demonstrated for some stopped organ pipes with the stopper in place and then removed. For one situation, the ends are the same (either fixed–fixed or free–free), and for the other, they must be opposite (fixed–free or free–fixed). As an alternative, use a short flute or whistle made from plastic plumbing pipe (with no finger holes), with the far end open and with a hand over the end.[7]

Although not a string at all, a device known as a *Shive wave machine*[8] (Figure 6.16) can be quite useful for demonstrating wave motion for a variety of end conditions. The motion for such a machine is also much slower than that for a typical string, so what is happening is readily visible to the eye. The wave machine is constructed using a large number of long rods attached parallel to each other and

---

[7] Whether an open end of a pipe should be considered fixed or free is a subject for later consideration. Whichever the case for the open end, it seems reasonable that a closed end should be the other.

[8] Developed in the 1950s by John N. Shive of Bell Labs.

Figure 6.16 A Shive wave machine slows down wave motion for the sake of demonstrations. The wave machine shown is just under 1 m long.

perpendicular to a wire. The wire is often made from spring steel.[9] The wire rests on supports but is free to rotate. The long rods provide a large rotational inertia and the wire provides a relatively small torsional return force, so the wave frequency is quite low compared to a typical string. Videos of these wave machines in action are easy to find online if one cannot be observed in person.

Quality Factor

A real string will have losses, of course, as was discussed for the driven harmonic oscillator. Each of the string resonances can be characterized with a quality factor, Q. For a typical string used in a musical application, that quality factor is frequency dependent. For strings used on musical instruments, it should not be surprising that the maximum Q is often found near where human hearing is most sensitive (1–3 kHz).

The fact that each string resonance has a different Q also means that each mode will have a different decay rate. After a string is plucked, the relative amplitudes of the overtones can be expected to change with time, and hence the perceived tone quality may also change with time. This change in tone can usually be heard by carefully listening to the tone, for example, from a guitar, and can certainly be seen in the changing waveform of the sound as observed on an oscilloscope.[10]

The decay of the string vibration arises from energy losses. Adding up all the loss rates contributing to Q can be very complicated; however, the dominant terms will include various powers of the frequency, at least approximately. Remembering that $1/Q$ is related to the loss rate per oscillation cycle (which is why *more* loss corresponds to a *smaller Q*), the frequency (f) dependence will usually look something like

$$\frac{1}{Q} = C_{-1}\frac{1}{f} + C_0 + C_2 f^2, \tag{6.12}$$

---

[9] Some designs use two closely spaced parallel wires or strings.

[10] The signal might be from a microphone or an electric guitar pickup.

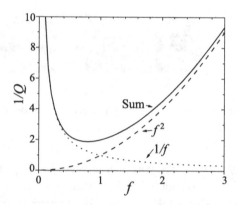

Figure 6.17 When loss rates proportional to $1/f$ and f^2 are added, the $1/f$ term dominates at lower frequencies, and the f^2 term dominates at higher frequencies. The minimum loss (maximum Q) occurs in the middle.

where the C_n are constants that depend on the particular circumstances. For a typical (metal) string used for music, the first term is largely due to air viscosity. Since $1/f$ gets very large when the frequency gets small, air viscosity dominates the loss rate at the lowest frequencies. The constant term, C_0, is largely due to friction, for example, from the interactions between windings on a wound string. This should be small for a properly wound musical string. The last term is largely due to **viscoelastic** energy losses. These arise within the material of the string. As the string is bent, some atoms have to move and can lose energy (to heat) in the process. As an example, if you repeatedly bend a small metal wire, such as from a paper clip, the wire will get hot – some of the energy you put in gets lost as heat. Since f^2 gets very large for large values of f, the squared term will dominate at higher frequencies. Somewhere between "low" and "high" frequency is a minimum loss per cycle and hence a maximum in Q. This is illustrated in Figure 6.17.

To gain a full understanding of all the nuances of string behavior as found in musical instruments would necessarily include the role of nonlinear interactions, such as time-dependent tension in the string, which is important for larger motions. The motion of the real string is also not limited to one plane. The motion can be up and down or side to side. For real strings, a rotation of the string (*torsional notion*), particularly when it is bowed, can also become important. These additional effects are beyond the scope here.

Summary

The fundamental vibrational frequency for an ideal string of length L, fixed at both ends and under tension F_T, was derived (Equation [6.2]), starting with the results for the mass on a spring. Additional vibrations occur at integer multiples, that is, harmonics, of that fundamental frequency.

When driven by a periodic sinusoidal source, string resonances can occur at the fundamental and harmonic frequencies, resulting in standing waves that vary sinusoidally in both time and position along the string. For these standing waves, the product of frequency and wavelength is a constant with units of speed.

Any free vibration of the string can be described as a sum of the standing sinusoidal waves with appropriate amplitudes and phases. The amplitudes of the harmonics will be strongly dependent on the initial conditions.

When losses are included, each vibration can be described using a quality factor, Q. For a given string, the quality factor for very low and very high frequencies will be low, with a maximum somewhere in between. For strings used for music, the maximum Q is usually near the middle of the audible range.

Changing the conditions at the ends of the string can result in a change in the set of vibrational frequencies that may occur.

ADDITIONAL READING

Giordano, N. "The Physics of Vibrating Strings." *Computers in Physics* 12, no. 2 (1998): 138–145.

Politzer, D. "The Plucked String: An Example of Non-Normal Dynamics." *American Journal of Physics* 83, no. 5 (2015): 395–401.

Valette, C. "The Mechanics of Vibrating Strings." In *Mechanics of Musical Instruments*, A. Hirshberg, J. Kergomard, and G. Weinreich, eds. Springer-Verlag, 1995.

Valette, C., and C. Cuesta. *Mécanique de la corde vibrante*. Hérmes, 1993.

PROBLEMS

6.1 A standing wave is described by

$$y(x, t) = 0.15 \sin(25x) \cos(66t),$$

where all values are in appropriate SI units. What is the amplitude, wavelength, and frequency (in hertz) of this wave?

6.2 Write an expression for a standing sinusoidal wave with a wavelength of 0.30 m, a frequency of 10 Hz, and an amplitude of 1 cm.

6.3 A string 30 cm long, fixed at both ends, is vibrating at its third harmonic frequency. What is the wavelength of the standing wave?

6.4 The vibrating portion of a violin string is 13 in. (33 cm). The violin A string is tuned to 440 Hz. If the tension is 49 N, what is the mass of the vibrating portion of the string?

6.5 A rule for placing frets on a guitar is sometimes referred to as the *rule of 18*. To place the first fret, divide the total length of the string, from the bridge to the nut, by 18 (or more precisely, 17.8). See Figure P6.5. That is then the distance from the nut for the first fret. Now the remaining distance, from the bridge to the first fret, is divided by 18 to locate the second fret, and so on. Show how that rule is related to the frequencies of the equal-tempered scale.

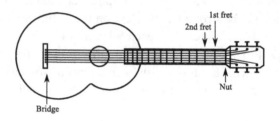

Figure P6.5

6.6 The adjacent strings on a violin are tuned a musical fifth apart (e.g., a factor of 3/2 in frequency). When the strings are "open" (no fingers used), the vibrating part of each string will have the same length, L.
   a. Why might you expect that all four violin strings should have (nearly) the same tension?
   b. Assuming they have the same tension, how do the masses of the length L of two adjacent strings compare?
   c. Assuming they have the same tension, how do the masses of the vibrating portion of two adjacent strings compare when they are made to sound the same note (e.g., by placing your finger at the appropriate point on the lower string to match the open higher string)?
   d. When the strings produce the same note, would you expect the tone quality, the timbre, of the notes to be the same? Why or why not?

6.7 Find at least three different positions on a string where it can be plucked, such as is done on a harpsichord, so that the sound from the seventh harmonic is minimal. Why might this be important for an instrument maker to know? Find a harpsichord (or piano) and see where the strings are plucked (or struck) to see if this practice is followed.

6.8 A string, fixed at both ends, is driven at resonance, and the wavelength of the standing wave is 30 cm. The frequency is then gradually increased to the next resonance, where it is observed that the wavelength is now 24 cm. How long is the string?

6.9 Consider the lowest-frequency mode – a half sinusoid – for a string under tension oscillating in the vertical direction. An obstacle is placed so that the string effectively shortens its length as it passes through horizontal, as shown in Figure P6.9. Estimate the average frequency for small oscillations of this vibrating system compared to the frequency in the absence of the obstacle.

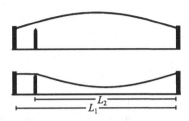

Figure P6.9

7   Normal Modes

The harmonic oscillator – the mass on a spring introduced in Chapter 5 – was the starting point for the discussion of vibrating systems. The harmonic oscillator has a single natural frequency that is both the oscillation frequency during the recovery after an initial displacement and the resonant frequency when the system is driven sinusoidally. The basic physics for the vibrating (ideal) string was found to be similar, although with multiple natural frequencies. For the string, each natural frequency is correlated with motion that is sinusoidal with distance along the string, as well as sinusoidal with time. The general solution for the vibrating string is the sum of those motions. For the string, the vibration frequencies that can occur are determined by the end conditions and, for the ideal string, are harmonically related.

The purpose of this chapter is to generalize some of these results to other, more complicated vibrating systems, including chimes, drumheads, and others. In particular, one task here is to look at what is the same as was found for the ideal string and what is different. What will be seen is that a very general description of the resulting motion has many similarities, although the mathematics rapidly gets somewhat abstract, and the details will vary. In each case, however, the general solution will be the sum of solutions, each of which has a sinusoidal time dependence at a corresponding natural frequency. In each case, the natural frequencies that occur are related to boundary conditions. And although the amplitude of the motion may not be sinusoidal with position, there are some common general features to the motion for all such vibrating systems. To start, the solutions for the string, from Chapter 6, are briefly reexamined.

Indexing the Modes

The solutions for the general motion of an ideal string are formed from the sum of standing waves. Not all standing waves are allowed since each must match the condition enforced at the end of the string. That is, if the string is tied down at both ends, each of the allowed standing waves must be zero at both ends. Those standing-wave solutions are the building blocks for any solution that can exist and are referred to as **normal modes**. Each of these string modes, considered separately, acts like a harmonic oscillator.

For the ideal string, each of the component standing waves – each of the modes – was found to obey $\lambda f = c$, where c is a constant with units of velocity. In addition, if f_1 is the lowest frequency of them all (the fundamental), then the remaining mode frequencies are all integer multiples of f_1. If the standing waves for a string (fixed at both ends) are ordered by their natural frequencies, from low to high, then the nth standing wave will have a natural frequency $f_n = nf_1$, where n is a positive integer. The wavelength corresponding to the solution with the frequency, f_1, is $2L$, where L is the length of the string, so there is one half wave along the length of the string. Hence, the wavelength for the nth mode solution is $2L/n$, and there are n half waves along the string.

Notice that in this discussion, n has (once again) been used in different ways. In particular, it is a label, or index, identifying which of the solutions is being considered, and at the same time, it is used as a numerical value for computation. The solutions could have been identified alphabetically, such as with a, b, c, and so forth, so that one would refer to f_a, f_b, and so on, or using "primes," such as f', f'', f''', and so on. In addition, there is no requirement that the labels go in any particular order. The convenience of using numerical digits as a label is that the relationship between the number to be used in the calculation and the label is often quite obvious. This is, perhaps, somewhat of a subtle point, but it is necessary to the understanding of the nomenclature.

Now consider the first few standing waves for a string tied down at the ends, as shown in Figure 7.1. This is a picture frozen in time, so the frequency of the motion for each mode is not obvious in the figure. However, it is straightforward to label these modes by counting the number of half wavelengths. Alternatively, the modes could be labeled using the number of **nodes**, locations that are always zero, either including or excluding the ends.

As another example, consider a **hanging chain** or rope, fixed at the top and free at the bottom. Here the tension in the chain is due to the length of chain being supported below. Hence, the tension at the top is the total weight of the chain, and the tension at the bottom is zero. Experimentally, it is easy to find the first few modes for small oscillations.[1] The motion, frozen in time, is illustrated for the first four modes in Figure 7.2. It is straightforward to count the nodes. If the node at the top is included, these modes can be labeled using the number of nodes,

---

[1] As discussed in Chapter 5, this can be accomplished by holding the rope at the top and moving your hand periodically in the horizontal direction. As a word of caution, the behavior for larger motions can become nonlinear, and in that case, this type of "modal analysis" is not as useful.

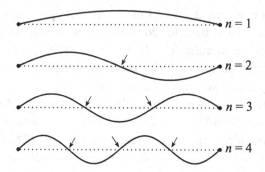

Figure 7.1 The first four modes for the vibrating string fixed at both ends. Nodes are identified with arrows. In this case, each mode can be indexed using the number of half wavelengths between the ends (n). The number of nodes, including the ends, is $(n + 1)$.

Figure 7.2 A hanging chain or rope has modes that are not sinusoidal in space but can still be identified using the number of nodes. ★

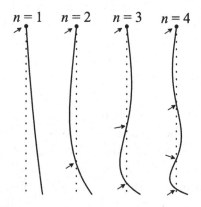

starting from 1. That labeling is easily accomplished, and the basic pattern is the same as was seen for the ideal string.

Computing the vibrational frequencies and the shapes for the hanging rope is, however, quite complicated. For reference,

$$f_n = \frac{c_n}{4\pi} \sqrt{\frac{g}{L}}; \ c_1 = 2.4048, \ c_2 = 5.5201, \ c_3 = 8.6537, \text{ and so forth}, \quad (7.1)$$

which looks like the frequency found for the simple pendulum, except with an extra series of seemingly random constants ($\{c_n\}$) out front. Those constants come from the "nth zeros of the zero-order Bessel functions of the first kind." To visualize the Bessel functions in this case, think about dropping a pebble into a still pond. The resulting waves, which decrease in amplitude with distance as they spread out from the center, are related to Bessel functions. Bessel functions will show up again in later chapters.

The solutions for the hanging rope are not harmonically related – they are **inharmonic**. Although labeling these modes was quite easy, finding the resulting vibrational frequencies and the shape of the motion is much more of a challenge. In addition, although a relationship exists between the index and the frequency, it is no longer simple. It is still the case, however, that as the number of nodes goes up, so also goes the frequency. This is often the case for many other systems as well, although there are exceptions. The point here is that it is easy to *visualize* what the solutions must look like, at least approximately, even if it is difficult to solve for the details.

Chimes

There are many types of chimes that can be modeled as transverse motion of a one-dimensional system – similar to the string. These include orchestral chimes, wind chimes, grandfather clock chimes, the small vibrating bars in a music box, and wooden bars or rods such as are used in xylophones and marimbas. The major distinction between a chime and a string is the origin of the return force. For a string, the return force arises from the tension. For a chime, the return force comes from the material properties of the chime – in particular, its response to bending motion.

Perhaps the simplest chime is just a long bar or rod. By "long," one means that it is significantly longer than it is wide. The oscillations considered here are perpendicular (i.e., transverse) to the length of the bar or rod. Such a bar or rod can also have compression waves that are longitudinal – that is, the compression is along the long direction of the bar or rod. These compression waves in bars and rods are less important for musical applications and will not be discussed here.

The return force for the oscillation is the force that resists bending. That force will depend on the material used and the shape of the bar or rod – a thicker bar is harder to bend. The return force per unit displacement will also depend on the length of the bar or rod – a longer bar is easier to bend. Think about supporting a plank at the ends and standing in the center. How far the center of the plank moves (downward) for a given weight is a measure of the return force for bending. The inertial portion for vibrational motion will depend on the mass density and mass distribution (i.e., shape) of the bar or rod.

As a rod vibrates, not all of the bar moves the same amount. Some portions move very little and some much more. The inertial term will depend more on those regions that move than on those than do not. Remember, inertia is the tendency to keep moving once set in motion. That tendency is less important for any portion of a bar that does not move very much in the first place. For a bar or rod, it is difficult to change the inertial portion without also changing the return force. The two properties both depend on shape and are thus interconnected.

The general solution for the transverse modes on a *uniform* rod or bar is known. Using x to represent position along the bar and $y(x,t)$ to describe transverse motion along the bar at a time t, the solution for each normal mode is given by

$$y(x,t) = \Big(A\cos(kx) + B\sin(kx) + C\cosh(kx) + D\sinh(kx)\Big)\cos(2\pi f t), \qquad (7.2)$$

where $A, B, C,$ and D are constants that depend on the particular situation, and k, in this context, is a constant that depends on the material and on the square root of the frequency, f. For simplicity, the time dependence is written simply using a cosine; however, in general, a phase factor might also be present. If unfamiliar, **cosh** and **sinh** refer to the **hyperbolic cosine** and **sine** functions, respectively.

The sine and cosine functions are often defined in terms of a unit circle. If you draw a circle with radius 1, a "unit circle," using an x–y graph, the points on that circle will satisfy the equation $x^2 + y^2 = 1$. Drawing a line from the origin to one of those points, and calling the angle between that line and the x-axis θ, then $x = \cos\theta$ and $y = \sin\theta$. A unit hyperbola consists of the points that satisfy the equation $x^2 - y^2 = 1$, which differs from the circle only by a minus sign. The hyperbolic functions can be similarly defined using a unit hyperbola, although the "angle" used is a bit more abstract. The hyperbolic trigonometric functions are cousins of the sine and cosine, and ultimately, the difference goes back to that simple minus sign. For reference, the hyperbolic functions are compared to the sine and cosine in Figure 7.3. All these functions are mathematically related and are part of the exponential family of functions.

The conditions at the ends of the bar or rod, the boundary conditions, are used to determine the constants $A, B, C,$ and D and the set of frequencies that will work. For rods, there are three simple idealized models for end conditions: **free**, **hinged**,

Figure 7.3
Hyperbolic sine and cosine functions compared to sine and cosine.

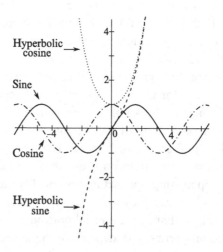

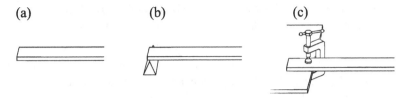

Figure 7.4 Simple models for the end condition for a bar: (a) a free end, (b) a hinged or simply supported end, and (c) a clamped end.

and **clamped**, as illustrated in Figure 7.4. The hinged end, also referred to as a simply supported end, is like that seen for the string. The magnitude of the motion is restricted to zero at the end, but the bar can come in at any angle – like a hinged door. A clamped end is more restrictive and is similar to what would be expected for a long board with one end firmly clamped to the edge of a table. The motion at the end must be zero, and also, the angle made at the end is zero. Since a bar or rod has two ends, there are a large number of possible combinations of end conditions.

The mode frequencies for three musically relevant cases are given *approximately* by the following:

$$\text{Hinged–free:}\quad f_n = \frac{1}{4}\frac{\pi K}{8L^2}\sqrt{\frac{E}{\rho}}\,(4n+1)^2;$$

$$\text{Free–free:}\quad f_n = \frac{\pi K}{8L^2}\sqrt{\frac{E}{\rho}}\,(2n+1)^2; \tag{7.3}$$

$$\text{Clamped–free:}\quad f_n = \frac{\pi K}{8L^2}\sqrt{\frac{E}{\rho}}\,(2n-1)^2.$$

Here, L is the length of the vibrating rod; E and ρ are, respectively, the Young's modulus[2] and mass density for the material used; K is a factor that depends on the thickness and shape of the rod; and n is any positive integer. The exact numerical values that appear in the parentheses are being approximated here using integers.[3] As seen in Chapter 6 for the string, **the series of overtones that occur depends strongly on the conditions at the ends.**

The **Young's modulus**, E, is a measure of the springiness of the material – if you put a weight on a piece of the material, the Young's modulus is a measure of how much the material will compress. It is assumed that the compression is reversed

[2] Some authors use the symbol Y for the Young's modulus.

[3] For more precise theoretical values, replace the integer values in parentheses for the smaller values of n as follows: for the hinged–free case, for $n = 1$, use 4.99951 instead of 5; for the free–free case, for $n = 1$ and 2, use 3.01124 and 4.99950 instead of 3 and 5, respectively; and for the clamped–free case, for $n = 1, 2,$ and 3, use 1.19373, 2.98835, and 5.00049 instead of 1, 3, and 5, respectively. For larger values of n, the values obtained using the integer values are already very accurate.

when the weight is removed. If too much weight is used, the compression will not recover, and the behavior is no longer like that of an ideal spring. Based on the simple harmonic oscillator result, it should not be surprising that the oscillation frequency of the bars depends on the square root of the springiness of the material divided by the material's mass density – the square root of the return force per displacement divided by the inertial term (see Equation [5.8]).

The factor K that appears is a geometric constant, known as the **radius of gyration**, and it has a value roughly equal to the thickness of the rod or bar, multiplied by a factor near 1, which depends on shape. As specific examples, for a solid circular rod of radius a, $K = a/2$. For a hollow circular rod with inside radius b and outside radius a, $K = \sqrt{a^2 + b^2}/2$. Finally, for a bar with a rectangular cross section, $K = h/\sqrt{12}$, where h is the thickness in the transverse direction (the direction of the wave motion). An interesting result for such a rectangular bar is that the frequency for transverse motion along the length of the bar does not depend on the width of the bar.

The hinged–free case is a model for a **hanging rod** that approximates the chimes used in many grandfather clocks. Those chimes are usually uniform rods, except that they become narrow at the attachment point – the top. Since the bending strength for a rod varies as the radius to the fourth power, if the width at the top of the rod is reduced by, for example, a factor of 3, it is then 81 times easier to bend at that point. The result is not exactly a hinge, but it can be modeled using a hinge.

The lowest mode for the hanging rod is a pendulum-like mode that has a very low frequency (ideally zero), is not musically relevant, and is no longer considered in this discussion. The lowest three modes that are musically relevant are illustrated in Figure 7.5.

Grandfather chimes are normally struck with a hammer or mallet near the top (near the "hinge"). Striking the chimes will excite the lowest mode ($n = 1$) and also higher-frequency modes as well. Those higher-frequency modes give rise to the overtones. For the hanging rod, the ratios of the lower overtone frequencies to

Figure 7.5 First few modes for a hanging rod. The lowest mode, not shown, is a pendulum-like mode that is not important for musical chimes.

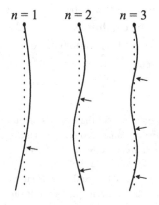

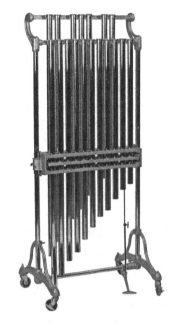

Figure 7.6 Orchestral chimes, also known as *tubular bells*, are examples of vibrating rods that are free at both ends. These chimes are roughly as tall as a person. (Photo by Stockbyte, Getty Images.)

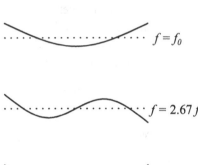

$f = f_0$

$f = 2.67 f_0$

$f = 5.41 f_0$

Figure 7.7 The lowest-frequency modes for a uniform free–free bar or rod, starting with the mode with two nodes. The motion is greatly exaggerated for the sake of illustration. Note that the nodes closest to the ends are in almost the same place for all three.

the lowest are close to an octave plus a minor second and an octave plus a minor sixth. That is, one might expect a "minor" sound. In fact, that is roughly what is produced by grandfather clock chimes. Interestingly, traditional church bells typically have overtones that correspond to minor intervals (although those intervals are not the same as for the grandfather clock chimes). The fact that both include minor intervals makes them sound somewhat similar.

An **orchestral chime** (also called a *tubular bell*; Figure 7.6) and many **wind chimes** can be modeled by a rod that is free at both ends. Some of the lowest of the free–free modes are illustrated in Figure 7.7. In this case, the modes with zero and one node are not important musically and are not shown.[4] Thus, in this case, when referring to the lowest-frequency mode, that refers to a mode with two nodes.

Of course, if a rod or bar had no support at all, then it would fall to the ground. In practice, the "free–free" rod is (gently) supported not at the ends but at one or both of the nodes of the lowest-frequency mode. For a uniform rod, those nodes are close to 22 percent of the length of the rod from each end. Since there is no motion at a node, a simple support there should not cause a problem. The outer two nodes for many of the higher-frequency modes are not that far away, so that position provides a satisfactory result for them as well. So, for example, the strings used to support wind chimes will usually be attached at roughly 22 percent of the length from the end.

Wind chimes are usually struck near their center, so the tone heard is (usually) dominated by the lowest-frequency mode. Orchestral chimes are typically struck on the end, exciting many higher-frequency modes. In fact, the perceived tone for an orchestral chime comes mostly from the relatively large fourth, fifth, and sixth overtones, which have frequencies in the ratio of $9^2:11^2:13^2$, which is close to 2:3:4. The sounds from these modes are perceived as being the first three (harmonic) overtones of a missing fundamental (see Chapter 4). The perceived pitch corresponds to that missing fundamental, one octave below the frequency of the fourth mode. See Figure 7.8. Since the tuning is approximate, some beating will often be present in the sound from these chimes.

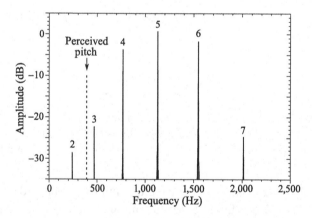

Figure 7.8 Measured spectrum for an orchestral chime used for the note G4 near 392 Hz. The perceived pitch is the (approximate) missing fundamental from the three largest overtones, which happen to have frequencies approximately in the ratio of 2:3:4.

---

[4] The solution with no nodes corresponds to translation at uniform speed. The solution with one node corresponds to rotation at uniform angular speed. In both cases, there is no return force from the bar and hence no oscillation.

(a)

(b)

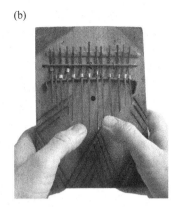

Figure 7.9 Music box chimes (*left*) and the thumb piano (kalimba; *right*) are musical examples that use vibrating cantilevers.

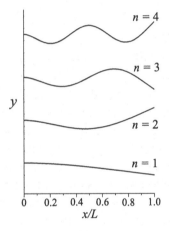

Figure 7.10 The lowest-frequency modes for a clamped–free bar, or "cantilever." The motion is greatly exaggerated for clarity.

Note that wind chimes and orchestral chimes are normally made from a length of pipe. Unlike organ pipes, however, the resonances of the air inside the pipe are not important contributors to the sound for these chimes.

The chimes in a typical music box and for a "thumb piano" (e.g., kalimba or mbira) are examples of clamped–free bars, or **cantilevers**. See Figure 7.9. Harmonicas rely on air-driven cantilevers for their sound. Some electric pianos also use cantilevers (tines) instead of strings. Figure 7.10 shows the first few cantilever modes. The pitch perceived is usually that of the lowest-frequency mode.

Cantilevers can also be used as a model for the reeds on clarinets, saxophones, and similar instruments, although only the lowest-frequency mode is usually relevant. Small cantilevers also find use in many nonmusical applications as well, including the accelerometers and gyroscopes used in cell phones and the atomic force and scanning tunneling microscopes used in the lab. Those microscopes are imaging systems that, under good circumstances, can resolve individual atoms on surfaces.

Note that for all these examples involving long bars or rods, the frequency depends on the inverse of the length *squared*. Hence, making a rod $\sqrt{2} = 1.414$ longer decreases the frequency by a factor of 2, one octave. In comparison, for the vibrating string, a change in length by a factor of 2 causes a change of one octave. For these chimes, a change in length by a factor of 2 causes a two-octave change, a factor of 4 in frequency.

The frequencies for all these bars also depend linearly on the thickness (through K). The material-dependent term, $(E/\rho)^{1/2}$, equals the speed of sound (for longitudinal or compression waves) within the material, which can often be measured independently. Hence, separate measurements of E and ρ are usually not necessary. In fact, one way to determine E is by measuring the speed of sound within, and the density of, the material and then computing E from those values.

Real Piano Strings

A brief revisit of vibrating strings is appropriate at this point. The idealized model for the vibration of a string results in mode frequencies that are harmonics of the fundamental. That is, the mode frequencies are integer multiples of the fundamental. The restoring force for the string vibrations is assumed to be solely due to the tension. A real piano string, however, like a chime, will also have some return force associated with bending. In fact, some of the lowest strings on a piano are thicker than the rods used for grandfather clock chimes. The highest-frequency strings on the piano are very short and have a width that, compared to their length, is also not unlike that of rods used for chimes. For a chime fixed at both ends – a hinged–hinged chime – the frequencies for the modes are (approximately) proportional to the square of integers, similar to the previous examples involving rods (Equation [7.3]).

A real piano string will be used under tension. If the return force for a string due to tension is much, much larger than the return force due to chime-like behavior, then one can hope that the string will behave more like an ideal string than a chime. Recalling that a 1 percent change in frequency is quite significant when tuning a musical instrument, even a small return force from bending can have a noticeable effect. In fact, the resulting inharmonicity is routinely observed for the tuning of pianos, making the overtones sharp (i.e., too high in pitch). Thus, in order to make the overtones of the lowest notes of a piano match the notes above them, those lowest notes must be tuned a bit low. Likewise, to make the highest notes match the overtones of the notes below them, they must be tuned a bit high. Compared to equal-tempered tuning, typical values found "by ear" range from about 3 percent low on the low end and 2 percent high on the high

Figure 7.11 A heavy string is wound around a central core to increase the mass of the string without a large increase in the bending return force.

end. This "mistuning" to make it sound good to the ear is quite significant. If the piano is tuned so that the fundamental frequencies of the strings simply match equal-tempered values, the piano will not sound as well in tune.

Piano strings used in the lower octaves, as well as those used for many other instruments, are often constructed using a wire that is spiral-wound on top of a solid wire core (Figure 7.11). In some cases, there may be more than one layer of spiral-wound wire. This is to provide mass with a minimal change in the bending strength and hence reduce the inharmonicity due to the chime-like return force. The wound wire will contribute much less to a bending return force, but it will certainly contribute to the mass of the wire – the inertial term. There is an art to making such strings, however. For example, if wound too tightly, the windings can rub against each other during vibration, which can cause a reduction in the quality factor due to the extra frictional losses.

It is easy to hear the inharmonicity in a metal wire "string" by tapping on a long length of solid piano wire that is under very little tension. While not very musical, such a string provides an interesting sound effect. Such a "string" with virtually no tension is better modeled as a very long, skinny chime. For an infinitely long chime, the hyperbolic sine and cosine terms must be zero, so the solutions are sinusoidal with distance, like waves on an ideal string. However, the relationship between wavelength and frequency for these sinusoidal waves is now frequency dependent. For the ideal string, it was found that $\lambda f = c$ (Equation [6.7]), independent of frequency. For a long string-like chime, $\lambda f^{1/2} =$ constant, so $\lambda f = c/\lambda$, which is what gives rise to the unusual overtones and the interesting sound effect. When the product, λf, depends on frequency (or wavelength), there is "dispersion." Dispersion is discussed more in Chapter 8.

Xylophone and Marimba Bars

While xylophone and marimba bars are perhaps not always thought of as being chimes, the basic physics for these bars is the same as that for chimes. A xylophone is any percussive instrument where the "chime" is made from wood (*xylo-* is derived from the Greek term meaning "wood"). Typically, there is one wooden chime for each note of the scale. Technically, a marimba is also a xylophone because it is made from wood, although in common usage (for modern Western music), both terms refer to distinct instruments. A marimba

Figure 7.12 A side view of a rectangular bar that has been "undercut."

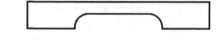

will have a significantly wider range than a xylophone. Similar instruments with bars made from metal are the glockenspiel and vibraphone.

The free–free analysis discussed previously can be used for a simple wooden bar with a rectangular cross section. That analysis showed that the overtones were not particularly musical. In addition, to obtain a wider range of notes, especially for the marimba, the relative lengths of the bars from low to high can become impractical. Both of these issues are addressed by tuning the bars using **undercutting**, as shown in Figure 7.12. While xylophones of various sorts have been around for a very long time, this tuning used for orchestral instruments is relatively modern, becoming prominent in the first half of the twentieth century.

The theory for the uniform bar no longer applies for undercut bars, of course, but the uniform bar can be used as a starting point. Recall that the vibrational frequencies come from the square root of a return force divided by the inertial term. The undercut bar will be much easier to bend about the center and hence has a smaller return force. The ends of the bar stay relatively straight but move significantly during vibration. Hence, the ends will still contribute most of the inertial term. Thus, undercutting will lower the frequency compared to the uniform bar. In addition, a comparison of the modes of the free–free bar (Figure 7.7) suggests that the higher-frequency modes may be affected differently than the fundamental. In fact, the depth and length (and other details) of the undercutting can be adjusted so that at least the first few overtones can be tuned to be a harmonic of the fundamental.

It has become tradition to undercut xylophone bars so that the first overtone is at three or four times the fundamental (an octave plus a fifth or two octaves), depending on local tradition, and for marimbas, that overtone is tuned to four times the fundamental (two octaves). Higher overtones that are still in the audible range can also be tuned. Since each piece of wood is different, the fine-tuning of these bars, which is done for higher-quality instruments, is a bit different for each bar. Less expensive instruments use imitation ("synthetic") wood bars, which can be manufactured with a much higher degree of uniformity.

Both xylophones and marimbas typically include a **resonator** under each bar that consists of a pipe that is open on top and closed farther down. The air in the pipe resonates, like a string fixed at one end and free at the other. That air resonance can be tuned by adjusting the position of the closed end.[5] When

---

[5] For decorative reasons, some pipes near the higher-frequency bars may be much longer than necessary. The closed "end" is near the top, and the pipe below that closure does not participate in the resonance. Pipes for the lowest frequencies may be bent to avoid hitting the floor.

tuned to match the pipe's fundamental, the bar's sound is reinforced. Some say the sound is "amplified" by the pipe, although that is probably a bad description. The sound is louder, but no extra energy is added by the pipe (after all, the resonator has no power source). The resonator creates a better coupling to the air, making the sound louder, and at the same time, it causes the bar to lose its energy faster – the quality factor is lowered. A vibraphone (or "the vibes"), which typically uses aluminum bars, has motor-driven plates that act as valves at the top of the resonator pipes. The plates rotate with time, a few times a second, causing the coupling to the pipes, and hence the volume, to oscillate in time (i.e., a **tremolo**). The oscillations will be at twice the plate-rotation frequency since the pipe is opened and closed twice for each full rotation of the plates.

Vibrations in Two Dimensions

Two-dimensional objects, such as drumheads, can also undergo vibrations. As was the case for the vibrating string, as long as the vibrations are small enough in amplitude, any vibration can be considered to be the sum of vibrational normal modes. **Each vibrational mode has a characteristic natural frequency and oscillates sinusoidally in time.** The computation of mode frequencies can be quite complicated; however, the basic description of each mode in terms of its nodes is still possible. In this case, nodes are described with **node lines** rather than points.[6] **The series of overtones that occur depends strongly on the shape and the conditions at the boundaries.**

The **ideal drumhead** is the two-dimensional version of the ideal string. The ideal drumhead has a return force that is due to a uniform tension, has a uniform mass density, and is fixed around the boundary. There are a number of mode calculators readily available, many with quality graphic interfaces, that allow users to visualize the modes. It is worth consulting one of these calculators to help visualize what is happening.

As a first example, consider a rectangular ideal drumhead, a **membrane**, that is fixed around the edges. While there are no real rectangular drumheads, this is a useful model to use as a starting point. The modes are simply described as the combination of sinusoidal oscillations both across and along the surface. Thus, a rectangular membrane that extends from 0 to L_x in the x-direction and 0 to L_y in the y-direction will have vibrational solutions that look like

$$z(x, y, t) = A \sin\left(\frac{n\pi}{L_x}x\right) \sin\left(\frac{m\pi}{L_y}y\right) \cos\left(\omega_{n,m}t\right), \qquad (7.4)$$

where z describes the wave displacement, and n and m are positive integers. Those integers are also used to identify the mode. In each case, there is an integer

---

[6] For an n-dimensional system, the nodes will be $(n-1)$-dimensional, so a three-dimensional oscillator would have node planes.

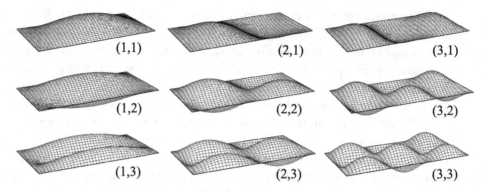

Figure 7.13 The first few modes for a rectangular membrane.

Figure 7.14
Simplified view of some of the normal modes for a rectangular membrane.

$+$ (1,1)	$+$ $-$ (2,1)	$-$ $+$ $-$ (3,1)

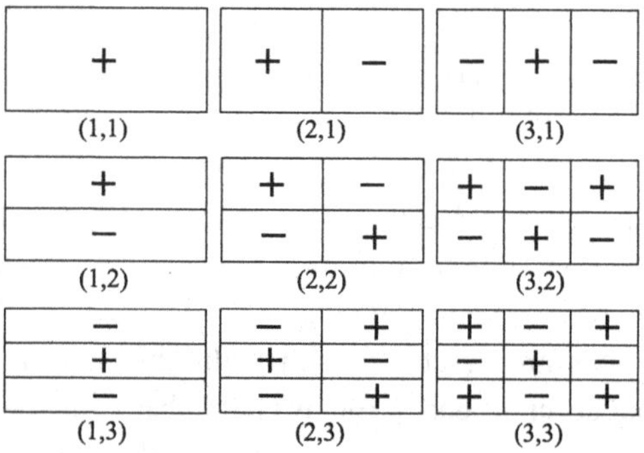

number of half wavelengths in both the x-direction and the y-direction. The natural frequencies of the modes are given by

$$f_{n,m} = \frac{c}{2} \sqrt{\left(\frac{n}{L_x}\right)^2 + \left(\frac{m}{L_y}\right)^2} , \qquad (7.5)$$

where c has units of speed and depends on the square root of the tension divided by the mass density – just like what was found for the vibrating string.

A picture, frozen in time, of the first few modes of a rectangular membrane, fixed around the edges, is shown in Figure 7.13 for the case where $L_x = 2L_y$. Figure 7.14 shows a simplified two-dimensional plot illustrating the node lines. In that simplified picture, "+" and "−" are used to describe the relative phases of the motion. That is, if you imagine the "+" regions as moving toward you, then the "−" regions are moving away. Of course, at a later time, the situation will be reversed since it is an oscillation. For such a drawing, at any fixed time, "+" must change to "−," or vice versa, when crossing any single node line. Because it is much easier to do so, vibrations of two-dimensional objects are often presented

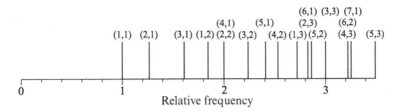

Figure 7.15 Spectrum for the rectangular membrane shown in Figures 7.13 and 7.14.

using these simplified drawings of the node lines. The "+" and "−" symbols can often be omitted.

A graphic illustrating the spectrum of natural vibration frequencies for the case when $L_x = 2L_y$ is shown in Figure 7.15. It is clear that adding a second dimension has made for a very complicated spectrum. Another interesting feature, which was not seen before, is that some very different modes happen to have the same natural frequency. That new feature will be addressed shortly.

As another example, consider a circular drumhead that is fixed around the edges (i.e., a circular membrane with a hinged edge). When there are oscillations involving an object with cylindrical symmetry, the solutions are not easily expressed using sines and cosines or, for that matter, x- and y-coordinates. The location on the membrane is more naturally described using cylindrical coordinates – a distance from the center, r, and an angle, θ, around the circle from a reference line. The solution for the radial direction involves Bessel functions. These are the same special functions that showed up for the hanging chain.

For a circular membrane fixed on the edges, the appropriate Bessel functions are the "mth-order Bessel functions of the first kind," usually designated J_m. To ensure that the edge is fixed, only those Bessel functions that are zero at the edge need be considered. Around the angular direction, the solution is sinusoidal with angle. So that the solution matches up with itself, only those sinusoids that have an integer number of full repeats can be included. That boundary condition would be expressed as saying that the motion at $\theta = 0$ must match the motion at $\theta = 2\pi$. Thus, the solution looks like

$$z(r, \theta, t) = A\, J_m\left(\lambda_{n,m} r\right) \cos(m\theta) \cos\left(\omega_{n,m} t\right), \tag{7.6}$$

where z describes the motion perpendicular to the membrane, m is an integer (in this case, including zero), and $\lambda_{n,m}$ is a value chosen so that the wave at the rim is zero. A few modes are illustrated in Figures 7.16 and 7.17, and the spectrum of natural frequencies is shown in Figure 7.18. Note that the modes are classified according to the number of node lines in the angular and radial directions. For the latter, the number of diameters is used for counting – each diameter is experienced twice as you go around the circle but is only counted once.

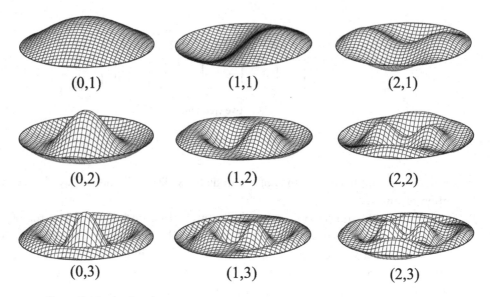

Figure 7.16 The first few modes for a circular membrane.

Figure 7.17
Simplified view of some
of the modes for
a circular membrane.

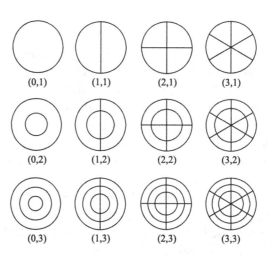

For membranes that cover an enclosed air space, such as the timpani ("kettle drum"), the extra return force from the enclosed air will also need to be taken into account. The so-called "timpani mode," (1,1), corresponds to a "sloshing mode" for the air, which involves less air compression than does, for example, the mode that is zero only around the edge, (0,1).

Another two-dimensional object that can vibrate is a **plate**, which is similar to a simple chime – that is, a plate with free edges, where the return force is due to the material properties. For example, a flat, uniform circular plate is a simple model for a cymbal and some gongs. As was found for the drumhead, the modes can be

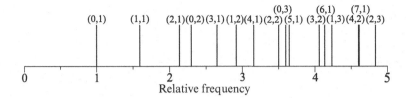

Figure 7.18 Spectrum for the circular membrane of Figures 7.16 and 7.17.

classified based on the number of node lines, and as found for the chimes, the mode frequencies scale with the speed of sound in the material, the thickness, and the inverse of an appropriate measure of the breadth of the plate, squared (e.g., *breadth* would refer to the diameter for circular plates, the length of the side for a square plate, and an appropriate average of the lengths of the sides for a rectangle).

The solutions for circular plates are a generalized version of the previously given solution for rods and bars, and they rapidly become mathematically complicated. The solutions involve a sum of Bessel functions and "modified Bessel functions." Modified Bessel functions are related to Bessel functions in the same way hyperbolic sines and cosines are related to sines and cosines. The details of these plate modes are summarized by Leissa (1969). The lower-frequency modes for a circular plate with free edges are shown in Figure 7.19, and the spectrum is shown in Figure 7.20.[7]

Degeneracies

For higher dimensions, especially with high symmetry, it may be that there are multiple modes that have the same oscillation frequency. In other cases, two modes may have very nearly the same frequency just by coincidence. The rectangular membrane in Figure 7.13 is an example, where the modes (4,1) and (2,2) have the same natural frequency. **Modes that have the same frequency of vibration are referred to as being *degenerate*.** If two modes have the same frequency, any combination of those two modes will also have that same frequency.

For a circular plate or membrane, the modes will always consist of doublets related by the rotational symmetry – the nodal diameters of one rest on the antinodes of the other, and vice versa. For a uniform square plate or membrane, doublets are related by a rotation of 90°. When the vibrations are set in motion, it is often difficult to excite just one of a doublet pair.

---

[7] For plates, a second material parameter, known as the *Poisson ratio*, is also required. Here that ratio is 0.30, a typical value.

Figure 7.19 Node lines for the lowest-frequency modes for a uniform circular plate with a free edge.

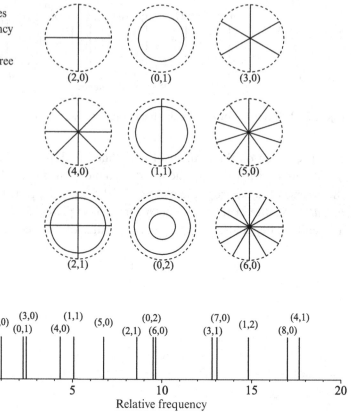

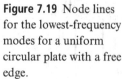

Figure 7.20 Spectrum of mode frequencies for a circular plate with a free edge.

Examples showing two degenerate modes for a square membrane and some combinations of those modes are shown in Figure 7.21. Additional examples can be found in Kang and Wei (2015). Many different node patterns may be possible, although they are not really new modes and can be written as a sum of simpler modes.

If degenerate modes have an additional coupling between them, even if that coupling is small, two different normal modes, with slightly different frequencies, will result. Musically, this means there may be beats in the sound.

To understand where the beats come from, consider a system consisting of two identical pendulums – call them pendulum 1 and pendulum 2 – to represent two modes. To start, the modes for the system can be considered to be the motions of each pendulum separately. Since they are identical, the modes are degenerate. To simultaneously describe the motion of both pendulums mathematically, that is, to describe the whole system at once, it is convenient to define the somewhat abstract symbols $\hat{x}_1$ and $\hat{x}_2$ to indicate the motions of pendulum 1 and 2, respectively, and $\mathbf{X}$ to symbolically refer to the motion of the entire system. This notation can be expanded to include any number of pendulums, although two pendulums are sufficient here.

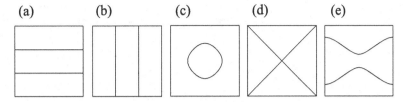

Figure 7.21 The (0,2) and (2,0) degenerate modes for a square membrane fixed around the edges are shown in (a) and (b). The sum of these modes is shown in (c), the difference in (d), and the (0,2) mode minus 2/3 times the (2,0) mode in (e). All of these oscillate with the same frequency. When possible, it is most convenient to use the modes shown in (a) and (b) as the normal modes, although this choice is not unique.

Example 7.1

If pendulum 1 is swinging with amplitude of 3.1 cm and pendulum 2 is stationary, the motion of the system is

$$\mathbf{X} = 3.1\ \hat{x}_1 \cos(\omega_0 t)\ \text{cm},$$

where ω_0 is the natural angular frequency of the pendulums.

If both pendulums are swinging together with an amplitude of 1.5 cm, the motion is

$$\mathbf{X} = 1.5\ (\hat{x}_1 + \hat{x}_2) \cos(\omega_0 t)\ \text{cm}.$$

If pendulum 1 is swinging with an amplitude of 1 cm and pendulum 2 is swinging with an amplitude of 4 cm in the opposite sense (i.e., pendulum 2 is going left when pendulum 1 is going right, and vice versa), the motion is

$$\mathbf{X} = (\hat{x}_1 - 4\hat{x}_2) \cos(\omega_0 t)\ \text{cm}.$$

Now add a small interaction between two identical pendulums, each of which, by itself, has a natural frequency ω_0. This can be modeled using a weak (and massless) spring, as shown in Figure 7.22. "Small" here means that the extra return force per unit displacement due to the spring is small compared to that due to gravity. For the motion shown in Figure 7.22a, the spring is not stretched or compressed, so the return forces experienced by the pendulums are unchanged, and hence the frequency is unchanged. However, for the motion shown in Figure 7.22b, the spring adds additional return force, so the frequency will be increased. Call the increase $\Delta\omega$. Then the motions shown can be described by

a. $\mathbf{X_a} = (\hat{x}_1 + \hat{x}_2)\cos(\omega_0 t) = \hat{x}_+ \cos(\omega_0 t)$ and

b. $\mathbf{X_b} = (\hat{x}_1 - \hat{x}_2)\cos\Big((\omega_0 + \Delta\omega)t\Big) = \hat{x}_- \cos\Big((\omega_0 + \Delta\omega)t\Big).$ (7.7)

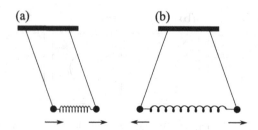

Figure 7.22 Coupled pendulums, where the coupling is modeled by a spring. For the motion in (a), the spring is not stretched or compressed, whereas in (b), the spring will add to the return force.

The new, even more abstract symbols $\hat{x}_+$ and $\hat{x}_-$ refer to the combined motions shown in Figures 7.22a and 7.22b, respectively. Since each of those motions has a single frequency, they can be used as normal modes. The motions of either of the two pendulums separately can no longer be described with a single frequency, and hence those motions cannot be normal modes.

Example 7.2

If the coupled pendulums are started from rest at $t = 0$ so that pendulum 1 has an initial amplitude of 1 cm and pendulum 2 starts at 0, what is the motion after they are released?

At any time, the system can be written as a combination of normal mode motions.[8] That is, for some constants A and B,

$$\mathbf{X} = A\hat{x}_+ \cos(\omega t) + B\hat{x}_- \cos\left((\omega + \Delta\omega)t\right).$$

At $t = 0$, this is

$$\begin{aligned}\mathbf{X}(t = 0) &= A\hat{x}_+ + B\hat{x}_- \\ &= A(\hat{x}_1 + \hat{x}_2) + B(\hat{x}_+ - \hat{x}_-) \\ &= (A + B)\hat{x}_1 + (A - B)\hat{x}_2.\end{aligned}$$

If there is no initial displacement of pendulum 2, then it must be the case that $A = B$. Then, to get the initial 1-cm displacement of pendulum 1, it must be that $A = B = 0.5$ cm.

Putting these values into the equation for the system motion yields

$$\begin{aligned}\mathbf{X} &= 0.5 \text{ cm } \left(\hat{x}_+ \cos(\omega t) + \hat{x}_- \cos\left((\omega + \Delta\omega)t\right)\right) \\ &= 0.5 \text{ cm } \left[\hat{x}_1\left(\cos(\omega t) + \cos\left((\omega + \Delta\omega)t\right)\right) + \hat{x}_2\left(\cos(\omega t) - \cos\left((\omega + \Delta\omega)t\right)\right)\right].\end{aligned}$$

The individual motions of pendulums 1 and 2 will exhibit a beat at frequency $\Delta\omega$.

[8] In general, the constants A and B need to also include phase-shift information.

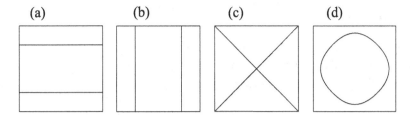

Figure 7.23 The degenerate (0,2) and (2,0) modes for a square plate, free on the edges, shown in (a) and (b), are coupled, resulting in normal modes that are approximately the difference (c) and sum (d) of the simpler modes. Due to the coupling in this case, the sum mode (d) has a higher natural frequency than the difference mode (c).

For vibrating systems, beating between modes (that would normally be expected to be degenerate) will often occur naturally and may be due to a number of smaller terms normally considered negligible and/or a result of imperfections in construction. For example, it is difficult to make a chime from a cylindrical bar or pipe that does not have some beating in the sound due to imperfections – the bar will not be exactly cylindrical. If a drumhead does not have uniform tension, the mode doublets may split and have different frequencies. For transverse vibrations of plates, bending forces naturally result in couplings between some otherwise degenerate modes. For example, the (0,2) and (2,0) modes of an idealized square plate (with free edges) will interact,[9] like the pendulum modes described previously, to give new modes that are (approximately) the sum and difference of the original idealized modes.[10] The node lines for such a combination are shown in Figure 7.23. Even the vibrating string, which has degenerate modes for vibrations in the vertical and horizontal transverse directions, can experience some couplings resulting in a circular (or elliptical), rather than back-and-forth, motion for the string.

When degenerate modes are excited simultaneously and/or if otherwise degenerate modes are coupled, interpreting the resulting node lines can become problematic. It may no longer be a simple matter to determine an appropriate index to use by counting the nodes.

The notation for the system motion, described earlier, can be used more generally for two (or more) oscillators (or modes) with different frequencies. In the general case, phase differences, other than in phase ("+") and opposite phase ("−"), can also be specified. This can be done by specifying a phase angle in addition to an amplitude, but it is more commonly accomplished using complex numbers – numbers that have a real and imaginary part. Complex numbers can be described using a magnitude and phase angle, so they are a natural for this use. Further consideration of complex numbers will, however, be deferred until Chapter 14.

---

[9] For a square plate, modes (n, m) and (m, n) will have a significant coupling if n and m differ by any even integer.

[10] There may be a coupling to other modes as well, although if they differ in frequency, their contribution to the normal mode will be small.

Example 7.3

Pendulum 1 oscillates with an amplitude of 1.2 cm and a natural frequency of ω_1, and pendulum 2 oscillates with an amplitude of 2.1 cm and a different frequency, ω_2. The motion of the system can be described as

$$\mathbf{X} = [1.2 \, \hat{x}_1 \cos(\omega_1 t) + 2.1 \, \hat{x}_2 \cos(\omega_2 t)] \text{ cm},$$

where, for simplicity here, the oscillators are taken to be in phase at $t = 0$.

Those familiar with the notation used to describe "vector quantities," quantities that have a magnitude and a direction, such as position, velocity, and similar measures, may notice the close similarities in notation. For example, in the case of the two pendulums, there is a direct connection to a two-dimensional motion of an object in a plane. In fact, much of the mathematics used for these vector quantities also works for, and is often used for, the description of vibrational modes. A vibrating string has an infinite number of modes, and so with such a notation, there are an infinite number of oscillators, $\hat{x}_n$. The notation expands to handle this situation, corresponding to an infinite-dimensional system. While such higher-dimensional systems may be difficult to visualize, they are simple to describe mathematically.[11]

If all of the normal modes of a system, $\{\hat{x}_n\}$, are known, then any motion of the system can be written as a combination of those modes.[12] That is, any motion of the system can be written

$$\mathbf{X} = \sum_n a_n \, \hat{x}_n \cos(\omega_n t + \varphi_n), \tag{7.8}$$

where ω_n is the normal mode frequency of mode n, and a_n and φ_n are appropriately chosen constants for the amplitude and phase (if needed) for the mode. In general, the summation includes all of the modes and is referred to as a **superposition** of the modes. While this notation may be very abstract and symbolic, when the description of the individual modes becomes complicated, this simplified representation can be useful. Notice the similarity to the Fourier series discussed in Chapter 4.

One musical application where coupled modes play a very significant role is for the acoustic string instruments, such as guitars and the violin family. As will be seen in more detail later (Chapter 15), the vibrating string does not produce much

---

[11] The notation shown is for a system with discrete ("countable") modes, which can be indexed using integers. The notation can be further generalized to describe a continuous system of modes, which would be indexed using real numbers. Such a continuous system is even harder to visualize.

[12] This might get more complicated if losses, such as from friction, are significant.

sound by itself. The string is thin and simply does not push on the air very effectively. However, the vibrations from the string are coupled to the body of the instrument, which is usually made of wood, and the body will have, by design, many resonances of its own. In addition, the air enclosed can have a Helmholtz resonator–like mode (Chapter 5), referred to as the *air mode*, that also plays a role. The coupling between the string and the body is through the bridge. Since the instrument's body has much more surface area, even relatively small vibrations can cause significant vibrations in the air. Hence, this coupling is essential for the instrument to produce a significant sound volume. On the other hand, couplings between the string and a particularly strong body resonance that is nearby in frequency can create an undesirable beat in the sound, analogous to what happens for the coupled pendulums. This results in an undesirable sound referred to as a *wolf tone*. The solution is to change the resonant frequency and/or coupling a bit, typically by adding a small weight (i.e., adding inertia).

Bells and Other Shapes

The vibrational modes of two-dimensional shapes that are bent into the third dimension, such as cymbals and bells, can be understood by starting with a corresponding two-dimensional shape (e.g., a circular plate). The node lines may move somewhat, due to the curvature of the material; however, the lines will look qualitatively similar. For example, the resulting modes for a bell will look something like those shown in Figure 7.24.

When there is cylindrical symmetry, the bell modes occur in degenerate doublets. Imperfections and/or couplings will result in a splitting of the doublet, and as discussed earlier, beats will result – a tremolo. Such a tremolo can be heard for many bells. Figure 7.25 illustrates slow beats for a 1-m-diameter church bell with

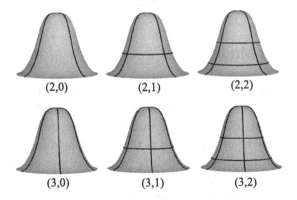

(2,0)   (2,1)   (2,2)

(3,0)   (3,1)   (3,2)

Figure 7.24 Approximate node lines for several of the lowest modes for a simple bell with cylindrical symmetry. Note that the locations in the vertical direction are estimates. Each of these modes is one part of a degenerate doublet.

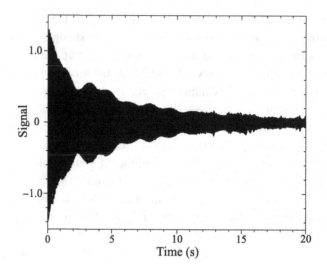

Figure 7.25 Recorded signal for a 1-m-diameter church bell. This bell has a fundamental at 228 Hz and prominent overtones at 399, 501, 589, and 832 Hz. The slow beats are due to imperfect tuning, in this case, mostly from the degenerate modes near 501 Hz.

a 228-Hz fundamental. Some bells are intentionally created without cylindrical symmetry and so naturally produce two different tones.

Church bells will be thicker at the top, where the bell is supported, and around the rim at the bottom, where it is struck. The computation of mode frequencies for bells, cymbals, and similar instruments is generally done numerically using finite-element methods. The *strike tone*, the perceived pitch when the bell is struck, is at the frequency of the first overtone. That first overtone is referred to as *prime*. The higher-frequency modes decay faster, so after a while, only the lowest mode (the fundamental) is heard. That mode is referred to as *hum*, and it may be as much as an octave below the strike tone. The next time there are church bells ringing nearby, listen carefully to the changes in the tone with time.

Torsional Modes

The discussion so far has been motivated by the transverse waves on a string and has considered transverse motions of more general objects. For these objects, there may be other types of motion. In particular, torsional, or rotational, motions may be present, such as those illustrated in Figure 7.26. If such modes occur at frequencies in the audible range, they may contribute to the sound.

The bars for the lowest notes on a marimba can have appreciable torsional motion that contributes to the sound in the audible range. If a uniform bar is struck in the exact center, torsional motion will not be excited. To ensure good sound in other cases, the bars can also be tuned so that the torsional motion is at

(a) 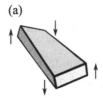   (b)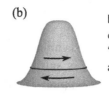

Figure 7.26 Examples of torsional, or "twisting," modes for a bar and a bell.

a harmonic of the fundamental. To do that, the undercut bar can also be shaped across the width of the bar. For the higher-frequency bars, the frequency of such motion is usually beyond that of human hearing and need not be tuned.

Bells can also have torsional motion, though that motion, by itself, does not produce much sound – such a motion does not effectively push the air. Those torsional motions may occur in combination with transverse motions, making tuning more challenging.

Visualizing Node Lines

The node lines on flat plates and membranes can often be visualized by orienting the objects horizontally, sprinkling sand on them, and exciting the resonant modes. The sand will move from locations where the vibration amplitude is significant – it is bounced away – and will settle in those regions with little or no motion. These are referred to as **Chladni plates and membranes**,[13] and this method is often used for demonstrations. The resonant modes may be excited with an external vibrator or by bowing, such as with a violin bow, on the edge.

Objects that are not flat, such as bells, can be investigated in a couple of ways. If the object is metal, a small electrical probe held near the object will form a capacitor (discussed more in Chapter 18) that is sensitive to local vibration. That small probe needs to be scanned over the surface to identify the nodes. Any motion sensor that is sensitive only to local motion can be used in a similar manner.

Another technique for curved objects is to use laser holography. Laser holography is mostly associated with a technique to record three-dimensional images of objects. Laser holography works by recording the phase differences between two coherent waves of (laser) light. One wave travels directly to a recording film. The other bounces off the target object on the way to the film. The combination of the two waves creates a record of the relative phases of the two waves at the film. The wave-interference effects involved are discussed more in Chapter 16. If the object is vibrating with an amplitude that is significant compared to the wavelength of the light (which is less than 1 μm), then a sum of many different phases will be recorded, with the details depending on the amplitude of the vibration

---

[13] Named for the German scientist Ernst Chladni (1756–1827), although he was not the first to use this technique.

compared to a wavelength. The resulting images show significant brightness variations, allowing a visualization of the motion and, in particular, of the node lines where there is no motion.

Sometimes it is sufficient to know mode frequencies. Those can often be determined by tapping on the object and observing the spectrum of the sound. There are many readily available tools ("apps") that can be used for this purpose.

Summary

Many properties seen for vibrating strings are also seen for other vibrating systems; however, the natural vibrating frequencies may not be harmonic and may be difficult to compute. The fundamental vibrational motion is described based on normal modes. In particular, the vibrational modes can be visualized and labeled by identifying the location of nodes, points where there is no motion for one-dimensional models of objects, node lines for two-dimensional and some three-dimensional models. The total vibrational motion of an object can be expressed as a sum of the motion of the normal modes. Some specific models for vibrating systems of musical relevance are presented as examples.

Some normal modes may have identical vibrational frequencies due to symmetry or coincidence. Such modes are referred to as being *degenerate*. In practice, degenerate modes can lead to beats due to small couplings between the modes or due to small imperfections in the manufacturing process.

For higher-dimensional objects, there may be many different types of motions, which may also exhibit modes, that may or may not contribute acoustically.

One way to study vibrational modes is to display the nodes. Experimentally, nodes can be visualized for flat objects using sand, which will move away from regions with high motion and will remain where the motion is minimal. More generally, some node lines have been visualized with the use of electrical couplings to conducting objects, laser holographic techniques, and other methods that are sensitive to local vibrational motion.

ADDITIONAL READING

Bhakta, H. C., V. K. Choday, and W. H. Grover. "Musical Instruments as Sensors." *ACS Omega* 3, no. 9 (2018): 11026–11032.

Fletcher, N. H., and T. D. Rossing. *The Physics of Musical Instruments*. Springer-Verlag, 1991.

Gordon, C., and D. Webb. "You Can't Hear the Shape of a Drum." *American Scientist* 84, no. 1 (1996): 46–55.

Kang, J-H., and B. Wei. "Degenerate Mode Shapes for Rectangular Membranes and Simply Supported Rectangular Plates." *Journal of Vibration and Control* 21, no. 8 (2015): 1633–1638.

Leissa, A. W. *Vibration of Plates*, NASA SP-160. National Aeronautics and Space Administration, 1969.

Martin, D. W., and W. D. Ward. "Subjective Evaluation of Musical Scale Temperament in Pianos." *Journal of the Acoustical Society of America* 33 (1961): 582–585.

Murray, C. J., and S. B. Whitfield. "Inharmonicity in Plucked Guitar Strings." *American Journal of Physics* 90, no. 7 (2022): 487–493.

Perrin, R., T. Charnley, and J. de Pont. "Normal Modes of the Modern English Church Bell." *Journal of Sound and Vibration* 90, no. 1 (1983): 29–49.

Rossing, T. D. "The Acoustics of Bells." *American Scientist* 72, no. 5 (1984): 440–447.

Suits, B. H. "Basic Physics of Xylophone and Marimba Bars." *American Journal of Physics* 69, no. 7 (2001): 743–750.

PROBLEMS

7.1 For the "free–free" chime, it was stated that the modes with zero and one nodes are not important musically. Illustrate the type of motion associated with those modes.

7.2 If possible, how might you strike a uniformly stretched, two-dimensional circular membrane (a drumhead) to most effectively excite vibrations while having the node lines illustrated? If it is not possible, explain why. What are some other modes that will also be excited if you strike those spots?

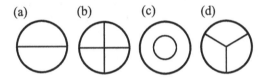

Figure P7.2

7.3 For each of the bells illustrated in Figure 7.24, choose one area between node lines and label it "+." Now label all the other (visible) regions based on that choice.

7.4 For a rectangular bar of length L, thickness h, and width w, with simple (hinged) supports at the ends (Figure P7.4), an added weight mg placed in the center of the bar will cause the center of the bar to sink a distance y, where

$$y = \frac{mgL^3}{4wh^3E}.$$

Using a convenient bar (such as a ruler or meterstick) and a known weight,[14] determine the Young's modulus, E, for the bar. Look up or determine the dimensions of the bar and the bar's mass and use them to determine the bar's density. Now predict the bar's lowest oscillating frequency when one end is

---

[14] A 1-L bottle of water is 1 kg plus the mass of the container – about 35 g for a typical one-time-use plastic bottle.

clamped to a table (without the added mass). How does theory compare to observation?

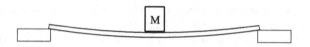

Figure P7.4

7.5 If a circular plate of radius 20 cm produces middle C (C_4) when it is struck near the center, what set of radii should be used to make a set of circular plates that will produce a major scale going up from middle C (C_4) to the C above middle C (C_5)? Assume the same plate material and thickness is used for all.

7.6 Three grandfather clock chimes are constructed with identical dimensions. One each is made using aluminum, steel, and bronze. The aluminum chime produces a sound with a frequency of 440 Hz. Predict the frequencies for steel and bronze. You will need to look up typical values for the material properties.

7.7 A series expression describing a string plucked in the center was shown in Example 6.2. Write out the first few terms in the series and compare to the notation of Equation (7.8). What are a_n and $\hat{x}_n$ for the plucked string?

8 Traveling Waves

Chapter 6, String Theory, dealt largely with wave motion on a string that is fixed at both ends and that is either plucked or driven sinusoidally. Another type of motion that can be observed on longer strings or ropes is that of a traveling pulse. Like waves coming onto a beach, such a traveling pulse has a sense of direction to its motion – it is not "standing." For a hypothetical ideal string or rope that is infinitely long, a wave pulse will travel indefinitely. To model many real systems involving waves, it will often prove convenient to divide the problem into two: a description based on traveling waves for an (ideal) infinite system, followed by a separate consideration of what happens when the disturbance gets to the boundaries of the system (e.g., the end of a finite string).

The motion of a pulse traveling along a string and the vibrational motion following a displacement (a pluck) seem to be very different types of motion. However, there is a common underlying description for both. A result that will be demonstrated in this chapter is that all waves on a string can be constructed out of the sinusoidal standing waves previously discussed, including traveling pulses. This works in reverse as well – all standing waves can be constructed as the sum of traveling waves.[1]

In some circumstances, it will be much easier to think about a problem in terms of standing waves, and in others, traveling waves will be the better choice. Why try to describe the waves coming into the beach using standing waves if traveling waves can be used and are a better match to the situation? In the same way, why use traveling waves for a violin string when there is no sense of travel? Traveling waves will often prove most useful for waves that move in more than one dimension – everyday examples include sound and light waves. To start, however, consider the simpler motion of a pulse traveling along an ideal string or rope.

---

[1] This statement applies to linear systems, including the ideal string, low-amplitude water waves, low-amplitude sound waves, and the electromagnetic waves that make up light and radio waves.

Motion of a Pulse

Figure 8.1 illustrates the creation of a single pulse at one end of a long rope of length L that is under tension. The rope is tied down at the other end. The pulse can be observed to travel along the rope, then reflect off the far end and return. With more tension, the pulse travels faster, and with less tension, slower. Aside from some decay in amplitude due to frictional losses, the shape of the pulse remains unaltered as it travels. The speed of the pulse can be measured by measuring the time, T, it takes to travel the full length and back, a total distance of $2L$. If friction were negligible, the pulse would continue to reflect at both of the ends with this repeat time T. This corresponds to a frequency $f = 1/T$. It is easily confirmed by measurement that this frequency is the same as the fundamental oscillation frequency for a string of length L, derived in Chapter 6, even though the motion seems quite different. More evidence that this is the case will be presented in what follows.

If the speed of a pulse on the string is c, and the total distance traveled to the far end and back is $2L$, then the total travel time is $T = 2L/c$, so $f = 1/T = c/2L$. Comparing this to the expression for the fundamental frequency for a string tied down at both ends (see Chapter 6), **the speed of travel is** $c = \sqrt{F_T/\rho}$. That is, the mysterious speed that was discovered for standing waves (Equation [6.8]) is equal to the speed for a traveling pulse on that same string. This is a first clue that there is a connection between these two types of string motion.

Figure 8.1 A pulse travels down a long rope with constant speed. The pulse will be seen to reflect back at the end of the rope. ★

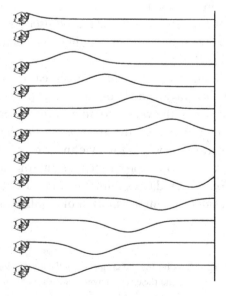

Mathematical Description of Traveling Waves

The general description of traveling waves includes short-duration signals, or pulses; longer-duration periodic signals; and more general shapes that may include elements of both. The important feature is that there is some shape to the wave, and that shape travels with time.

Traveling-Wave Pulses

In the absence of friction, a traveling pulse on an (infinite) ideal string will maintain its shape. That shape will simply slide along with time. Let that shape be described by a mathematical function, $F(s)$. That function depends on one variable, s. That is, a single numerical value is plugged in to find the result. The function F does not change with time, so every time you plug in a value, for example, the value 3, you get the same result, $F(3)$. The shape function, F, can be any valid mathematical function. The function could describe a single pulse, multiple pulses traveling together, or even a much more complicated shape.

To describe a traveling pulse, which certainly does change in time, the single variable s is made to depend on time and position. For a wave traveling along the x-direction with constant speed c, the wave motion at a position x and time t is given by $F(s)$, where $s = x - ct$. Two values, x and t, are combined to determine the appropriate single value, s, to put into the shape function, F. Hence, **the displacement of the traveling pulse on the string, y, at any position along the string, x, and time, t, can always be written as**

$$y(x, t) = F(x - ct).\tag{8.1}$$

The reverse is true as well. Any pulse that can be put in this form is a traveling pulse.

Example 8.1

Consider the particular shape function

$$F(s) = e^{-s^2/9}\cos(5s/2),$$

which is illustrated in Figure 8.2a. For reference, two particular values are identified in the figure: $F(0) = 1$ and $F(3) = 0.12752$.

To make this shape travel along x with speed $c = 2$, we need $s = x - 2t$, or equivalently, $x = s + 2t$. Hence, when $t = 0$, $x = s$, and when $t = 0.5$, $x = s + 1$, and so on. That is, when $t = 0$, the value $F(0)$ is observed at $x = 0$, and the value $F(3)$ is observed at $x = 3$. However, when $t = 0.5$, $F(0)$ is observed at $x = 1$, and $F(3)$ is observed at $x = 4$. The entire wave has moved 1 unit to the right during this time. Similarly, the wave is 1 unit to the left when $t = -0.5$. The motion of this wave pulse is illustrated in Figures 8.2b–8.2d.

Figure 8.2 (a) The
example shape function
$F(s)$.
(b)–(d) A traveling wave
based on $F(s)$ that is
traveling along x for
three different times. The
two dots on each curve
represent the locations of
$F(0)$ and $F(3)$ on the
respective curves.

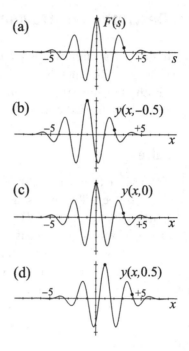

To make a pulse travel toward $-x$, use $s = x + ct$ instead of $s = x - ct$. That is the same as changing the sign on the speed, c; however, speeds are normally kept as positive values. Thus, a minus sign in the definition of s results in a pulse moving toward positive x, and the plus sign in the definition results in a pulse moving toward negative x.

It is common to see traveling pulses written mathematically without explicitly writing out a shape function, F, and variable s. For example, using the wave shape from the previous example, if the pulse were traveling toward negative x with speed c, it could be written directly as

$$y(x,t) = e^{-(x+ct)^2/9} \cos\left(5(x+ct)/2\right),\qquad(8.2)$$

and the numerical constants (in this case, 9 in the exponential and 5/2 in the cosine) will have units associated with them that cancel any units associated with $s = x + ct$. As written here, s will have units of length, for example, meters for Standard International (SI) units. Hence, if the intent is to use SI units for x and t, then the constant 9 in the exponential must have units of m², and the constant 5/2 in the cosine must have units of $1/m = m^{-1}$.

It is not always obvious whether or not an expression represents a traveling pulse. Consider the following examples.

Example 8.2

Consider the expression

$$y(x,t) = \frac{3x}{4x^3 - 16x^2t + 16xt^2 + 5x},$$

where the units (not shown) are appropriate for x in meters and t in seconds. Is this a traveling pulse, and if so, what is its speed and direction?

To answer this question, try to put it into the form of a shape function that depends on a single value, $s = x \pm ct$. In this case, first divide the top and bottom by x, and then note that the quadratic in the denominator can be factored to give

$$y(x,t) = \frac{3}{4(x^2 - 4xt + 4t^2) + 5} = \frac{3}{4(x - 2t)^2 + 5}.$$

Since the wave depends only on the combination $x - 2t$ and constants, this is a traveling pulse with speed $c = 2$ m/s traveling in the positive x-direction. The point here is not to get into a lot of algebra. What is important is the fact that whether or not an expression describes a traveling pulse depends on whether it *can* be written in a form involving a shape function and $s = x \pm ct$, not whether it *is* written in that form.

Example 8.3

Consider

$$y(x,t) = \frac{2x + 6}{x^2 - 16t^2 + 6x + 9},$$

where, again, x and t are in SI units. Can this be described in terms of traveling pulses? Sparing the details of the math, this expression is equivalent to

$$y(x,t) = \frac{1}{(x - 4t) + 3} + \frac{1}{(x + 4t) + 3},$$

so this looks like the combination of two traveling pulses, each traveling at 4 m/s; however, one is going toward positive x, and the other is going toward negative x.

The latter example, of course, could not be of a wave on a string since at certain times and positions, x and t, the displacement, y, becomes infinite. A string cannot do that. In fact, this function may not describe a physical situation at all. However, it is an example showing that rather complicated mathematical expressions might have a simpler physical interpretation if they are written in the right way – even if they do not directly represent a physical situation. Knowing that can sometimes help to visualize a function's behavior – in this case, there are two pieces that move (like pulses) in opposite directions with time.

Sinusoidal Traveling Waves

In the language of physics, the traveling pulses just described are all referred to as **traveling waves**, even if there is no "waviness" to them. If a traveling wave has a sinusoidal shape function, that is, a sine or cosine, then it is a **sinusoidal traveling wave**. Motivated by the ideas from Chapter 4, it is worth considering such waves, even though they have the unrealistic property that they extend to infinity in both directions. Mathematically, a sinusoidal wave will look like

$$y(x, t) = A \cos\left(k(x - ct) + \varphi\right) = A \cos(kx - \omega t + \varphi), \tag{8.3}$$

where A, k, c, and φ are numerical constants with appropriate units, and the cosine was chosen for this example, although a sine function can also be used. These constants are as follows:

A = the amplitude of the wave;
k = the "wave number" for the wave (and not a spring constant);
c = the speed of the wave;
$\omega = kc$ = the angular frequency (in radians per second) of the wave; and
φ = a "phase factor," usually in radians.

Many of these constants were already defined for standing waves on a string. However, it is worth reviewing them here in this new context.

The units for the amplitude A depend on the type of wave. For waves on a string, $y(x, t)$ describes a displacement, so A will be in distance units (e.g., meters). Be aware that it is common to place the numerical value associated with A ("A's magnitude") at the beginning and the units of A at the end of the expression, even though they go together.

The wave number k is a scaling factor that will have units of 1/length. The wave number converts distances into the units that the cosine function requires (radians). The speed of the wave, c, was already described. The phase factor is just a simple value, in radians, that generally must be included, and its effect is equivalent to an offset in time. Hence, the phase factor will depend on when the clock (used to measure t) was started. For many traveling-wave problems and many of the examples that follow, it is okay to assume, for simplicity's sake, that the phase factor is zero. As will be seen later (Chapter 16), when there are two (or more) waves present simultaneously, such an assumption may not always be possible.

Sinusoidal traveling waves are one example of periodic traveling waves. A periodic wave frozen in time will repeat exactly after a distance, λ, the wavelength, and at any fixed location with a time, T, the period. That is, for all integer values of n and m,

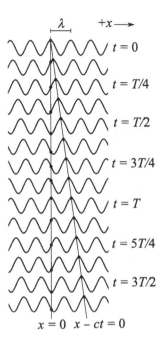

Figure 8.3 A sine wave traveling toward $+x$. For fixed time, the wave looks identical if you move by a wavelength. As the wave evolves in time, it will also look identical after one period, T, two periods, $2T$, and so forth.

$$y(x + n\lambda,\ t + mT) = y(x, t), \tag{8.4}$$

and the speed of this wave will be given by $c = \lambda/T = \lambda f$. This is illustrated for a sinusoidal wave in Figure 8.3.

Traveling sinusoidal waves are often described by specifying their frequency, f (in Hz), and their wavelength, λ (in meters). As has already been shown, the relation between angular frequency (radians/second) and frequency (repetitions/second) comes from the fact that the sinusoidal functions repeat for each 2π radians. That is, 2π radians = 1 repetition. Hence, $\omega = 2\pi f$, or equivalently, $f = \omega/(2\pi) = 1/T$, where T is the repeat time for (i.e., the period of) the motion at a fixed position on the wave. The wavelength, λ, is the repeat distance for the wave at a fixed time.

Keeping time fixed, the sinusoidal function will repeat when the product kx changes by 2π. Hence, $k\lambda = 2\pi$, or equivalently, $\lambda = k/(2\pi)$. Furthermore, $\omega = kc$ leads directly to $\lambda f = c$, a result derived previously for standing waves – a relation that is such a fundamental part of wave motion that it would be good to commit it to memory.

Thinking about units may make the relationships less difficult to remember. The speed must be in meters per second. The wavelength is meters per repetition (at fixed time), and the frequency is repetitions per second (at fixed position). The only way to combine those to get the units of speed is $\lambda f = c$:

$$\frac{\text{meters}}{\text{repetition}} \times \frac{\text{repetitions}}{\text{second}} = \frac{\text{repetitions}}{\text{repetition}} \times \frac{\text{meters}}{\text{second}} = \frac{\text{meters}}{\text{second}}. \tag{8.5}$$

The sinusoidal traveling wave may also be written

$$y(x,t) = A\cos\left(\frac{2\pi}{\lambda}x - \frac{2\pi}{T}t + \varphi\right),$$ (8.6)

where the "conversion factors" to get from distances (x) and times (t) to the radians that the cosine wants to see are "2π radians per repeat length" and "2π radians per repeat time." And of course, recall that frequency, f, is the inverse of the repeat time, T: $f = 1/T$.

It is worth looking at some specific examples. In each case, think about the repeat time for fixed position and the repeat distance for fixed time, along with the factor of 2π that converts "repeats" to "radians." Thinking in these generalities, rather than trying to memorize each of the various relationships, is strongly encouraged and is probably more fruitful in the long run.

Example 8.4

Consider the wave described by

$$y(x,t) = 2.3\cos(72x - 95t) \text{ cm},$$

where the units are such that x and t are in meters and seconds, respectively. What are the amplitude, wavelength, frequency, and speed for this wave, and in which direction is the wave traveling?

The amplitude, A, is the overall scale factor that is usually out front, along with its units, often at the end. Hence, $A = 2.3$ cm. Note that the "cm" specified at the end goes with the value 2.3 at the beginning and *none of the other values in between*.

The value 72 converts the distance x, which is in meters, to radians. Hence, $72 = 2\pi/\lambda$ or $\lambda = 2\pi/72 = 0.0873$ m $= 8.73$ cm. The value 95 converts the time into radians so that $95 = 2\pi/T = 2\pi f$, so $f = 95/(2\pi) = 15.1$ Hz. The speed is found by multiplying the wavelength and frequency: $c = 0.0873 \times 15.1$ m/s $= 1.32$ m/s. Note that the speed can be found a bit faster working directly from the two factors ω and k – the values of 2π used earlier in the intermediate steps will simply cancel. Since the factor $k = 72$ must have units of 1/m ("per meter") and the factor of $\omega = 95$ units of 1/s ("per second"), meters per second is found from $c = \omega/k = 95/72 \; (1/\text{s})/(1/\text{m}) = 1.32$ m/s.

There is a minus-sign difference between the two factors (72 and 95), and the spatial variable is x, so this wave is traveling in the $+x$-direction.

Example 8.5

Write down the mathematical description for a sinusoidal ultrasonic wave with wavelength 0.85 cm and frequency 40 kHz, traveling along the x-direction toward $-x$, if it has an amplitude of 1.2 pascals (Pa).

Ultrasonic? Pascals? It does not matter if these terms are (as-yet) unknown. The solution will be:

$$y(x,t) = 1.2 \cos\left(\frac{2\pi}{0.0085\text{m}}x + \frac{2\pi}{1/40,000 \text{ Hz}}t\right)\text{Pa} = 1.2\cos(739\,x + 251,000t)\,\text{Pa},$$

where x is in meters, and t is in seconds. A cosine function with a phase of zero was chosen. A sine function and/or the addition of a phase factor would have led to an equally valid answer. The information given does not specify a phase factor, so any convenient value can be chosen. By the way, the speed of this wave is $c = \lambda f = 0.85$ cm $\times$ 40 kHz $= (0.0085 \times 40,000)$ m/s $= 340$ m/s.

Traveling and Standing Waves

It is straightforward to show that a traveling sinusoidal wave can be written in terms of sinusoidal standing waves, and vice versa. That connection results from the same mathematical identities discussed earlier in connection with "beats" (Chapter 3). Those identities are summarized in Appendix A.

Example 8.6

Consider the traveling wave

$$y(x,t) = 8.4\cos(3.2x - 78t) \text{ cm},$$

where x and t are assumed in SI units. The wave has an amplitude of 8.4 cm and is traveling toward the $+x$-direction with speed $78/3.2 = 24.4$ m/s. Using the identity for the cosine of a difference (see Appendix A),

$$\cos(A - B) = \cos A \cos B + \sin A \sin B,$$

the traveling wave can be written, substituting $A = 3.2x$ and $B = 78t$,

$$y(x,t) = 8.4\cos(3.2x)\cos(78t) \text{ cm} + 8.4\sin(3.2x)\sin(78t) \text{ cm},$$

which has products of sinusoidal functions involving space (x) and time (t). Those are just two standing waves with the same wavelength, frequency, and amplitude as the original traveling wave, and they have no sense of travel separately. The fact that one standing wave involves cosines and the other involves sines is, in this case, of crucial importance for the combination to exhibit travel toward $+x$.

Example 8.7

Consider the standing wave

$$y(x,t) = 8.4\cos(3.2x)\cos(78t) \text{ cm}.$$

A relationship for any values A and B (see Appendix A) is

$$\cos A \, \cos B = \frac{1}{2}\left(\cos(A+B) + \cos(A-B)\right),$$

so that, substituting $A = 3.2x$ and $B = 78t$,

$$y(x,t) = 8.4\cos(3.2)\cos(78t) \text{ cm} = 4.2\cos(3.2x + 78t) \text{ cm} + 4.2\cos(3.2x - 78t) \text{ cm}.$$

The standing wave is equal to the sum of two traveling waves moving in opposite directions, each with the same wavelength and frequency, although one-half of the amplitude, of the standing wave.

In the same way that standing and traveling sinusoidal waves are related, so, too, are more general disturbances, such as the pulses discussed at the beginning of this chapter. That is, consider a string tied down at both ends, with a pulse traveling back and forth. Such a pulse can be described using two waves consisting of an infinite series of pulses that are simple traveling waves. Mathematically, those traveling waves extend beyond the end of the string; however, the solution is valid only over the length of the string. This is illustrated in Figure 8.4. The conclusion is that **all standing waves can be described as a sum of traveling waves**.

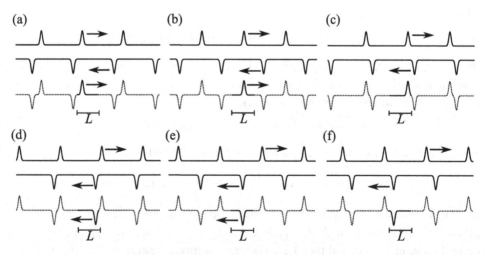

Figure 8.4 A sequence, (a)–(f), showing that the sum of two traveling pulse trains going in opposite directions describes the motion of a pulse traveling on a string tied down at both ends. ★

The pulse traveling between the two ends of the string can also be written in terms of sinusoidal standing waves. Fourier's theorem was previously used in Chapter 4 to look at functions that depend on time. The same theorem works for functions of position. Fourier's theorem says that any function can be written as a sum of sinusoidal functions. For a string tied down at both ends, one need only include the sinusoidal functions that are always zero at those ends. Each of those functions corresponds to a standing wave that, for the ideal string, has a known behavior. So if the motion of the string is known at some specific time, say, $t = 0$, that initial motion can be described as a sum of sinusoidal standing waves. The subsequent motion is found by adding the appropriate frequency dependence for each of those initial sinusoids. This is illustrated in Figure 8.5, where a pulse traveling back and forth is constructed from the sum of sinusoidal standing waves. The same mathematical analysis can be used for a string that is incredibly long so that in practice, the pulse never gets to the end. Thus, **all traveling waves can be written as the sum of standing waves**.

Of course, if the pulse can be constructed from sinusoidal standing waves, and each of those standing waves can be described as the sum of two traveling

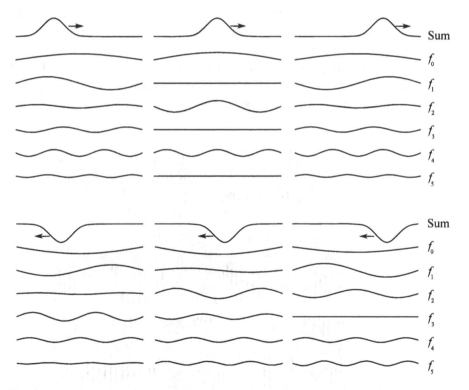

Figure 8.5 A pulse traveling back and forth, reading clockwise, can also be constructed using the sum of harmonic sinusoidal standing waves. Only the first few harmonics are shown in the figure. ★

sinusoidal waves, then the original pulse can also be constructed from a sum of sinusoidal traveling waves, and vice versa.

Polarization

Thus far only transverse waves, such as those seen for a string, have been used as examples. Here, *transverse* refers to the fact that what is waving – in this case, the motion associated with the wave – is perpendicular, that is, "sideways," compared to the direction of travel for the traveling wave. For an ideal string, the wave motion of the string is always sideways as the wave moves along the direction of the string. There are an infinite number of directions that are perpendicular to the string – for a horizontal string, transverse waves can be in the vertical direction, the horizontal direction, or at any angle in between. Only two distinct transverse directions are necessary, however, since all the "angles in between" can be constructed using the sum of two waves, one along each of the two distinct directions chosen. For example, thinking in terms of an x-, y-, z-coordinate system, if the string is along x, then wave motions in the y- and z-directions are both transverse to the direction of travel, and any transverse motion in the y–z plane can be considered to be a combination of separate motions in the y- and z-directions. Any such wave is referred to as having **transverse polarization**.

A long spring can exhibit another type of wave motion that is along the direction of travel. A compression at one end will travel down the spring – see Figure 8.6. Such a wave is referred to as having **longitudinal polarization**. A long

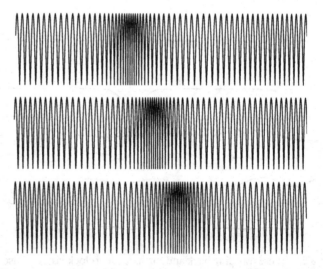

Figure 8.6 Compression waves on a long spring are examples of longitudinal waves, where the motion associated with the wave is along the same direction as wave travel. ★

spring is, of course, also capable of transverse wave motion. A compression in a rod or bar is also an example of a longitudinal wave.

The coordinates sometimes used to describe longitudinal waves may be confusing. The position along the spring, in the absence of the wave, is designated with one coordinate, say, x, and the time-dependent wave motion from that point, which is also along the x-direction, may be designated by another, say, $y(x, t)$. This use of the variables has the advantage that transverse and longitudinal waves are treated the same, mathematically, but at the same time, it creates the odd circumstance that both x and y refer to the same spatial direction – one being time independent and one being time dependent.

All types of waves will have a polarization. Electromagnetic waves in air or vacuum (i.e., light and radio waves) can only be transverse. In that case, nothing is physically moving, but "what is waving" (the amplitude of electric and magnetic fields, each of which is in a direction perpendicular to the travel direction) is perpendicular to the motion. Such waves are often characterized in terms of vertical and horizontal polarization. Polarized glasses will block one sense of polarization. Sunlight includes both polarizations in a random way. Due to the physics of the reflection process, sunlight that has been reflected off flat surfaces is predominantly polarized parallel to the reflecting surface. For example, for a pool of water, the surface of which is horizontal, the reflected light will be predominantly horizontally polarized. Hence, if polarized glasses allow only vertically polarized light but block horizontal, less reflected light – that is, less "glare" – will be visible. Radio waves behave in a similar manner, where, in addition to any reflections along the way, the polarization sent and received depends on the type and orientation of the sending and receiving antennas, respectively.

In contrast, as will be discussed in more detail in Chapter 12, **sound waves in air** (or more generally, sound waves in "ideal fluids") ***can only be longitudinal***. They are compression waves. Thus, there is no way to use polarization to selectively block reflections of sound waves.

Traveling Waves Transport Energy

For a pulse on a string, the pulse travels along the string; however, the string itself does not travel. Each piece of the string will move as the pulse goes past, but then it will return to its original position. While the pulse does not cause a net transfer of material, it can be used to transfer information (e.g., that a pulse had occurred) and energy. Some effort was necessary to start the pulse, and that effort, so to speak, travels down the string.

For a sinusoidal wave, the energy in the wave will be proportional to the amplitude of the wave squared. For an infinite string, which is simply a model used for mathematical convenience, the total energy of a sinusoidal wave would be infinite, and thus total energy is not a useful concept for these models. However, the energy per unit length – an energy density, which remains finite – is meaningful. The rate at which energy is delivered or "used up" is called **power**. Power is energy per unit time. In SI units, power is measured in watts.[2] For a traveling wave, the power transported (neglecting friction), P, depends on the amplitude squared and the speed of travel:

$$P \propto A^2 c. \tag{8.7}$$

The constant of proportionality will depend on the specific circumstance. For a sinusoidal wave with angular frequency, ω, on an ideal string that has mass per unit length, ρ,

$$P = \frac{1}{2}\rho\omega^2 A^2 c. \tag{8.8}$$

This, of course, will be the same as the power (the "effort") necessary to create the wave in the first place. The relations $c = \lambda f = \sqrt{F_T/\rho}$ can be used to write Equation (8.8) in many alternative forms, if desired.

Other types of traveling waves also carry energy, including sound, light, and radio waves; water waves; and so forth. A notable exception is the "matter waves" of quantum mechanics. They involve energy but have a more complicated interpretation.

Dispersion

The foregoing discussion is for waves on an ideal string and other systems that behave like an ideal string. The ideal-string model assumes that the return force arises solely from the tension. As discussed in Chapter 7, real strings, including piano strings, will have some tendency to straighten, even in the absence of tension. This "imperfection" compared to the ideal-string model can give rise to **dispersion**. For sinusoidal traveling waves, dispersion means that the speed is not a single constant value but depends, at least somewhat, on frequency (or equivalently, wavelength). That is, λf is not a single, constant value.

When a system exhibits dispersion, a pulse will not, in general, maintain its shape as it travels. The usual description of this effect first appeals, once again, to Fourier's theorem. The traveling pulse can be considered to be the sum of sinusoidal traveling waves. Each of those sinusoidal waves has a wave speed. For the ideal string, that speed is a constant, so the sinusoidal waves all move together. When that speed depends on the wave's frequency, the pulse will change

---

[2] Named for the Scottish inventor and engineer James Watt (1736–1819). Based on considerations basic to thermodynamics, Watt made fundamental improvements to the performance of the steam engine.

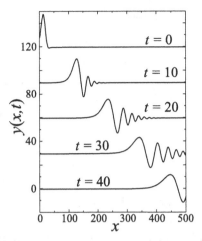

Figure 8.7 An illustration of the behavior of a simple pulse when dispersion is present. In this example, the higher-frequency components travel faster than the lower-frequency components.

shape with time because the sinusoidal waves do not move together, and hence how they sum will depend on time. An example of the evolution of a pulse with time when there is dispersion is shown in Figure 8.7.

Dispersion for visible light waves as they travel through water droplets and reflect back toward the source is what gives rise to rainbows. Dispersion is also observed for shallow-water waves. Dispersion for sound waves along long metal cables ("long chimes"; see Chapter 7) can give rise to some interesting sound effects.

Summary

A transverse disturbance will travel along an ideal string under tension. Such motion can be used as a model for other one-dimensional systems. The disturbance on a string can be described mathematically using a shape function that moves along the string with constant speed, c. For one-dimensional motion, the description relies on a shape function that depends on one variable, s, that is in turn determined by two variables describing position, x, and time, t, such that $s = x \pm ct$. A special type of traveling wave is based on an (infinite) sinusoidal shape function.

Standing waves can be described as the sum of traveling waves, and traveling waves can be described as the sum of standing waves. More complicated waves can always be described as the sum of sinusoidal waves.

Waves have a polarization. The waves on a string are examples of transverse waves, where the wave motion is perpendicular to the direction of travel. A long spring can also exhibit wave motion along the direction of travel, referred to as *longitudinal polarization*. Sound in air will always be longitudinally polarized.

Traveling waves can transport energy. Systems where the wave speed depends on the frequency of the wave exhibit dispersion, where a simple time-independent shape function cannot be used.

ADDITIONAL READING

Many introductory physics texts cover traveling-wave motion in simple systems. Consult any such text. Examples include:

Knight, R. D. *Physics for Scientists and Engineers*, 3rd ed. Pearson, 2013. (See Chapter 20.)

Urone, P. P., and R. Hinrichs. *College Physics*. OpenStax, Rice University, 2020. (See Chapter 16.)

PROBLEMS

8.1 Sounds in the range of human hearing are generally considered to be from 20 Hz to 20 kHz. If the speed of sound in air is 340 m/s, what are the corresponding wavelengths for traveling waves in air at 20 Hz and 20 kHz?

8.2 Observe, either directly or on video, surface-water waves coming in toward a beach. Estimate the wavelength, frequency, and wave speed. What is the polarization of these waves? If a sinusoidal traveling wave were used to model those waves mathematically, how well would it work? That is, which behaviors would such a model describe well, and which would it describe poorly? How about surface waves farther out, away from the beach?

8.3 A traveling wave is described by

$$y(x,t) = 7.6\cos(32x + 96t) \text{ mm},$$

where x and t are in SI units.

  a. What are the units associated with the constants 32 and 96?
  b. What is the amplitude of the wave?
  c. What are the wavelength and frequency of the wave?
  d. What are the speed and travel direction for this wave?

8.4 The speed of sound in air is about 340 m/s, whereas the speed of sound in water is about 1,500 m/s. Compare the wavelengths for a 440-Hz sound wave in air and in water.

8.5 The power transported by a traveling sinusoidal wave on a string is given by $P = \rho\omega^2 A^2 c/2$.

  a. How much energy is contained in each wavelength, λ, of such a traveling wave? Recall that power is energy per second.
  b. Ignoring friction, how much energy would it take to create the lowest-frequency *standing* wave, with a frequency of 440 Hz and an amplitude of 1.0 mm, on a string with a length of 33 cm and a tension of 12 N, tied down at both ends?
  c. Compare the values from part (b) to the power used by various electronic devices.

8.6 A string that is along the z-direction has transverse waves that can be described using polarizations in the x- and y-directions, where x and y are perpendicular to z and to each other. Compare and contrast these traveling waves on a string described by the sum of two sinusoidal waves, one polarized along x and the other along y, with displacements given by the following:

a. $x(z,t) = \sin(2\pi(z-t)); \quad y(z,t) = \sin(2\pi(z-t))$.
b. $x(z,t) = \sin(2\pi(z-t)); \quad y(z,t) = \cos(2\pi(z-t))$.

These waves have an amplitude of 1 unit, a wavelength of 1 unit, and a period of 1 unit.

(Hint: Start by making an x–y plot for the motion at a fixed value of z, say, $z = 1/4$, for several equally spaced times between 0 and 1. Then consider a different value of z, say, $z = 3/4$, using the same times.)

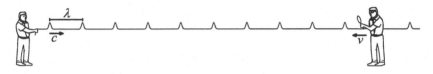

Figure P8.7 If a sender sends N pulses per second, how many pulses per second does a moving observer measure? The size of the pulses has been exaggerated for clarity.

8.7 Pulses are sent down a (hypothetical) infinitely long (ideal) string. One pulse is sent every T seconds, and the pulses travel with speed c. An observer located away from the source counts the pulses as they pass by. The observer is walking toward the source with speed v. A snapshot of the situation at a fixed time is illustrated in Figure P8.7. Consider the following questions based on this information:

a. What is the spacing between the traveling pulses on the string, λ?
b. What is the frequency, f, of the pulses that are sent down the string by the source – that is, how many pulses per second are sent? Is it true that $c = \lambda f$ for these pulses?
c. What is the frequency of pulses observed by a stationary observer (i.e., if $v = 0$)? How many pulses are observed during a time t (assume $t > T$)?
d. If the observer is moving with speed v, how far does the observer move toward the source during a time t?
e. How many pulses are counted during the time t by the moving observer compared to a stationary observer? (Hint: Assuming pulses start from the same location at $t = 0$, consider how many pulses the moving observer will have seen that a stationary observer will not yet have seen during the time t. Assume t is significantly larger than T when thinking about this.)

f. How does the frequency of the pulses observed by the moving observer, f', compare to the frequency of the pulses that are being sent by the source? For what real circumstances will a similar effect be observed?

g. What happens if the observer is moving away from the source with speed v?

9   The Uncertainty Principle

So far, sinusoidal standing and traveling waves have been considered without much concern about what happens when they start and stop. The purpose of this chapter is to take a bit of time to consider this issue. The finite duration of waves that start and stop directly affects the spectrum through what is called the *uncertainty principle*.

One of the perplexing results that arises from the very successful theory of quantum mechanics is known as *Heisenberg's uncertainty principle*. Basically, the principle states that you cannot simultaneously know the position and momentum (e.g., speed) of a particle to an arbitrarily high degree of precision. Furthermore, the more accurately you know the particle's position, the less accurately you know its momentum, and vice versa. This principle arises due to the wave nature of particles and, in practice, is only of consequence if the particles are quite small on our everyday scale. An alternative version of the principle states that a similar relationship exists between energy and time. That is, a measurement that determines energy quite precisely will take a long time, and an event that lasts for only a short time will necessarily have a large uncertainty in its energy. *Uncertainty* in this case means that if you repeat the experiment many times, you will not get the same answer every time, no matter how good your equipment and measurement technique may be.

It is not the purpose here to explore the world of quantum mechanics but to look at another situation involving waves – the sounds associated with music. The basic mathematics is the same for all waves, so it should not be surprising that a similar phenomenon occurs for sounds. For sounds, there is a relationship between the frequency and duration of the sound that will influence the perception of pitch. Generally, the relationship is sometimes referred to as the *Fourier uncertainty principle*, and when applied specifically to sound perception, it is called the *Gabor uncertainty principle*.

Wave Pulses

With the use of modern electronics, it is not difficult to produce a nearly perfect waveform of any desired shape. For example, near-perfect sinusoidal signals, but of finite time duration, can be used as a signal source that may be sent through an amplifier to a speaker. Trials with such signals show that a single cycle from a sine wave with a nominal frequency of, say, 440 Hz, will be described as a click.[1] The pitch of such a sound is difficult to identify.[2] Even with several cycles present, it may not be clear to the listener what frequency is being used. Only when many cycles are used – that is, the tone is left on for a longer time – can a pitch can be easily identified.

The basic phenomenon can be understood by referring back to beats (Chapter 3). Recall that when two sinusoidal signals are present simultaneously but with different frequencies, there will be a "beat" at the difference frequency. Hence, if two sinusoidal signals are at frequencies that are 10 Hz apart, there will be a beat at 10 Hz. On the other hand, if both signals are at the same frequency, no beats are heard. Thus, in order to hear whether two different frequencies are present in a sound, the signals should last long enough so that, if present, the beat can be heard. If two signals differ by 10 Hz, so there is one beat each 0.1 s, those signals must last for 0.1 s for a full beat to occur. Note that since the perception of a beat, or not, is a gradual phenomenon, the best that can be said about the *detection* of such a beat is "about 0.1 s." A precise time cannot be specified. For many listeners, it may not be necessary to hear an entire beat to know that one is there. Also, if the signals are restricted to being to sine functions, the listener has additional knowledge of the signal, allowing even shorter time comparisons than for composite signals.

It was shown before that sound perception relies heavily on the spectrum, and the spectrum is a description of the sound "in the frequency domain." Hence, it is worth looking at some short-duration signals, that is, "pulses," in the frequency domain. A few examples are shown in Table 9.1. In these examples, T is the basic unit of time used (e.g., seconds, tenths of seconds, milliseconds, etc.).

A time duration, Δt, can be defined for all these signals. There is not a unique choice for Δt. The goal is to encompass at least most of the signal during that time duration. Likewise, a width in frequency, Δf, can be defined to encompass most of the signal when viewed in the frequency domain (i.e., in the spectrum). Possible choices for these widths are included in Table 9.1. Any detailed discussion involving these widths must start with a definition of how they are chosen. Note also that here Δt and Δf refer to measured properties of a signal and not

---

[1] The US National Institute for Science and Technology (NIST) currently broadcasts a time standard on several frequencies in the short-wave radio bands using the radio station WWV. The basic "clock tick" heard there is actually a short-duration sinusoidal signal (e.g., 5 cycles of a 1,000-Hz tone).

[2] When a train of such signals is used, and when compared to a similar train using a different center frequency, the relative pitch ordering – for example, which is higher pitched– is easier to hear.

Table 9.1 Mathematical descriptions of some pulses.

	Time Domain	Frequency Domain	$\Delta f \cdot \Delta t$
Rectangular pulse			2
Trapezoidal pulse			8/3
Gaussian pulse			3

to the perceived duration and range of pitches. Perceived duration and pitch perception can vary somewhat from person to person and for different circumstances. Understanding perception is a more complicated endeavor.

Recall that the signal in the frequency domain is a representation of the amplitudes of the terms in a Fourier series for that signal (Chapter 4). For these pulses, there will be a large number of terms in the series that have frequencies that are all within the range Δf. It is important to understand that the inability to hear a pitch for short pulses is not solely a matter of perception. The signal is equivalent to the sum of many signals with frequencies throughout that range, so there is no single frequency to be heard but rather a distribution of frequencies.

Also shown in Table 9.1 is the product of the two widths, $\Delta f \cdot \Delta t$, for each case. Obviously, that product will depend in detail on the choices made to define those widths. What is interesting is that for any reasonable choice used to define the widths, **the product $\Delta f \cdot \Delta t$ is always a value close to 1** – typically within a factor of about 2 or 3 but certainly closer than a factor of 10. A physicist would say the product is "of order 1" – that is, using a log scale (see Chapter 2), the power of 10 that is near to these results is $10° = 1$.

For a given pulse shape, the product $\Delta f \cdot \Delta t$ does not depend separately on either the time duration or the width in frequency. So, for example, if the time duration is made shorter, the width in the frequency domain will get larger, so the product remains the same. This is illustrated in Figure 9.1 for gaussian pulses. Gaussian (bell-shaped) functions have the interesting property that a gaussian amplitude in time transforms into a gaussian amplitude with frequency, and vice versa.

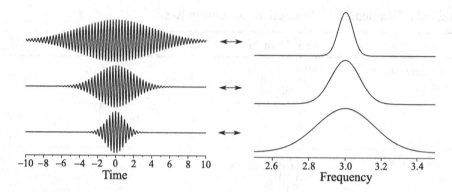

Figure 9.1 As a pulse's time duration is reduced, the width of its spectrum becomes proportionately longer.

For the sake of making estimates in cases where the details of the pulse shape are unknown, it is reasonably safe to use the equation

$$\Delta f \cdot \Delta t = 1 \tag{9.1}$$

to estimate the range of frequencies, Δf, given the duration of the signal in time, Δt, or vice versa. Note that Δf is the full width in the frequency domain about an appropriate center frequency, f_0. Hence, most of the signal will be in the frequency range from $f_0 - \Delta f/2$ to $f_0 + \Delta f/2$.

Example 9.1

How short in duration should you play middle C (262 Hz) so that the width in the frequency domain is 50 Hz?

We have,
$$50 \text{ Hz} = 1/\Delta t \rightarrow \Delta t = 1/50 \text{ s} = 0.02 \text{ s} = 20 \text{ ms}.$$

Thus, playing middle C for 20 ms puts most of the sound energy in a frequency range from about $262 - 25$ Hz = 237 Hz to $262 + 25$ Hz = 287 Hz, or about 240 to 280 Hz. Note that the notes adjacent to middle C are nominally B at 247 Hz and $C^{\#}$ at 277 Hz. For a 20-ms duration, there is overlap of the spectrum of this note C with its neighbors. For fast playing, the notes quite literally get blurred into one another because the frequency is "uncertain."

The examples so far all involve a single sinusoidal signal that has a short time duration. The spectrum involves a single peak that has been broadened. If a complex sound, including many overtones, has a short time duration, the spectrum will have many peaks, each of which is broadened.

A pulse that is extremely short in duration, a very short click, can be useful for making some types of measurements. Such a pulse includes a large range of frequency components all at once. So, for example, if the frequency-dependent reflection from a particular wall is desired, a short pulse can be transmitted and the reflected pulse recorded. The spectrum of the received pulse, compared to the transmitted pulse, will have the desired information. Such a technique has the advantage that data from reflected pulses from more distant (or closer) walls, or other objects, can be eliminated based on the speed of sound and the arrival time. In contrast, if steady tones were used, one frequency component at a time, it would be difficult to eliminate signals reflected from other objects.

Attack and Release

Any real signal will have a start and a stop. For music, these are referred to as the **attack** and **release** (or decay) of a sound. The attack and release are known to provide a listener with very important clues as to the origin of the sound, and in some cases, those clues are no less important than is the spectrum.

The short pulses already discussed have both their start and stop within a short time span, and the resulting sound could be understood to some extent by looking in the frequency domain. It is not obvious that it is that simple for sustained sounds. Some examples of starting transients are shown in Figure 9.2. However, for a sound that is turned on at some time and then is left on, perhaps even for minutes or hours, a simple spectrum of the entire signal cannot be expected to be of similar help. The listener will soon forget about the start of the signal and will hear it as a continuous tone. Likewise, the signal in the frequency domain will be dominated by the long time for which the signal is steady, rather than transients, and so the information about the start and stop will become overwhelmed by the rest of the signal.

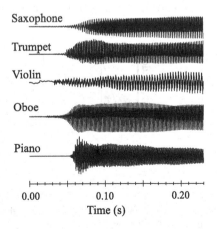

Figure 9.2 Examples showing the onset of a signal from several musical instruments.

If a sound turns on very gradually, over a period of many seconds, it is heard simply as a gradual increase in the volume of a steady tone. On the other hand, if a sound is started quickly, there is an extra sound – almost a click – that occurs at the beginning of the sound. For musical sounds and human listeners, the difference between slow and fast seems to be somewhere around 50 ms (0.05 s). Exactly what that value is depends on the listener and conditions, but 50 ms seems to be a reasonable value, with variations by factors of 2 being common but by factors of 10 being rare. It seems as though, at least in part, a person hears and interprets sounds in a running window of time that is about 50 ms long.

ADSR – a Model

A simple model for sounds, used in some forms of sound synthesis, is called **ADSR**. The signal is divided up into four regions: **attack (A), decay (D), sustain (S), and release (R)**. The model describes the envelope (i.e., the sound amplitude as a function of time) for a signal and can be specified with a small number of values. In its simplest form (see Figure 9.3), three amplitudes and four times are specified. ADSR will be discussed more later on (Chapter 20); however, the modest success of the model suggests that it contains many of the elements that are important for musical sounds.

Spectrographs

Signals that are changing in time can be visualized using various forms of combined time–frequency analysis. A common form is known as a **spectrograph**.[3] The modern spectrograph digitally records the data and produces a two-dimensional plot. Time is used as the horizontal axis and frequency as the vertical axis. The spectrum displayed along the vertical axis is obtained from signals during a short time interval near the corresponding time specified on the horizontal axis.

Figure 9.3 The ADSR model used as a simple approximation for the time dependence of signals.

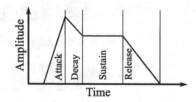

---

[3] Other names include *sonogram* when used for sound, although *sonogram* has other meanings, and *voice prints* when used specifically for voice analysis.

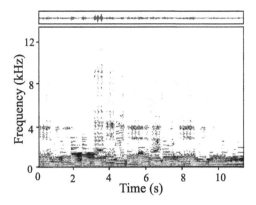

Figure 9.4 An example of a spectrograph, which shows the spectrum as a function of time. For comparison, a representation of the total input signal is shown across the top.

Often, the square of the signal is used, which is proportional to the signal power, and thus all phase information is lost in such a plot. The intensity is displayed using brightness and/or false colors. An example is shown in Figure 9.4. Since the frequency spectrum is obtained using data collected over a short time, Δt, the time resolution of the plot is no better than Δt, and the frequency resolution will be limited to about $\Delta f = 1/\Delta t$.

Phase Shifts and Transients

As was mentioned previously (Chapter 4), any signal can be written using a Fourier series that includes amplitude and phase coefficients. *For sustained periodic signals*, the phase constants seem relatively inconsequential when it comes to how the sound is perceived, at least at lower volume levels. The same cannot be said for sounds that start and stop. This can be easily proven using one simple example – playing some sounds backward. It is a simple matter to try this with modern sound-editing software, and clearly, any sounds that change in time, such as speech, do not sound the same forward and backward. On the other hand, it is difficult to hear a difference for a sustained note, say, from a clarinet, when played forward and backward.[4] What is left to show is that this reversal in time is equivalent to a change in the phase constants in a Fourier series.

Previously, the Fourier series was written using only the cosine function with an amplitude and phase constant for each harmonic. An alternate form uses both sine and cosine functions without phase constants,

---

[4] The starting and ending of the tone should be trimmed or faded so that only the sustained portion is heard.

$$S(t) = \left(A_0 + A_1 \cos(\omega_1 t) + A_2 \cos(\omega_2 t) + \dots\right)$$
$$+ \left(B_1 \sin(\omega_1 t) + B_2 \sin(\omega_2 t) + \dots\right). \tag{9.2}$$

There are still two constants per harmonic, although now both are amplitudes (A_n and B_n). This works because a sine function is equivalent to a cosine function with a phase shift.[5] Now the direction of time can be reversed by simply replacing t with $-t$. Using the facts that

$$\sin(-x) = -\sin(x) = \sin(x + \pi) \text{ and } \cos(-x) = \cos(x), \tag{9.3}$$

then

$$S(-t) = \left(A_0 + A_1 \cos(\omega_1 t) + A_2 \cos(\omega_2 t) + \dots\right)$$
$$+ \left(B_1 \sin(\omega_1 t + \pi) + B_2 \sin(\omega_2 t + \pi) + \dots\right), \tag{9.4}$$

which is just the original series but with some phase shifts added. **Reversing time –** playing a sound backward – **is equivalent to a particular set of phase shifts**.

Another interesting experiment is to take a recording of a signal with lots of starts and stops, such as a recording of speech, and alter the spectral information while retaining the phases. For example, make all the (significant) coefficients in the Fourier series have equal amplitude, but keep the phase constants as they were. For such a modified sound, the content of the sound is often still intelligible. In that case, for perception, the phase factors may be more important than the spectrum.

Hence, one can conclude that at least **some phase shifts will affect the perception of sounds that are changing in time**. For music played and enjoyed acoustically – without the aid of electronics – there are generally few, if any, deleterious phase shifts to worry about. However, much of the music heard today has gone through some electronics, perhaps with significant signal processing along the way. This processing may include microphones, cables and/or wireless systems, amplifiers, recording devices, speakers, and so forth, as well as computer processing. Each of those steps can add a phase shift to the signal. Some of these phase shifts may change the perceived sound, even if no change can be heard for sustained tones. Thus, for good sound reproduction, some attention to potential phase shifts, in addition to amplitude shifts, is justified.

Summary

A very short-duration signal will be perceived as a click, without a definite frequency. Finite-duration signals or sounds will have a width in time, Δt, and the spectrum of that signal will have a width in frequency, Δf. For any reasonable

---

[5] Refer to Appendix A for properties of these functions.

definitions of those widths, $\Delta f \cdot \Delta t \approx 1$. Thus, a very short signal or sound will include components with a wide range of frequencies, and vice versa.

A signal that starts and then later stops, or releases, will be perceived differently at the beginning (the attack) and at the end (the release), even if the signal is sustained for a long duration in between. A simple model for the changes in the sound is called *ADSR*, standing for *attack*, *decay*, *sustain*, and *release*. Signals that evolve with time can be displayed using a spectrograph, a method to show the spectrum as a function of time.

Playing a recorded signal backward is equivalent to a certain set of phase shifts. Unlike sustained tones, sounds that change with time, such as talking, do not sound the same when played backward, showing that, at least sometimes, phase shifts in the Fourier representation can be very important for perception.

ADDITIONAL READING

Gabor, D. "Theory of Communication." *Journal of the Institute of Electrical Engineering* 93, no. 26 (1946): 429–457.

Hsieh, I. H., and K. Saberi. "Imperfect Pitch: Gabor's Uncertainty Principle and the Pitch of Extremely Brief Sounds." *Psychonomic Bulletin & Review* 23, no. 1 (2016): 163–171.

Jeon, J. Y., and F. R. Fricke. "Duration of Perceived and Performed Sounds." *Psychology of Music* 25, no. 1 (1997): 70–83.

Oppenheim, J. N., and M. O. Magnasco. "Human Time-Frequency Acuity Beats the Fourier Uncertainty Principle." *Physical Review Letters* 110, no. 4 (2013): 044301.

PROBLEMS

9.1 (a) If middle C is played for 10 ms, approximately how many other notes will be overlapped by its spectrum? What about for 10-ms notes played (b) two octaves above or (c) two octaves below middle C? How might these results relate to the ability of bass instruments, such as a tuba, to clearly play fast runs that are routine on higher-pitched instruments, such as a piccolo?

9.2 Multiple musicians play a note near 4,500 Hz. If the note is played for 0.2 s, approximately how close in frequency do they need to play so that no beat is heard? Express the result as a percentage. How about two musicians playing a note near 45 Hz for 0.2 s? Recall that neighboring notes differ in frequency by about 6 percent, regardless of whether it is a high or low note. What does this suggest about the different tuning challenges for groups of musicians who play piccolos compared to tubas?

9.3 Consider the recorded signal shown in Figure P9.3, which consists of one complete cycle of a sinusoid. How accurately can you determine the frequency of that sinusoid from the graphic? How does that compare to the accuracy predicted from the uncertainty relation (Equation [9.1])? What is different about those two determinations that yield very different results?

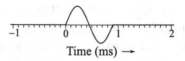

Time (ms) →

Figure P9.3

9.4 Use an audio editor and recordings of both music and speech to investigate the sound that results after the following manipulations: (a) greatly over-amplify the signal so that it is strongly "clipped," and then rescale it back to a normal volume; (b) explore the effects of high-pass and low-pass filters using very low and very high cutoff frequencies, respectively, adjusting the volume as necessary; and (c) if available, take the Fourier transform, overamplify that, and then transform back (or take another Fourier transform). If the transform is not available in your audio editor, it may be accomplished by exporting the sound to a spreadsheet that can do the transform, then import-ing that result back into the editor (details will depend on the software used).

9.5 Use a Fourier series to demonstrate that if a phase shift, φ, is applied to all the partials in the series, and those phase shifts depend linearly on the frequency of the partial (e.g., $\varphi = \alpha\omega$, where α is a constant), the result is equivalent to a time delay. That is, such a phase shift will change the time at which the signal occurs but not what it sounds like. Can you think of a physical situation where this occurs?

10   Nonlinear Physics

The complexities of even very simple nonlinear physics problems go well beyond the scope of this text. However, nonlinear physics is essential for the operation of all bowed and blown musical instruments and thus cannot be avoided entirely. In what follows, only that which is essential to give a flavor of what nonlinear physics is, and why it is important for music, will be considered. To understand what a nonlinear problem is, it is first important to define a linear problem.

Linear Problems

The driven damped (ideal) harmonic oscillator was presented earlier (Chapter 5). It is an example of a linear physics problem. The basic equation governing the motion for the simple case of a sinusoidal driving force is

$$ma + cv + kx = F \cos(\omega t + \varphi), \tag{10.1}$$

and although the solutions, $x(t)$, are somewhat complicated to write down, they are known (Equations [5.14] and [5.15]). In general, the time-dependent driving force may not be a simple sinusoidal function. However, any real time-dependent force can always be expressed using a Fourier series expansion, so the damped harmonic oscillator with *any* time-dependent driving force will be governed by an equation equivalent to

$$ma + cv + kx = F_1 \cos(\omega_1 t + \varphi_1) + F_2 \cos(\omega_2 t + \varphi_2)$$
$$+ F_3 \cos(\omega_3 t + \varphi_3) + \dots, \tag{10.2}$$

where as many "driving terms" as are necessary are included on the right-hand side.

This equation is linear in the mathematical sense, which means that a "divide-and-conquer" approach can be used to solve it. That is, **the total solution is simply the sum of each of the separate solutions** found using just one sinusoidal term at a time (plus zero for the transient part of the solution). Each of those separate solutions is known already. In particular, solutions, $x_n(t)$, are found to all the *separate* equations,

$$\left.\begin{array}{l} ma_0 + cv_0 + kx_0 = 0, \\ ma_1 + cv_1 + kx_1 = F_1 \cos(\omega_1 t + \varphi_1), \\ ma_2 + cv_2 + kx_2 = F_2 \cos(\omega_2 t + \varphi_2), \\ \text{and so forth,} \end{array}\right\} \tag{10.3}$$

and then the solution to the original problem is simply given by

$$x(t) = x_0 + x_1 + x_2 + \ldots = \sum_n x_n. \tag{10.4}$$

A similar situation occurs for the vibrational modes of a string (Chapter 6). Each mode behaves like a separate harmonic oscillator, and the total motion of the string can be found as the sum of the separate harmonic oscillator solutions. The driven string can be solved as the combination of a large number of oscillators, each experiencing a large number of sinusoidal driving terms. For these linear problems with many separate terms, the total motion may appear to be extremely complicated; however, it is easily treated as the sum of relatively simple motions.

What Makes a Problem Nonlinear?

A nonlinear problem will include some term in the equation(s) that causes the divide-and-conquer approach to fail, and in general, **the solution cannot be simply written as the sum of solutions to simpler problems**. Since many equation-solving techniques rely directly or indirectly on the presence of linearity, nonlinear problems, where these techniques fail, tend to be much more difficult to solve. Whole careers are built around this very issue.

There are some nonlinear physics problems where the basic behavior is very similar to a linear problem but where the solution simply needs to be adjusted. These are the less interesting nonlinear problems. There are also nonlinear physics problems that have a solution that just is not possible, even approximately, with a completely linear system. The latter includes the physics of the blown and bowed musical instruments. In addition, there are nonlinear systems with very interesting behavior, even though they are not normally associated with music or music production. One such nonmusical example of interest gives rise to behavior known as *chaos*.

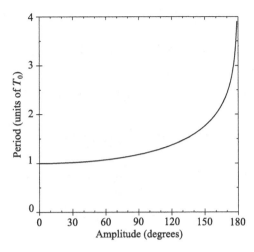

Figure 10.1 The period of a simple pendulum for larger swing amplitudes in units of T_0, the period in the small-angle limit. The period approaches an infinite duration for an amplitude approaching 180°.

The simple pendulum is an example of a nonlinear problem. That problem was treated earlier (Chapter 5) using an approximation, appropriate for small-angle swings, that turned the problem into a linear problem – a harmonic oscillator. For larger angles, the approximation is not valid, and the problem is more complicated to solve. However, the basic behavior of the pendulum does not change for larger angles. It swings to and fro as before, although the frequency of oscillation and the position as a function of time are not exactly the same as those of a harmonic oscillator. The only extra behavior for the real pendulum is that it can be made to spin in vertical circles, a nonoscillatory motion.

The simple pendulum problem for large angles (without friction) is solvable in terms of functions known as *elliptic integrals*. There may not be a calculator button for these functions, but there are well-known methods for computing their values. Figure 10.1 illustrates the period of a simple pendulum for different oscillation amplitudes, in terms of the small-angle result. While the frequency of oscillation depends on amplitude, the problem can be thought of as being, more or less, the (ideal) simple pendulum with a correction factor. Note that for larger angles, particularly for those larger than 90°, the pendulum's "string" must be considered to be rigid (i.e., a rod) so that the "string" remains straight.

The motion of the pendulum is periodic but is not quite sinusoidal. Hence, the time dependence can be described using a Fourier series with a fundamental frequency and harmonic overtones (see Chapter 4). The overtone content depends on the nonlinearity; the larger the angle of swing, the larger will be the amplitudes of those harmonics.

There are a few musical instruments that rely on this simple type of nonlinear physics. A type of "opera gong" (Figure 10.2) uses the nonlinear properties of curved vibrating metal plates to produce a sound that can either rise or fall in pitch after a strong strike. The frequency depends on amplitude, as for the pendulum, and as the amplitude decays due to friction, the frequency changes.

Figure 10.2 A simple gong with nonlinear properties gives rise to a pitch that changes with time if it is struck strongly enough in the right location.

Some other instruments use the nonlinear properties in a way so that one motion can control the properties of another. That is, the total motion consists of two parts, one of which is larger and controlled by the player, and the other is a small time-dependent oscillation. Due to the nonlinear interactions, the solution depends on both at once. One such instrument, used for various effects, is called a *flexatone* (or sometimes, *flexitone*). It includes small spring-loaded hammers that bang on a bent piece of metal as the instrument is shaken. There is a small time-dependent bend in the metal due to the vibrations and a larger bend in the metal that is under the control of the player. The player can change the pitch by changing the bend angle. One of the most melodic instruments that uses this same idea is the musical saw. The musical saw, originally based on standard wood saws, can be struck but is normally bowed, and it will have resonant frequencies for small-amplitude vibrational (bending) modes that depend on how much the saw is bent (curved) by the player's hand.

The bowed and blown instruments exhibit a more complicated behavior. You can bow or blow with a constant speed, and yet time-dependent oscillations result. This behavior arises from the coupling between an oscillator, similar to a mass on a spring or a string under tension; a source of energy; and a nonlinear force. **A simple everyday nonlinear force is friction**, which, via the so-called "stick–slip" mechanism, is essential for the operation of bowed instruments.

Friction

One of the most common nonlinear forces in physics is friction. Simple mechanical friction arises when two surfaces are in contact. **The frictional force tends to reduce to zero, or maintain at zero, the relative velocity of the two surfaces**. It is a velocity-dependent force. Everyday observation, perhaps while rubbing one's hands together, also shows that the size of the force can be increased if the two surfaces are pressed harder together and reduced if they are pressed together less hard. Careful measurements show that friction is a highly complicated process that exhibits considerable variability from one situation to another.

Fortunately, there is a simple model that can be used for smooth surfaces and "dry friction." *Dry* in this context means there is no lubricating layer. What is meant by *smooth* is left a bit vague, but there should be no jagged edges or small bumps that can get caught or hooked on other similar shapes. In any event, the model works reasonably well for many real situations and is quite sufficient for the purposes here.

The model says that the force of friction for moving surfaces is always in a direction opposite to the relative velocity of the surfaces and that it is simply proportional to the contact force between the surfaces. For two surfaces that are stationary with respect to one another, the model says the force will be in the direction to oppose any change from zero velocity, and it will take on whatever value is necessary to keep the velocity at zero, up to some maximum possible value, after which the surfaces break free and begin to move. Note that for this model, the force does not depend on the speed, other than the difference between zero and nonzero values, and it does not depend on the area of contact. This simplified model for friction is variously attributed to Leonardo da Vinci (1452–1519), Guillaume Amontons (1663–1705), and Charles-Augustin de Coulomb (1736–1806).

The force of contact represents how hard the two surfaces are being pressed together and will always be in a direction perpendicular to the surfaces. Sometimes that force is referred to as the **normal force**, where here the word *normal* has one of its mathematical definitions, meaning "perpendicular to the surface," rather than its everyday definition of "usual" or "the opposite of abnormal." It does not matter whether this is the force of surface 1 on surface 2 or the force of surface 2 on surface 1 since only the magnitude matters, and by Newton's third law, those two forces must have the same magnitude.

Example 10.1

If a box of mass m is sitting on a horizontal tabletop, as shown in Figure 10.3, the box is being pulled downward due to the force of gravity with a force mg, and the table exerts an upward force of equal magnitude on the box (otherwise, the box would fall or rise), so the magnitude of the force of contact is mg.

If the force of friction is given by F_f and the magnitude of the force of contact between the two surfaces by F_c, then the model says that for a relative velocity v,

Figure 10.3 The force of friction opposes the velocity and depends on the force of contact.

if $|v| > 0$, $F_f = \mu_k F_c$ in a direction opposite to the velocity, and

if $v = 0$, $F_f \leq \mu_s F_c$ with a magnitude and direction to oppose all other applied
 forces.

Here, μ_k and μ_s are constants known as **coefficients of friction**, and they depend on
the particular combination of objects used. The subscript k stands for *kinetic*,
indicating that the surfaces are moving, whereas the subscript s stands for *static*.
It is usually the case that $\mu_s > \mu_k$, meaning that it takes more force to start an
object moving from zero than it does to keep it moving. Note that even though the
force of kinetic friction does not depend on speed, it is velocity dependent because
the direction of the force depends on the direction of the velocity.

   It is easy to demonstrate that this model for dry friction gives rise to nonlinear
problems. Imagine a large piece of furniture that needs to be moved. It is large
enough so that one person can push with only about three-quarters of the force
necessary to overcome static friction. Hence, the solution when one person
pushes is that there is no motion. A second person pushing by themselves also
produces a solution with no motion. But if the two people push in concert, the
furniture moves. The solution with two people pushing is *not* the sum of the
solutions for each of them separately, which would be no motion at all, and hence
the situation must be nonlinear.

The Stick–Slip Mechanism

Imagine a box that is being pulled across the floor using an elastic cord. This
situation can be modeled as a mass (the box) on a spring (the cord) with dry
friction. For simplicity, assume the cord is being pulled by a person walking with
a constant speed, v_0. The force applied to the box includes the force of friction
with the floor and the force due to the cord. If the box is stationary, the force of
friction exactly counters the force due to the cord up to a maximum magnitude,
$F_1 = \mu_s mg$. Once the box begins sliding, the force of friction is a constant value,
$F_2 = \mu_k mg$, with $F_2 < F_1$, opposite to the motion. The force from the cord will
depend on how much the cord is stretched. If you start with the cord unstretched
and the mass stationary, the mass will remain stationary until the cord is stretched
enough to overcome static friction, F_1. Once the mass begins moving, it will
remain in motion, reducing the stretch in the cord. If the force from the cord

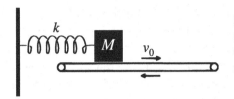

Figure 10.4 A box on a conveyor belt as an example of stick–slip motion. ★

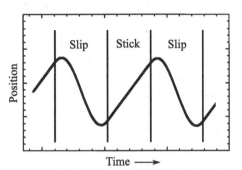

Figure 10.5 The solution to the simple model for the stick–slip behavior has sinusoidal and straight-line behavior, patched together.

drops below that due to kinetic friction, F_2, the velocity of the mass will decrease until it comes to a stop, and then the process can repeat.

To clearly see the oscillatory part of the motion and to be more quantitative, it is useful to look at the problem in another way. Consider that one end of the spring (cord) is fixed, the other is attached to the mass, and the mass is on a conveyor belt that moves with constant speed, v_0, to the right (Figure 10.4). If the mass starts with the spring unstretched, the initial motion will simply be at speed v_0 to the right, the position will then be $x = v_0 t$, and the force from the spring will be $kx = k v_0 t$ to the left. As long as $kx < F_1$, the force of static friction will exactly counter the force due to the spring, and the box will continue at a constant velocity. When $kx > F_1$, the mass begins to slide, and the force of friction becomes a (smaller) constant value, F_2, still to the right, until the velocity of the mass again matches the velocity of the conveyor belt. Hence, the equations governing the motion are $ma + kx = F_2$ as long as $v \neq v_0$, and when $v = v_0$, the velocities stay equal as long as $kx < F_1$.

The solution to the first equation will be that of a simple harmonic oscillator but with an offset, valid only until $v = v_0$, at which point the second solution takes over. The final solution is obtained by patching together the straight-line solutions during the stick phase with the sinusoidal solutions during the slip phase. The general case involves some mathematical manipulation that will not be reproduced here. A representative solution is shown in Figure 10.5.

Note that this stick–slip solution is periodic but not sinusoidal. That means the solution can be written as a Fourier series with a fundamental frequency and harmonic overtones (see Chapter 4). The overtone content is, in a sense, "generated" by the nonlinear forces involved.

It should be recognized that the problem with the box pulled along the floor and the problem with the mass on the conveyor are really the same basic physics problem, seen from different points of view. A physicist would simply say that there has been a change in the **frame of reference**. In the first case, the floor is considered stationary while the end of the cord is pulled at speed v_0. In the second case, the end of the cord is considered stationary while the "floor" is moving with a speed v_0. Since only the relative speed matters, both descriptions are equally valid. The first description might be more appropriate for the start–stop motion when you drag something across the floor or for the motion of Earth's tectonic plates as they rub against each other, whereas the second is more appropriate for tone production from bowed musical instruments.

The stick–slip mechanism for a bowed string is a bit more complicated. The string has many modes of vibration, and the point of contact with the bow is only on one small portion of the string. When the slip occurs, a pulse is sent down the string and is reflected. While the pulse is traveling, the string near the bow rapidly re-sticks. As the pulse travels back to the bow, it shakes the string loose, causing another slip (Figure 10.6). The first technical description of this process is attributed to Hermann Helmholtz (1821–1894), and this motion is sometimes referred to as the **Helmholtz mechanism**. The string motion generated is periodic but not sinusoidal. The time dependence can be described using a sum of standing-wave solutions similar to those of the plucked string (see Chapter 6). The spectrum of vibrational frequencies will include a fundamental and the harmonics of that fundamental as necessary to satisfy the stick–slip process. That is, the relative overtone content can be traced back to the nonlinear interaction between the bow and string.

Back to the Swing

The general form for several musically related nonlinear systems, in simplified form, can be written as a harmonic oscillator with a driving term, but where the driving term depends directly on position and/or velocity (in a nonlinear way) rather than on a simple function of time. That is, the problem is best described by

$$ma + cv + kx = F(x, v) \tag{10.4}$$

(compare to Equation [5.11]). Of course, since x and v may depend on time, the force may change with time, but in this case, the dependence on time is indirect.

In the earlier discussion of driven harmonic motion (Chapter 5), the driving force was taken to have a sinusoidal time dependence, and this gave rise to resonant behavior, where a large response is observed when the driving frequency matches the natural frequency of the oscillator. An example was used, with

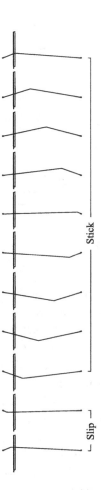

Figure 10.6 The simplified stick–slip motion of a violin string. ★

qualification, of pushing someone on a swing set – the biggest response occurs when the push frequency matches the natural frequency of the swing. Taking a closer look, however, suggests that that problem might be better described if the force were thought of not as a function of time but as a function of position and velocity.

When pushing someone on a swing set, the pushes naturally occur at a certain point during the motion when the swing is going in a certain direction. If the swing amplitude is constant, the amount of energy lost due to various forms of friction is compensated for by the extra energy added with each push. Under these conditions, the swing oscillates periodically, and hence the pushes are applied periodically. However, if the swing amplitude is changing, the period of oscillation will change with amplitude, as mentioned earlier for the pendulum; however, the pushes still occur at the same *place* when the swing is going in the correct direction. The time dependence of the force can thus become complicated, but when thought of as depending on position and velocity, it can be quite simple. A simple model would be a force that is either

Figure 10.7 A periodic pulse sent down a string with a flick of the wrist. The flick occurs at a particular time that is determined by the motion – that is, when the pulse returns. ★

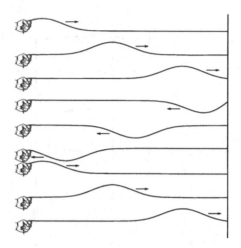

on or off depending on position and velocity – the mass m gets a little kick each time it goes by in the correct direction.

A more musically realistic model, and a situation that is easily replicated, would be a pulse sent down a string or rope (see Chapter 8) using a flip of the wrist (Figure 10.7). The pulse travels with velocity c, reflects off the far end, and returns inverted. If nothing else is done, when the pulse returns, it will get inverted again, so it is now back upright, and the process repeats. For a real string, there will be some friction, and the size of the pulse will diminish as it travels and with each reflection. If, on the other hand, each time the pulse returns, a new flip of the wrist adds back the energy lost, the system will continue in a periodic motion indefinitely. The new flip of the wrist occurs when the pulse returns – that is, when the pulse is at a certain position traveling in the correct direction. If the string is of length L, the periodic motion will be at a frequency $c/2L$. If the length of the string is changed, the periodic motion will automatically change to the new frequency since the flip of the wrist occurs only when the pulse has returned, no matter how long that takes.

The bowed and blown instruments exhibit the same basic behavior, in that the applied force is best described based on the current state of the system and not as a simple periodic function of time.

Reed Instruments Simplified

If you look closely at a clarinet mouthpiece, as shown in Figure 10.8, it consists of a rigid tubular piece, narrower at one end, with a rectangular opening on one side, and that opening is covered by a "reed" – basically a thin, flexible piece of cane or similar synthetic material. In its resting position, there is a small gap between the reed and the rest of the mouthpiece through which air can enter and leave.

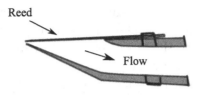

Reed

Flow

Figure 10.8 A clarinet reed controls the flow of air into the instrument.

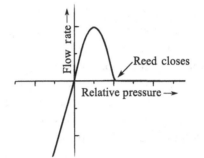

Flow rate →

Reed closes

Relative pressure →

Figure 10.9 Qualitative illustration of flow as a function of pressure for a clarinet.

When played, the narrower end of the mouthpiece is placed in the mouth, the lips make a seal around the circumference, and air is blown into the instrument.

The flow rate through the instrument will look something like that shown in Figure 10.9. At low blowing pressures, the flow increases with increasing pressure. As the blowing pressure increases, the reed will start to close, shutting off the flow. At very high pressures, the reed is entirely closed, and there is no flow. Of particular interest is the region just before it closes, where an *increase* in pressure causes a *decrease* in flow. Imagine pushing an object across the floor with friction – the harder the push, the faster it goes. On the other hand, what would happen if that were reversed, so the harder the push, the slower it goes?

Friction causes a harmonic oscillator, once started, to gradually reduce its amplitude to zero. Now if a force is added that does the opposite, at least for small changes in the motion, normal friction can be counteracted. Then, rather than decaying with time, oscillations will build up. The motion would look like a video of an oscillator running backward in time – starting with little motion, then gradually building up oscillations with time. By supplying a constant blowing pressure so that, on average, the operation is in the decreasing region of the curve in Figure 10.9, the oscillations will act as if there is negative friction. If attached to an acoustic oscillator, a tone may result. The constant blowing pressure and flow provide the energy (power) that leads to the increasing oscillation amplitude. **If a system looks like it has negative friction, there must be a power source somewhere.**

There are two types of forces that cause the reed to close. The rather obvious force is the blowing pressure that forces the reed closed. The other is due to the flow – more (faster) flow into the instrument tends to close the reed, reducing the flow. The latter is due to the Bernoulli, Coanda, and related effects. The effect of flow can be observed by hanging two sheets of paper parallel to each other, a few centimeters apart, then blowing between the two sheets. While it might be expected that the sheets will be forced apart, they actually go toward each other, narrowing the spacing.[1]

When a reed is initially blown, a pulse of air enters the instrument before the reed closes. That pulse travels down the instrument, and some of it reflects back to the reed. When it arrives at the reed, it forces the reed open, allowing another pulse of air to enter, and the process repeats. This is qualitatively similar to the pulse on a string mentioned earlier. Of importance when used with systems that have more than one mode is that the reed will be open or closed for all vibrational modes. It cannot be open for some and closed for others. This nonlinearity – the inability to treat the modes separately – couples the modes and determines the relative sizes of the overtones that will be produced. The reed motion couples best to the natural resonant modes of the instrument attached to it. More details of wave propagation in pipes will be considered in Chapter 14.

Additional details of reed instruments rapidly go beyond the discussion here. There are reed instruments where the blowing pressure tends to close the reed and others where it opens the reed. Other reed instruments may use a double reed – two back-to-back reeds. Reed organ pipes, accordions, and harmonicas typically use a thin strip of metal (e.g., brass) rather than cane. The basic physics is similarly nonlinear and is surprisingly complicated.

Lip Reeds

To a crude approximation, the brass instruments can also be treated as reed instruments, where the function of the reed is provided by the player's lips. The lips are drawn tight, to provide tension, and pressed together. As the player tries to blow, the flow is stopped by the closed lips, up to a point. Once the lips open and flow begins, the blowing pressure at the lips drops, and the lips again close. Since the lips have some inertia (mass), it takes some time for the lips to close. It is relatively easy to demonstrate such an air reed without an instrument, resulting in a very raspy sound.

The brass instruments will include a mouthpiece connected to a long tube. The mouthpiece has a cup and a constrained (small-diameter) connection to the tube. When held against the lips, the volume in the cup plus the constrained section of tube will act somewhat as a Helmholtz resonator. That is, in turn, connected to the length of tube. That tube will have multiple resonances of its own. The combined system will interact with the lip reed to produce a musical tone. The interaction is nonlinear since the lips are either open or closed and cannot be open for some modes and closed for others.

---

[1] Search for "Bernoulli effect" and/or "Coanda effect" to find many other simple demonstrations related to airflow.

While it is possible to produce a trumpet-like sound with just a tube, such as a short length of garden hose, only a limited number of notes can be produced.[2] The mouthpiece and the shape of the instrument are designed to provide resonances that are nearly harmonic that can work together to interact with the reed. The player adjusts the tension in their lips to match the frequency of their lip reed, and its many overtones, to interact with the resonances of the instrument. A harmonic set of tube resonances will allow the player to more easily produce tones with a fundamental frequency at any of the harmonic frequencies, which are in turn related to the musical scale (Chapter 3). A bugle is the simplest example of such an instrument and illustrates the importance of the nonlinear interaction. With no change in the instrument, the player can produce several different notes (typically five) by altering their (nonlinear) interaction with the instrument. By adding valves used to switch in short lengths of tubing, different sets of harmonics can be used.

Air Reeds

Flutes, some organ pipes, and similar instruments produce sound by blowing an air jet at an edge. The jet can be produced by the player's lips (such as for standard orchestral flutes) or by the design of the instrument (e.g., the "fipple" of a recorder). A cutaway of a wooden organ pipe, illustrating the airflow, is shown in Figure 10.10. To imagine an air jet, think of the flame that exits a rocket during launch. The jets used for music are less extreme, of course, but are a similar phenomenon. The sound produced can be thought of as being due to an air reed. The jet of air flexes back and forth in a manner similar to the reed of the clarinet. In this case, however, control is not due to pressure but to the flow rate.

If the jet is initially blowing into the instrument, the flow will go to the (effective) end of the instrument and reflect back, like the pulse on a string. When the flow returns, it blows the air jet outward, decreasing the flow. With reduced flow, the jet is again drawn back into the instrument. Hence, the jet is alternately drawn into the instrument and then pushed out of the instrument. The jet will be flowing in or out for all modes, and it cannot flow in for some and out for others.

Figure 10.10 A cutaway view of a wooden organ pipe showing the airflow. A jet of air is aimed at an edge and can either go into or out of the pipe. The jet interacts with the air within the pipe.

---

[2] Large seashells, hollowed-out animal horns or bones, and other such objects can also be used.

The nature of the nonlinear interaction with the instrument is a very important factor in determining the relative amplitudes of the overtones – the sound of the instrument – often much more so than the rest of the instrument. For example, if a clarinet mouthpiece is attached to the body of a flute, the result will sound much more like a clarinet than a flute, and if a flute head joint replaces the clarinet head joint, the result sounds much more like a flute.

Playing in Tune

When nonlinear interactions are driving oscillations, the frequency is determined, in part, by the source of the driving force(s). This is easy to observe with a simple whistle, recorder, harmonica, or bagpipes. The pitch will change quite noticeably when blowing harder and softer. For orchestral and similar instruments, the player's lips and mouth are part of the interaction, and the player can, and will, adjust to maintain intonation (the tuning) with changes in blowing pressure. Saxophone players can often be heard using this technique artistically, as an effect, to "bend the pitch" across many half steps. Another example can be heard in the clarinet glissando (gradual change in pitch) at the beginning of Gershwin's *Rhapsody in Blue*, where much of the last portion is often done solely by changing the way the instrument is blown. However, when this type of real-time pitch correction is not possible, the instrument can only play at one volume level and still be in tune.

Pipe organs are an example where the amplitude of the sound from each pipe is fixed. You cannot play any individual pipe louder or softer because it will go out of tune. In fact, part of the tuning of an organ pipe includes the adjustment of the volume of each pipe. However, a pipe organ will have multiple pipes for each note. The largest organs will have more than 10,000 pipes. These pipes will be of different types (open and closed ends, metal and wood, flute or reed mechanism, cylindrical and conical, etc.), each producing different levels of overtones, so that different sounds (timbres) can be produced. To enable a certain set of pipes, the player will "pull out the stop," which opens the appropriate air valve (unstops the flow) corresponding to that type of pipe. To increase the volume, multiple pipes, playing the same note, are used at the same time. To get the maximum volume, the player will "pull out all the stops." Some organs are also equipped with a set of doors that are between some of, or all of, the pipes and the listeners, and those doors are under the control of the player. These doors allow for more or less sound to get out to the listener, an additional method for volume control. Pipes contained behind such doors are referred to as *swell pipes*.

Some pipe organs may include a device known as a *tremulant* in the air supply for some pipes. The tremulant periodically varies the strength of the air supply to the pipes, causing a small periodic change in the volume (i.e., a tremolo) and, associated with the change in volume, the pitch (i.e., a vibrato) of the sound.

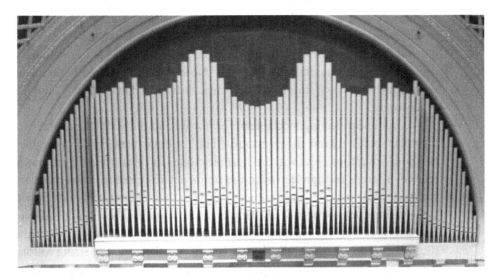

Figure 10.11 The organ pipes out front, the facade pipes, provide a decorative look but may not be functional. For this organ in Hill Auditorium at the University of Michigan, none of the visible pipes produces sound. There are more than 7,000 active pipes housed just behind the facade. The design of the facade pipes has changed somewhat since this photo was taken in the early 1970s.

Interestingly, for a typical pipe organ (e.g., a church organ), most of the pipes are not visible. They are located in a space behind the pipes you see, with a covering that is transparent to sound. Often some, and possibly all, of the visible pipes are nonfunctional. They are called *facade pipes* or *display pipes* and may be there only for looks. See Figure 10.11. Some entirely electronic organs may also include such display pipes that are clearly just for decoration.

Mode Locking

The *ideal* string will have mode frequencies that are harmonic. Likewise, the analogous vibrations of the air in an *ideal* pipe will have modes that are harmonic. For real strings and real pipes, the modes may be close to harmonic but will not be exactly harmonic. That is, the modes are not precisely in tune with each other. Each mode will also have some energy loss (e.g., friction), leading to a finite quality factor and, hence, a width in the frequency response.

If nonlinear interactions are present, a phenomenon known as **mode locking** can occur if two oscillators (or modes) are close in frequency or if their frequencies differ by a factor that is very close to an integer – for example, if they are an octave apart. The phenomenon appears to have been first documented by Christiaan Huygens (1629–1695) while he was working on pendulum clocks.

He noticed that two similar pendulum clocks mounted on a common support would synchronize. Since the clocks would not be expected to have the *exact* same natural frequency, the result was surprising. The effect is easily demonstrated using multiple mechanical metronomes.[3] Note that if the interaction is linear, the pendulum motion will simply exhibit beats, as discussed in Chapter 7, and will not synchronize.

The nonlinear driving mechanism for bowed and blown instruments results in a similar lock between the overtones. The resultant tone is then periodic with harmonic overtones, even if the normal modes are not precisely harmonically related. There is enough leeway in the frequency response for each mode so that the modes will be driven at a frequency close to their respective mode frequencies.

Referring back to the bowed string, as an example, the bow is applied over one very short section of string, and whether the string sticks or slips is determined by the sum of the modes, evaluated at the location of the bow. The string will stick or slip for all modes – it cannot stick for some and slip for others. Modes that are a bit low in frequency will experience the slip a little bit earlier than might be ideal for that mode by itself. Similarly, modes that are a bit high will see the slip delayed by a small amount. A compromise solution results that locks the oscillations together. A similar description applies to blown instruments, where the interaction at the mouthpiece is based on the sum of modes. Thus, even though the frequencies of the modes may not be exactly harmonic, the frequencies of the overtones produced, when bowed or blown, can be exactly harmonic due to the nonlinear interaction.

Mode locking can also occur between separate organ pipes that are closely spaced, a phenomenon known to organ builders for some time, and this is taken into account during organ design. Interestingly, the underlying physics that gives rise to mode locking can occur in some other, very different situations, such as the production of ultrashort laser pulses.

Chaos (Optional)

A new behavior that occurs for some nonlinear systems is referred to as *chaos*. Normally, the future state of a system can be predicted accurately from the starting conditions. If there is an inaccuracy in the knowledge of the starting conditions, the accuracy of the prediction degrades with time no worse than linearly. That is, the accuracy after a time $2t$ is no worse than about twice the accuracy after a time t. In a chaotic system, the accuracy degrades exponentially. The result is that even a very small difference in the initial conditions can create a huge difference in the result after even a relatively short time. That result shows the power of exponential growth. This effect is sometimes referred to as the *butterfly effect*.

---

[3] Search using the keywords "synchronizing metronomes" or similar.

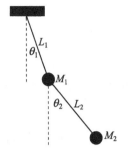

Figure 10.12 A double pendulum is a simple physical system that can exhibit chaotic motion.

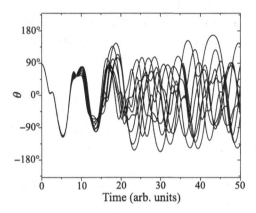

Figure 10.13 Results of a series of precision calculations for a double pendulum with $M_1 = M_2, L_1 = L_2$, started with $\theta_1 = 90°$ and $\theta_2 = 169°$ to $171°$ in $0.25°$ increments. Note how the solutions all start together but then rapidly diverge. This is a sign of chaotic motion.

The idea is that the subtle effects of the flight of a butterfly in one part of the world may have extreme effects on the weather somewhere else in the world.

A very simple physical system that exhibits chaos is the double pendulum – a pendulum hanging from a pendulum. See Figure 10.12. Recall that a pendulum has significant nonlinear behavior as the angle becomes larger than about 1 radian (~57°). Figure 10.13 shows the behavior for one of the angles of a double pendulum from several high-precision calculations but with slightly different starting conditions. The motions are similar for a short time, but then they rapidly diverge. In practice, results at longer times simply cannot be predicted.

While it is not likely that the formal theory of chaos plays a significant role in the physics of musical applications, some composers have incorporated chaos, along with related concepts such as **fractals**, into their musical compositions.

Summary

For linear physics problems, such as the ideal vibrating string, the problem can be divided into smaller parts, and the solution to the whole is the sum of the solutions to the smaller parts. For a nonlinear problem, a solution based on such a division is not possible.

Friction is an everyday example of a nonlinear force. A simple model of friction can be used to understand the stick–slip mechanism important for bowed instruments.

For reed instruments, the action of the reed leads to a nonlinear relationship between the blowing pressure and the flow rate. For some values of blowing pressure, increased pressure leads to a decrease in flow, which acts like a negative resistance that can result in oscillations. The average blowing pressure provides the energy for those oscillations. As a very simplified model, the brass instruments can be considered to have a lip reed, and the flute family an air reed, both of which are also nonlinear.

The nonlinear interaction is largely responsible for the relative amplitudes of the overtones produced and, hence, the sound of the instrument. The nonlinear interaction can also lock together oscillations of different modes at harmonic frequencies, even if the natural frequencies of those modes are not exactly harmonic.

Nonlinear forces can also result in a phenomenon known as *chaos*, where solutions at longer times simply cannot be predicted.

ADDITIONAL READING

Campbell, D. M. "Nonlinear Dynamics of Musical Reed and Brass Wind Instruments." *Contemporary Physics* 40, no. 6 (1999): 415–431.

Fletcher, N. H. "Mode Locking in Nonlinearly Excited Inharmonic Musical Oscillators." *Journal of the Acoustical Society of America* 64, no. 6 (1978): 1566–1569.

Fletcher, N. H. "The Nonlinear Physics of Musical Instruments." *Reports on Progress in Physics* 62, no. 5 (1999): 723–764.

Hirschberg, A., J. Kergomard, and G. Weinreich, eds. *Mechanics of Musical Instruments*. Springer-Verlag, 1995.

Newman, E. B., and S. S. Stevens. "The Nature and Origin of Aural Harmonics." *Journal of the Acoustical Society of America* 8, no. 3 (1937): 208.

Schumacher, R. T., and J. Woodhouse. "Computer Modelling of Violin Playing." *Contemporary Physics* 36, no. 2 (1995): 79–92.

PROBLEMS

10.1 Show that $x = 2$ and $x = 3$, each individually, satisfy the equation $3x^2 - 15x + 18 = 0$, but $x = 2 + 3 = 5$ does not, thus proving that the equation is not linear.

10.2 Real springs will not obey Hooke's law if they are compressed or stretched too much. A possible model for an undamped oscillator with a nonideal spring is given by

$$ma + kx + \beta x^3 = 0,$$

where β is a constant that describes the deviation from Hooke's law behavior.

a. Start by assuming $\beta = 0$, and show that the equation is solved by $x(t) = A\cos(\omega t)$ for any amplitude, A, for an appropriate choice for ω. See Equation (5.7) for the relation between position and acceleration for sinusoidal motions.

b. Put the solution from part (a) into the equation with $\beta \neq 0$ to see what happens. Use the identity

$$\cos^3 z = (\cos(3z) + 3 \cos(z))/4$$

to show that (i) the solution to this model nonideal oscillator can be expected to contain terms that oscillate at some of the harmonics of ω, and (ii) the solution will depend on amplitude A. In particular, if A is very small, the nonlinear term becomes negligible.

10.3 For a real string, in what ways would you expect the spectrum from a plucked string to differ from that of a bowed string?

10.4 For loud sounds, the nonlinear properties of the ear can cause sounds to be perceived that are not present in the original sound. These extra sounds are called *aural harmonics*. Use identities for sinusoidal functions to determine the perceived frequencies expected if the response of the ear is described by

$$H(t) = S(t) + a\, S^2(t),$$

where $H(t)$ represents the (subjective) sound that is heard, and (a) $S(t) = A\cos(\omega t)$ and (b) $S(t) = A\cos(\omega t) + B\cos(3\omega t/2)$ describe the sounds heard in the absence of the nonlinearity. Here, a is a constant that affects the amplitude of, but not the frequencies of, the aural harmonics.

10.5 Get a tour and demonstration of a pipe organ – the larger, the better. What physics did you see that was important for the organ's design?

11 Classical Gases

As was previously stated, it is relatively easy to show that the sounds we hear travel through air, a gas. In addition, sound production in the wind instruments – the woodwinds and brasses – necessarily involves the motion of air. Hence, it would seem to be useful to take a look at the behavior of air or, more generally, of gases.

Much of the understanding of the behavior of gases comes from studies in the area of physics known as *thermodynamics*. The scientific works that are generally considered to be the beginnings of thermodynamics date back to the mid-1600s. The winters in northern Europe were particularly cold at that time, giving rise to an increased interest in the measurement and interpretation of "temperature."[1] Many of the contemporaries of Isaac Newton were involved, including English scientists Robert Hooke (of Hooke's law for springs) and Robert Boyle (1627–1691). At the time, the components of air were largely unknown. Later, developments in the field of chemistry, in particular the development of the periodic table, such as that put forward in 1869 by the Russian chemist Dmitri Mendeleev (1834–1907), gave rise to the idea that a gas was made of atoms or molecules. Based on evidence from many scientific studies from around the world, the modern picture of the atom was put forward in 1913 by the Danish physicist Niels Bohr (1885–1962).

A simple picture of gases will be developed in this chapter in reverse historical order, starting with atoms from the periodic table and working backward. Such an approach for studying thermodynamic properties, performed in a more complete and general manner than what is presented here, is often referred to as *statistical mechanics*. Thermodynamics and statistical mechanics are both appropriate for the study of many systems composed of a large number of atoms. Here, only a basic understanding of the behavior of simple gases is the goal. Many other interesting results from thermodynamics and statistical mechanics are left to be

---

[1] This became of particular interest when their wine started freezing.

discovered elsewhere. The discussion of gas properties, in particular the speed of sound, will be continued in the next chapter.

The Elements

The periodic table (Figure 11.1) is used to display the known **elements** in a manner derived from, and useful for, chemistry. While some tables may show additional information, the basic information in all such tables includes the atomic number of the element, the symbol used for the element, and the atomic mass in atomic mass units (u). Since the symbol and atomic number go together, having both is actually redundant information. The layout may differ from one table to another, but often the atomic number is above the symbol, and the atomic mass is below the symbol.

Example 11.1

Sodium, which has the symbol Na (from the Latin name for the element, natrium), has, by definition, atomic number 11. The (average) mass of a sodium atom is 22.989 u.

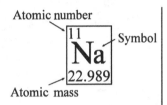

The basic structure of an atom is described by the **Bohr model**, named for Niels Bohr. Each atom consists of a central nucleus, containing most of the mass, and **electrons orbiting around the nucleus**. The classical (as opposed to quantum mechanical) picture is much like that of the planets orbiting the sun (Figure 11.2). **The nucleus of the atom consists of a combination of particles of roughly equal mass known as** *protons* **and** *neutrons*. Both are referred to generically as *nucleons*. The proton is positively charged, and the neutron has no charge. The **atomic number of an atom equals the number of protons in the nucleus**. Electrons are much less massive than are protons, and they are "negatively charged." In the normal state, the atom has no net charge, meaning that the number of electrons equals the number of protons. If the number of electrons is larger or smaller than that, then the atom is referred to as an **ion**. Since the electron orbits are largely determined by the number of protons, and the electrons are responsible for forming chemical bonds, the chemical behavior of an element depends in large part on the number of protons and very little on the number of neutrons.

It is quite impractical to draw a Bohr atom to scale, in the same way it is difficult to make a model of the solar system that is to scale. A typical atom is about 10^{-10} nm $= 0.1$ nm across, whereas a typical atomic nucleus is

Periodic Table of the Elements

1	2	3	4	5	6	7	8	9	10	11	12	13	14	15	16	17	18
1 H 1.0080																	2 He 4.0026
3 Li 6.939	4 Be 9.0122											5 B 10.811	6 C 12.011	7 N 14.007	8 O 15.999	9 F 18.998	10 Ne 20.183
11 Na 22.990	12 Mg 24.312											13 Al 26.982	14 Si 28.086	15 P 30.974	16 S 32.064	17 Cl 35.453	18 Ar 39.948
19 K 39.102	20 Ca 40.078	21 Sc 44.956	22 Ti 47.867	23 V 50.942	24 Cr 51.996	25 Mn 54.938	26 Fe 55.846	27 Co 58.933	28 Ni 58.690	29 Cu 63.546	30 Zn 65.37	31 Ga 69.723	32 Ge 72.630	33 As 74.922	34 Se 78.971	35 Br 79.904	36 Kr 83.798
37 Rb 85.468	38 Sr 87.620	39 Y 88.906	40 Zr 94.122	41 Nb 92.906	42 Mo 95.940	43 Tc (98)	44 Ru 101.070	45 Rh 102.906	46 Pd 106.420	47 Ag 107.868	48 Cd 112.411	49 In 114.818	50 Sn 118.710	51 Sb 121.760	52 Te 127.600	53 I 126.905	54 Xe 131.293
55 Cs 132.906	56 Ba 137.327	57 * La 138.906	72 Hf 178.490	73 Ta 180.948	74 W 183.840	75 Re 186.207	76 Os 190.230	77 Ir 192.217	78 Pt 195.078	79 Au 196.967	80 Hg 200.590	81 Tl 204.383	82 Pb 207.200	83 Bi 208.98	84 Po (209)	85 At (210)	86 Rn (222)
87 Fr (223)	88 Ra (226)	89 † Ac (227)	104 Rf (261)	105 Db (262)	106 Sg (266)	107 Bh (264)	108 Hs (277)	109 Mt (278)	110 Ds (281)	111 Rg (282)	112 Cn (285)	113 Nh (286)	114 Fl (289)	115 Mc (290)	116 Lv (293)	117 Ts (294)	118 Og (294)

*Lanthanides

58 Ce 140.116	59 Pr 140.908	60 Nd 144.240	61 Pm (145)	62 Sm 150.36	63 Eu 151.964	64 Gd 157.250	65 Tb 158.925	66 Dy 162.500	67 Ho 164.930	68 Er 167.259	69 Tm 168.934	70 Yb 173.040	71 Lu 174.967

†Actinides

90 Th 232.038	91 Pa 231.036	92 U 238.029	93 Np (237)	94 Pu (244)	95 Am (243)	96 Cm (247)	97 Bk (247)	98 Cf (251)	99 Es (252)	100 Fm (257)	101 Md (258)	102 No (259)	103 Lr (262)

Figure 11.1 The periodic table of elements.

Figure 11.2
Schematic to illustrate
the Bohr model of an
atom.

about 1–10 fm across (1 fm $= 1$ femtometer $= 10^{-15}$ m). The size of an electron is not known and is difficult to define, but by all accounts, it is certainly not much larger than the nucleus and may be more than 1,000 times smaller. Thus, an atom, like our solar system, is mostly empty space.

Since electrons have very little mass, the mass of an atom, the **atomic mass**, is largely determined by the combination of protons and neutrons. The atomic mass is usually given in unified atomic mass units (u) or, equivalently, daltons (Da),[2] where the mass of a proton, and hence also of a neutron, is close to 1. The small differences will be discussed later. Thus, if you have an atomic mass of about 23 u, and there are 11 protons, there must be 12 neutrons. To convert to Standard International (SI) units, use the fact that

$$1 \text{ u} = 1.66054 \times 10^{-27} \text{ kg}. \tag{11.1}$$

Clearly, the mass is small in everyday units. However, for N particles, each having a mass of 1 u, it is easy to compute how many make a total mass of 1 g:

$$N \times 1.66054 \times 10^{-27} \text{ kg} = 0.001 \text{ kg},$$
$$N = 0.001/1.66054 \times 10^{-27} = 6.02214 \times 10^{23}. \tag{11.2}$$

The value $6.02214076 \times 10^{23}$ is known as *Avogadro's number* or *Avogadro's constant*, often abbreviated N_A.[3] A **mole** is a quantity of particles equal in number to Avogadro's number. Hence, if an atom has an atomic mass of 23 u, then 1 mole of those atoms has a mass of 23 g. Thus, the atomic mass values on the periodic table can be considered to be in units of grams per mole rather than atomic mass units.

---

[2] Recently renamed for the English chemist John Dalton (1766–1844), although not yet universally adopted. Previously, *amu* was used to stand for "atomic mass units," although 1 amu is defined slightly differently than is 1 u.

[3] The 2019 redefinition of SI units defines N_A to be exactly this value.

Isotopes

Elements can exist as different **isotopes**. The number of protons establishes which element the atom is, and then the number of neutrons establishes which isotope of that element it is. Thus, all atoms with 11 protons in the nucleus are sodium, but there may be 12 neutrons or 11 neutrons, corresponding to different isotopes of sodium. Isotopes are often labeled with the elemental symbol and the total number of nucleons. For example, a sodium atom with 12 neutrons would be labeled Na-23 or ^{23}Na. Only a relatively small number of the known isotopes are stable. The lighter elements may have a few stable isotopes, and the rest are unstable, or "radioactive," and eventually will spontaneously break apart. Some elements do not have any stable isotopes; well-known examples include technetium (the lightest), uranium, and plutonium. The fraction of any given element you expect to find (on Earth) as a specific isotope is known as the **natural abundance** for that isotope. Small variations in natural abundance can occur, depending on the source of the material, and can sometimes be used to trace the origin of the material. For example, the ratio of Cr-54 to O-17 is sometimes used to classify meteorites. Larger variations can occur if one isotope is radioactive, such as is found for carbon. For carbon, such variations provide the basis for carbon dating.

Carbon Dating (Optional)

Since it was mentioned, a brief description of carbon dating seems in order. The nitrogen in the air, virtually all of which is N-14, is continuously bombarded by cosmic rays, and a small fraction is converted to C-14 during those collisions. C-14 is radioactive, with a half-life of about 5,730 years – that is, for each 5,730 years, one-half of the C-14 is lost, a multiplicative rate. When C-14 decays, it turns back into N-14. Stable carbon is mostly C-12. Since living things ingest or inhale carbon, in various forms, during their lifetimes, living things will incorporate both isotopes. After death, however, the C-14 will decay and will not be replaced. Hence, the ratio of C-14 to C-12 will decrease (exponentially) after the time of death.

Example 11.2

For each 5,730 years, you lose one-half of the C-14. What fraction is left after 10,000 years?

This is a problem involving an exponentially decreasing quantity, similar to the examples seen in Chapter 2. Number of half-lives = $10,000/5,730 = 1.75$; fraction remaining after one half-life = 0.5 (by definition). Hence, the fraction remaining after 10,000 years = $(0.5)^{1.75} = 0.30$, or 30 percent. The fraction lost is then 0.70 (or 70 percent).

Notice the similarity of this problem to that of the frequencies of the musical notes on a keyboard (Chapters 2 and 3). For each octave decrease, the frequency decreases by one-half. Hence, if the frequency of a particular note represents the initial quantity of a material, then the quantity left after n half-lives would be represented by the frequency of the note n octaves lower. They are both examples of an exponentially decreasing quantity.

Isotopes and Atomic Masses (Optional)

Different isotopes have different atomic masses. Since the periodic table is designed for chemistry, and the chemistry is principally determined only by the number of protons, the table will show an average mass for such an atom. If more than one isotope of an element occurs with a significant natural abundance, it is more difficult to determine the number of neutrons using the periodic table.

Example 11.3

Naturally occurring copper (Cu) is predominately composed of two isotopes, Cu-63 and Cu-65. Copper has atomic number 29, so these isotopes have 34 and 36 neutrons, respectively. A typical copper sample is 69 percent Cu-63 and 31 percent Cu-65. Hence, an average mass of about $0.69 \times 63 + 0.31 \times 65$ u $= 63.6$ u would be expected. This value is indeed close to what you find in the periodic table.

At the risk of getting quite far afield, it is interesting to take a closer look at the masses of atoms. The unified atomic mass unit, u, is currently defined to be exactly 1/12 the mass of a carbon-12 atom (C-12). A C-12 atom has 6 neutrons and 6 protons. If you look at He-4, which has 2 protons and 2 neutrons, you might expect to get 1/3 of that, or exactly 4 u. However, careful measurements show that it is 4.0026032 u. Similarly, Li-6, with 3 protons and 3 neutrons, is not 6 u, but 6.015122 u. While the discrepancy seems small, it is quite measurable, and for both of these, the mass is larger than expected. On the other hand, Si-28, which has 14 protons and 14 neutrons, has a mass of 27.97693 u, a bit less than 28 u, so it is less massive than expected. How can this be?

The origin of this mass difference was ultimately resolved by Albert Einstein following the development of his special theory of relativity. A basic starting point for that theory was that all observers measure the same value for the speed of light no matter how or where that light was created.[4] After working through the consequences of that assumption, one of the most famous results was

---

[4] More technically, this applies to all observers who are in "inertial reference frames." See Chapter 5.

$$E = mc^2 , \tag{11.3}$$

which says that mass, m, is associated with energy, E, and the amount of energy is found using the conversion factor c^2 (the speed of light squared). Hence, if three He-4 atoms could be combined to get one C-12 atom, there would be 0.00781 u of mass lost in the process, and that means energy (such as heat) must be released in the process. When several atoms are combined together to create a heavier atom, the process is called *fusion*.

If a very heavy nucleus, such as U-235 (atomic number 92), which has a mass of 235.043923 u, breaks up into Th-231, mass 231.036297 u, and He-4, mass 4.0026032 u, the total mass afterward is 235.0389 u, which is less than what was there initially. In this case, energy must also be released. The process where an atom is broken up into smaller parts is called *fission*. That this change in mass is important is evidenced by the fact that during the first nuclear bomb detonations, films of which can be viewed and show the devastating power involved, the energy released came from the loss of less than 1g of mass during a fission process.

The previous examples show that lighter elements can be combined to give off energy, and a heavier element can be broken apart to give off energy. In the middle of the periodic table are various isotopes of iron (Fe) and nickel (Ni) that cannot be combined or broken apart to give off energy. These isotopes are considered to be, in some sense, the most stable of the isotopes.

While this discussion has been a bit far away from music, it will be evident later that if you could somehow turn off special relativity, which would change the mass of the atoms, you would actually need to retune your organ pipes at the same time. The two topics are connected through the common laws of physics.

Molecules

A combination of two or more atoms can be chemically bound together to form a molecule. Some examples are represented in Figure 11.3. Many of the materials that are present in our everyday world consist primarily of molecules, including human bodies, violins, trumpets, and air.

Figure 11.3
Schematic representations of the principal molecules that are found in air.

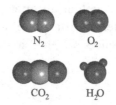

N_2 O_2

CO_2 H_2O

States of Matter

Matter is usually described as belonging to one of four states: solid, liquid, gas, and plasma. Most materials fit neatly into one of these classifications, although the dividing lines can become indistinct in some circumstances. When a material undergoes a change from one to another, such as ice melting to water, or water boiling to steam, it is referred to as a *phase change* of matter. In the following descriptions, the word *molecule* should be understood to also include atoms (a molecule with one atom).

Solids – Solids consist of molecules that are so tightly packed together that the molecules can be regarded as touching one another. The motions of molecules in a solid are restricted by bonds to their neighbors, so the packing is also quite rigid. That is, other than small vibrations made in place, an atom only moves if the entirety of the material moves. In crystalline solids, the atoms or molecules are arranged in a neatly ordered array. In glasses and amorphous solids, the arrangement is somewhat haphazard.

Liquids – Liquids also consist of molecules that are so tightly packed together that they can be regarded as touching one another; however, they are not strongly bound to their neighbors and can easily slide past one another.

Gases – Gases consist of molecules that are so loosely packed that they can move around quite freely. Gas molecules travel randomly through space, only occasionally bumping into a neighbor or the walls of their container. There is only negligible bonding between molecules in a gas.

Plasma – A plasma is like a gas; however, a significant fraction of the molecules have one or more electrons that have become separated from the molecule. That is, there is a significant fraction of ions present. Plasmas emit light when the electrons recombine with an atom or molecule. Fluorescent light bulbs, the interior of the sun, and interstellar space all contain plasmas.

For a given material, the same volume of a liquid or solid will have roughly the same mass. That is, the liquid and solid states have roughly the same mass per volume, or density. Gases (and plasmas), on the other hand, tend to fill the available space and so can have a very wide range of mass for a given volume.

As a round number, when a solid or liquid is converted to gas under normal room conditions, the *volume* expands by a factor of about 1,000. Thus, a typical *spacing* between molecules in a gas under normal room conditions is about 10 times the size of the molecule (Figure 11.4).

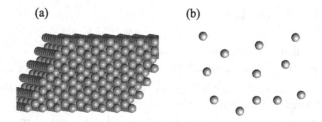

Figure 11.4 A schematic representation of the spacing between atoms in (a) a solid compared to (b) a typical gas.

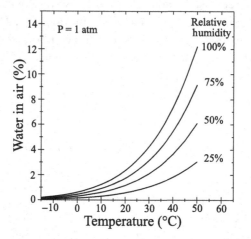

Figure 11.5 The temperature-dependent relationship between humidity and water content for air.

Air

Air is a gas consisting of many different molecules. Table 11.1 lists the major components of Earth's atmosphere, excluding water. The molecules were already illustrated schematically in Figure 11.3. Nitrogen and oxygen make up most of the air, and both come as diatomic (two-atom) molecules, abbreviated N_2 and O_2, respectively. Argon (Ar) is one of the noble gases (also referred to as *rare gases* or *inert gases*) and is present as an atom. Carbon dioxide, CO_2, which is very important for Earth's greenhouse effect, is actually only a small fraction of the air. It is an example illustrating that sometimes even a very small quantity can have a large effect on the overall properties. The amount of water in the air, typically reported using "relative humidity," is highly variable (see Figure 11.5) but may be up to about 5 percent of the air on a hot, humid day, in which case the remaining 95 percent will be the components of dry air.

Table 11.1 The major constituents of dry air.

Component	Fraction of Dry Air (%)
Nitrogen (N_2)	78.09
Oxygen (O_2)	20.95
Argon (Ar)	0.92
Carbon dioxide (CO_2)	0.03

Temperature and Kinetic Energy

The molecules in a gas move around randomly and undergo occasional collisions with other molecules or the walls of a container. The "amplitude" of this motion is the velocity (or momentum), and the energy associated with that motion is known as the **kinetic energy**, *KE*. As is often the case, energy goes like the amplitude squared. For a single particle of mass *m*, the kinetic energy is given by

$$KE = \frac{1}{2}mv^2. \tag{11.4}$$

In a gas, there is not one velocity for all molecules but a wide distribution of velocities that, in the absence of wind, average to zero. There are as many molecules going left as right, as many going up as down, and so forth. However, the velocity squared, which is equal to the speed squared, will always be a positive value and will not average to zero. Thus, it still makes sense to talk about the average kinetic energy for a molecule, even though the average velocity is zero.

A classical result from statistical mechanics, which will not be derived here, says that the energy for a simple gas of particles that is at equilibrium at a temperature, *T*, will be given by

$$E = N_d \times \left(\frac{1}{2}k_BT\right), \tag{11.5}$$

where *T* is the **absolute temperature**, k_B is a constant (the *Boltzmann constant*) that converts temperature units to energy units, and N_d is the number of degrees of freedom. The absolute temperature scale is the one where classical physics predicts all motion will stop when $T = 0$. In Standard International (SI) units, absolute temperature is measured in kelvin (K), and $k_B = 1.38066 \times 10^{-23}$ J/K. The temperature in kelvin can be found from the temperature in Celsius by adding 273.15.[5] The number of degrees of freedom for the simple translational motion of a single particle in three dimensions is just 3 (e.g., corresponding to motion in the x-, y-, and z-directions). If you have N particles, then the total number of degrees of freedom is 3 N. In this simple case, temperature is a measure of energy.[6]

Comparing these two results for the energy, an average speed for a molecule of mass m that is in a gas at absolute temperature T can be determined as follows:

$$KE_{ave} = \frac{1}{2} m \left(v^2 \right)_{ave} = \frac{3}{2} k_B T \rightarrow \left(v^2 \right)_{ave} = \frac{3 k_B T}{m} \rightarrow v_{rms} = \sqrt{\frac{3 k_B T}{m}}, \quad (11.6)$$

where **rms** stands for **root mean square**, which is the square root of the average of the square. While the average velocity is always zero because the direction is random, the root mean square velocity can be used to give an indication of a typical speed.

Example 11.4

Compute the typical speed of nitrogen *molecules* (N_2) in air at 20°C.

$$m = 2 \times 14.01 \times 1.6605 \times 10^{-27} \text{ kg} = 4.665 \times 10^{-26} \text{ kg},$$

$$T = 20 + 273.15 \text{ K} = 293 \text{ K},$$

$$v_{rms} = \sqrt{\frac{3 \times 1.38 \times 10^{-23} \times 293}{4.665 \times 10^{-26}}} \text{ m/s} = 510 \text{ m/s}.$$

and 510 m/s corresponds to about 1,800 km/hr or 1,100 miles per hour (mph). These molecules are certainly moving at high speed. In fact, the speed of sound in air at 20°C is typically about 340 m/s, about two-thirds of the average speed. Many of the molecules are "supersonic." It will be shown later that this relationship between the speed of sound and the average speed is not at all a coincidence.

[5] Note that it is proper to say "kelvin" and not "kelvins" or "degrees kelvin." The absolute scale based on Fahrenheit degrees is the Rankine scale. To convert Fahrenheit to Rankine, add 459.67.

[6] Although it is true in many other circumstances, one must be careful about extending this statement in general. For example, as water is boiled, energy is added, but the temperature is constant. Note also that this statement applies only if one uses absolute temperature since the kinetic energy cannot be negative.

Example 11.5

Estimate the typical speed of an argon atom at 100 K.

To compute the average speed for other circumstances, proceed as for nitrogen, or simply take the result for nitrogen at 20°C (293 K) and algebraically replace the mass and temperature with the new values. From the periodic table, the mass of an Ar atom is 40 u, whereas that of a nitrogen *molecule* is 28 u. Hence, for argon at 100 K,

$$v_{rms} = 510 \text{ m/s} \sqrt{\frac{28}{40}} \sqrt{\frac{100}{293}} = 250 \text{ m/s}.$$

In essence, the first square root replaces the mass of the nitrogen molecule with the mass of argon, and the second replaces the temperature 293 with 100. Remember that more mass and colder temperatures each correspond to slower motion.

Impulse

The molecules in air, all traveling in different directions at about 500 m/s, will undergo collisions with each other and with the walls of their container. A collision is a short-duration encounter between two objects. *Short* is a relative term, and one must ask, Short compared to what? The collision of a baseball with a bat occurs over about 1 ms, so it is short on our everyday scale of seconds. On the other hand, a collision of two galaxies will take place over a hundred million years or so, which seems like a long time but is short compared to the expected lifetime of most galaxies. In general, *short duration* might be defined as short compared to other times of interest. In the case of molecular collisions in a gas, *short* might reasonably be taken to mean "short compared to the time between collisions."

For a collision, it is often convenient to describe the net result of the collision without worrying about the details of the collision. For a simple collision of a particle of mass, m, that takes place over a time interval, Δt, the result of the collision can be described using the change in velocity of the particle, Δv. In terms of Newton's laws, but using the average acceleration during the collision, the average force, $\overline{F}$, during the collision is given by

$$\overline{F} = m(v_f - v_i)/\Delta t \quad \text{or} \quad \overline{F}\Delta t = m\,\Delta v, \tag{11.7}$$

where it must be remembered that the direction of the initial and final velocities (v_i and v_f, respectively) must be taken into account. The quantity $\overline{F}\,\Delta t$ is known as the **impulse** for the collision. If the collision is of short enough duration, only the product of $\overline{F}$ and Δt is important for determining what happens, and not either one individually. Note that an impulse, like velocities and forces, will have a direction. The direction of the impulse acting on the mass, m, will be in the direction of the (average) force that caused the change in the velocity of the mass.

Example 11.6

Suppose a 1,500-kg car is driven at 30 m/s ($= v_i$) southward toward a wall. Upon hitting the wall, the car comes to a stop ($v_f = 0$). The impulse on the car for this collision is 45,000 kg·m/s northward, directed away from the wall.

It is noted that while the time, Δt, may not be all that important for describing the "before and after" of the collision of a particle, it may be important in some real-life applications. In Example 11.6, the car could have been built very rigidly, in which case it would come to a stop very suddenly, or the car could have been built to crumple, drawing out the collision as long as is practical (although still obviously over a short time compared to what might be desirable). In both cases, the impulse would be the same. However, a longer collision time for a fixed impulse corresponds to a smaller average force. In the case of a car hitting a wall, a longer rather than shorter time can determine whether or not the driver survives the collision. A safer car is one that can crumple, provided it crumples around the driver, since that reduces the average force on the driver during the collision.

Collisions can be further classified as being elastic or inelastic. For an **elastic collision**, the total kinetic energy (*KE*) is the same before and after the collision. All other collisions are inelastic. When a single particle makes an elastic collision with a stationary rigid object, such as a wall or floor, the speed after the collision will be the same as the speed before the collision – only the direction changes. When the speed is smaller afterward, the collision is inelastic. For the special case where the particle sticks to the wall or floor, it has lost all its kinetic energy, and the collision is referred to as being *totally*, or *perfectly*, *inelastic*.

Consider now a single molecule inside a box (Figure 11.6a). For simplicity, imagine the molecule is moving horizontally with speed v_x, as shown, and the collisions with the wall are elastic. When the molecule strikes the right wall, the impulse on the particle is $2mv_x$, directed away from the wall. The molecule will then travel to the left wall, collide, and head back to the rightmost wall, where it will then make another collision. The total time between collisions with the

Figure 11.6 (a) A single molecule moving in a straight line in a box and (b) random motion of many molecules in a box.

(a)

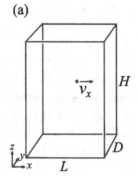

(b)

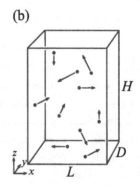

rightmost wall will be $2L/v$, the total distance traveled divided by the speed. Thus, using Newton's third law, the time-averaged force on the rightmost wall due to the collisions with the molecule will be

$$\overline{F} = \frac{2mv_x}{2L/v_x} = \frac{mv_x^2}{L},$$ (11.8)

The average speed squared for a molecule in a simple gas that is at (absolute) temperature, T, was found in Equation (11.6). That expression is, however, for a molecule moving in three dimensions (Figure 11.6b). The expression for the average force on a particular wall considers motion in one dimension only. Dividing the velocity associated with the random motion of molecules into three component speeds, corresponding to each of the three dimensions labeled x, y, and z, and squaring yields

$$v^2 = v_x^2 + v_y^2 + v_z^2.$$ (11.9)

Since the direction of the velocity is random, it must be that, on average, $v_x^2 = v_y^2 = v_z^2$, and so, on average, $v_x^2 = v^2/3$. Hence, the average force on the rightmost wall due to a molecule from a gas at temperature T can be expected to be

$$\overline{F}_{\text{ave}} = \frac{m}{L}\frac{1}{3}\frac{3k_{\text{B}}T}{m} = \frac{k_{\text{B}}T}{L}.$$ (11.10)

This result is already interesting in that the mass, m, has no effect. Hence, it is immaterial what type of molecule may be involved.

Pressure

The wall of the box described previously will experience many collisions from many different molecules, and those collisions can occur across the entire area of the wall. To remove the specific wall from the description, it is useful to talk about the total force provided by the gas for a specific area. Dividing the force pushing on the wall by the area of the wall yields the *force per unit area*, which does not depend on the specific wall used. In this context, **pressure is the force per unit area**. For SI units, pressure will be in N/m^2 = pascal (Pa).[7]

Returning to the force on the rightmost wall, the pressure on that wall will be

$$P = N\frac{\overline{F}_{\text{ave}}}{HD} = N\frac{k_{\text{B}}T}{HDL},$$ (11.11)

---

[7] Named for Blaise Pascal (1623–1662), a French mathematician, scientist, and philosopher.

where N is the number of gas molecules in the box. Recognizing that the total volume inside the box is HDL, this result can be rewritten

$$PV = Nk_\mathrm{B}T. \tag{11.12}$$

Since $N = nN_A$, where n is the number of moles of gas and N_A is the number of molecules per mole, this result will also be written

$$PV = nRT, \tag{11.13}$$

where the R is called the *molar gas constant*; $R = k_\mathrm{B} \times N_A = 8.3145$ in SI units. No matter which constant is used, this relationship between pressure, volume, and temperature is known as the **ideal gas law**. For gases at everyday temperatures and pressures, this simple law works surprisingly well.

It is important to remember that when using the ideal gas law in practice, most pressure gauges measure what is known as the **gauge pressure**. The gauge pressure is the *difference* in pressure from one side of the gauge to the other. Such a measurement is a straightforward measurement of the net force (e.g., on the walls of a container). For a simple container, the gauge pressure (inside) will be the inside pressure minus the outside pressure. Thus, when the pressure of a tire on a car is measured to be 30 pounds per square inch (psi), that refers to the difference between the pressure inside the tire compared to the pressure outside the tire.[8] If the gauge pressure is negative, that means that the pressure inside is smaller than the pressure outside. The actual pressure inside, sometimes referred to as the **absolute pressure**, can never be negative – an atom colliding with a wall cannot produce a negative impulse. A common error is the use of the gauge pressure in the ideal gas law. **When using the ideal gas law, *absolute pressure must be used***; the pressure "outside" is irrelevant. Some examples will come later.

Atmospheric Pressure

The pressure exerted by Earth's atmosphere – that is, by the many billions of billions of molecules that are constantly bombarding us at supersonic speeds each second – can be measured in a number of ways. The general principle is to balance one force against another, as was done for springs earlier (Chapter 5). Most of the time, the atmosphere exists all around, so the forces from the atmosphere balance themselves out. By removing the atmosphere from one side, to the extent possible, the force due to the remaining air can be determined.

A simple method used to measure differences in pressure utilizes a U-shaped tube containing a liquid. Such a device is called a **manometer** (Figure 11.7). On one side, there is a gas at a known pressure; on the other, a gas at the pressure to

---

[8] If you see the unit *psig* or *psia*, the extra letter is indicating whether it is gauge or absolute pressure. This practice is limited to pounds per square inch (psi) and is not used for other pressure units.

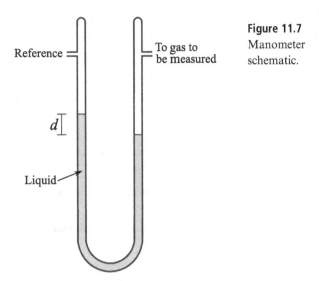

Figure 11.7
Manometer
schematic.

be measured. The pressure difference is proportional to the difference in the height of the fluid on one side compared to the other. The pressure difference gives rise to a difference in force, and that force is balanced by the weight of the extra liquid on one side. When used with an appropriate liquid, such as mercury, the reference side can be hooked to a vacuum pump in order to determine the absolute pressure. That is, **to measure the absolute pressure, the reference should be a vacuum.**

Measurements show that at sea level, the atmospheric pressure is typically

$$101 \, \text{kPa} = 1 \, \text{atm} = 14.7 \, \text{psi}. \tag{11.14}$$

To appreciate how large this value is, consider the force on the top of a typical desk. A typical desktop has a surface area of about 1 m^2, so the force is 101,000 N, or the weight of about 10,000 kg. A typical car has a mass of about 1,500 to 2,000 kg, so the desktop has a force on it equivalent to the weight of five to seven cars. Of course, a force of the same magnitude is being applied by the air on the underside of the desktop, so the air provides no *net* force.

For pressure measurements, there are several common non-SI units in use, particularly for lower pressures. Some of these are summarized in Table 11.2. It is a simple matter to convert from one to another.

While "1 atmosphere" is a defined quantity, the actual atmospheric pressure will depend significantly on the altitude above sea level and on the weather. Representative variations are shown in Figures 11.8 and 11.9. Instruments designed for measuring atmospheric pressure are referred to as *barometers*. In meteorology, the atmospheric pressure reading is often adjusted, using a simple equation, to remove the effects of the elevation of the barometer above sea level, and the result is referred to as the *pressure corrected to sea level*. If such pressure

Table 11.2 Some of the many units used for pressures.

Unit of Pressure	Definition/Origin	Equivalent
1 atm	Agreed-upon standard based on atmospheric measurements	101,325 Pa
1 cm of water	Pressure needed to raise a column of water by 1 cm	98.064 Pa
1 in. of water	Pressure needed to raise a column of water by 1 in.	249.08 Pa
1 cm of mercury (Hg)	Pressure needed to raise a column of mercury by 1 cm	1,333.2 Pa
1 in. of mercury (Hg)	Pressure needed to raise a column of mercury by 1 in.	3,386.4 Pa
1 bar	A convenient value (1 atm in Pa, rounded to the nearest power of 10)	100,000 Pa
1 torr	Same as 1 mm Hg	133.32 Pa
1 hPa	1 hectopascal = 100 Pa = 1 millibar (seen principally in meteorology)	100 Pa

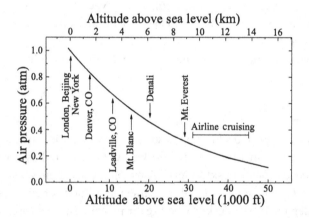

Figure 11.8 Representative variation of atmospheric pressure with height above sea level.

values are to be used in the ideal gas law, the effects due to elevation will need to be reintroduced.

Ideal Gas Law Examples

Now consider some simple examples that illustrate the use of the ideal gas law.

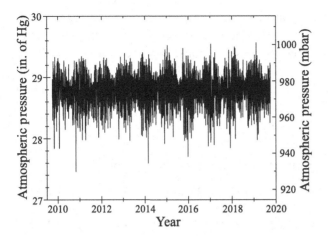

Figure 11.9
Variations in atmospheric pressure at one location associated with changes in the weather.

Example 11.7

A tire is filled until the tire gauge reads 30 psi when the temperature is 20°C. What will the pressure gauge read when the temperature rises to 35°C? It will be assumed that the tire is rigid enough that the change in the volume of the tire is negligible. Also, it is assumed that the tire is sealed, so the number of gas molecules inside the tire will not change.

First, to use the ideal gas law, the temperatures must be on an absolute scale, accomplished in this case by adding 273, and second, the pressure must also be converted to an absolute scale. A tire gauge will report "gauge pressure," which is the absolute pressure inside minus the absolute pressure outside. The pressure outside is presumed to be about 14.7 psi (1 atm), so the absolute pressure inside the tire is $(30 + 14.7)$ psi $= 44.7$ psi. Thus, the final absolute pressure in the tire is found from the ideal gas law:

$$\frac{P_f V}{P_i V} = \frac{Nk_B T_f}{Nk_B T_i} \rightarrow P_f = \frac{T_f}{T_i} P_i = \frac{308\,\text{K}}{293\,\text{K}} 44.7\,\text{psi} = 47.0\,\text{psi}.$$

Now convert back to a final gauge pressure, which is what was asked for, by subtracting the outside pressure of (about) 14.7 psi. Hence, after the temperature change, the gauge can be expected to read $47.0 - 14.7 = 32.2$ psi.

Example 11.8

A 2.0-L bottle, equipped with a pressure gauge, is opened at sea level when the temperature is 7°C. The bottle is then tightly sealed. The bottle and contents are then transported to Denver, Colorado, where the temperature is 25°C. What is the reading on the pressure gauge? Note that due to the altitude above sea level, the absolute air pressure in Denver is about 83 percent of that at sea level.

Once again, the volume and number of molecules do not change, and care must be taken to make sure absolute pressures and temperatures are used. The initial gauge pressure in the bottle is zero because the bottle was opened to the air. That means the initial absolute pressure is equal to the pressure at sea level, or about 1 atm (14.7 psi). The effect of the change in temperature is treated as in the previous example,

$$P_f = \frac{298}{280} 1 \text{ atm} = 1.064 \text{ atm},$$

which is converted to gauge pressure by subtracting the outside pressure in Denver,

$$1.064 - 0.83 \times 1 \text{ atm} = 0.23 \text{ atm} = 3.4 \text{ psi} = 24 \text{ kPa}.$$

In this case, most of the change in the gauge pressure is due to the change in the outside pressure.

The ideal gas law is a very successful rule for gases of all kinds, as long as the average spacing between molecules remains large compared to the size of the molecules. This law was derived earlier using a simple model. That model started with just one molecule traveling in one direction and that underwent a simple motion and an ideal elastic collision only with the wall. One is tempted to argue, from the success of the result, that reality must match the model. Such an extrapolation is always risky.

In the case of gases composed of many, many molecules, there is much going on that is described by values that average to zero. Hence, a model that does not include such effects in the first place can still be successful. For example, the model assumed that molecules make a simple elastic collision with the walls. Real gas molecules often stick to the surface of the wall for a while and then are released at a later time. With a large number of molecules, the odds are pretty good that anytime one molecule sticks, another is released – on average, with the same kinetic energy – and so the simpler model provides the same end result. Likewise, except at very low pressures, a molecule will also collide with many, many other molecules long before it will traverse the entire width of its container. However, some other molecule will, on the average, strike the opposite wall in its place. The basic atomic-level picture of a gas, a collection of many molecules flitting about in random directions at high speed, remains reasonably correct, however.

Pressure for Wind Instruments

Blowing pressure is what provides the force and, ultimately, the energy to excite gas flow and acoustic resonances in the wind instruments. Typical values are shown in Table 11.3. These are measured by running a small tube in through the

Table 11.3 Approximate range of blowing (mouth) pressures for wind instruments. (Averaged from various sources.)

Instrument	Pressure Range (kPa)
Flute	1–5
Oboe	4–7
Saxophone	3–6
Clarinet	3–8
Trumpet	2–20
French horn	2–11
Trombone	2–17

corner of a player's mouth. Note that these are gauge pressures, comparing the pressure inside the player's mouth to the outside, and are quite small compared to 1 atmosphere. There is a wide range for all the instruments when playing low/soft notes is compared to loud/high notes.

An interesting question arises at this point. It was shown previously that the absolute air pressure can vary with altitude and weather. The full range of atmospheric pressures experienced by the world's symphony orchestras is about 20 percent. Since a 6 percent change in frequency is a half step, such a large change in pressure could be problematic. How the absolute pressure influences, or does not influence, the instrument's pitch is an important consideration and will be addressed in the next chapter.

Summary

Atoms are the basic constituents of all matter. An atom has a nucleus consisting of protons and neutrons. The name of the atom (i.e., the element) depends solely on the number of protons. Many atoms can exist with different numbers of neutrons, called *isotopes*. The available atoms are often presented using a periodic table, which includes the atomic mass, based on the average of the naturally occurring isotopes. Atoms can bond to form molecules.

There are four types of matter: solids, liquids, gases, and plasmas. For solids and liquids, the atoms or molecules will be closely spaced. In gases and plasmas, there will be a lot of empty space between the molecules. Normal dry air consists mostly of nitrogen and oxygen molecules, with a small amount of argon and an even smaller amount of carbon dioxide. Water vapor is often present in varying amounts of up to 5 percent.

The kinetic energy of the molecules in a gas depends on the absolute temperature. From that, it is possible to deduce that the nitrogen molecules in air are, on average, moving with speeds near 500 m/s for normal room conditions. Heavier molecules will move slower, and molecules at colder temperatures move slower, both having a square root dependence.

Impulse can be used to characterize a collision in terms of an average force and time for the collision. The average force over an area, the absolute pressure, will be proportional to the number density of molecules present and the absolute temperature, expressed mathematically as the ideal gas law. Pressure gauges usually report a gauge pressure, which is the pressure difference between the inside and outside of a container. The gauge pressure must be converted to the absolute pressure before use in the gas law. Pressure differences (gauge pressures) are important for describing the net force due to air pressure.

ADDITIONAL READING

Almeidaa, A., D. George, J. Smith, and J. Wolfe. "The Clarinet: How Blowing Pressure, Lip Force, Lip Position and Reed 'Hardness' Affect Pitch, Sound Level, and Spectrum." *Journal of the Acoustical Society America* 134, no. 3 (2013): 2247–2255.
Fletcher, N. H., and A. Tarnopolsky. "Blowing Pressure, Power, and Spectrum in Trumpet Playing." *Journal of the Acoustical Society America* 105, no. 2 (1999): 874–881.

Any comprehensive introductory physics text will have a section on basic thermodynamics. Examples include:

Knight, R. D. *Physics for Scientists and Engineers*, 3rd ed. Pearson, 2013. (See Chapter 16.)
Urone, P. P., and R. Hinrichs. *College Physics*. OpenStax, Rice University, 2020. (See Chapters 11 and 13.)

PROBLEMS

11.1 What is the human population of Earth expressed in moles?

11.2 The mass of a typical person is about 70 kg. What is the mass of 1 mole of people? How does that compare to the mass of Earth?

11.3 Under normal room conditions, air has a density of about 1.2 kg/m^3. If you have a vertical column of air with a base that is 1 m^2, estimate how high the column would need to be to have a weight of 101,000 N. How does that compare with the height of Earth's atmosphere?

11.4 If you drive your car from sea level to visit the highest roads of the Rocky Mountains, should you worry about, or make any adjustments to, the air pressure in your tires? Be sure to state your assumptions and criteria.

11.5 Consider the situation in Example 11.8, where a bottle ends up in Denver and contains air at a gauge pressure of 0.23 atm = 3.4 psi. If the bottle is now opened, so the air inside equilibrates with the air outside, how many moles of gas exit or enter the bottle?

11.6 Potassium-40 (K-40) is a naturally occurring radioactive isotope. Unlike C-14, which is continuously created, K-40 is believed to be "primordial," meaning that all that now exists was created sometime in the distant past, for example, in a star or star explosion. The half-life for K-40 is 1.28 billion years. If your body contains 130 g of potassium, 0.012 percent of which is K-40, how many of those potassium atoms decay each second? Note: You can use the approximation that if ε is very small compared to 1, then $(1/2)^{\varepsilon} \approx 1 - \varepsilon \ln 2$.

11.7 If, with each breath, you convert 150 ml of oxygen gas into carbon dioxide gas, by how much will you change the carbon dioxide content of a small room, expressed as a percentage, after an hour? Assume the room is sealed and has no ventilation.

11.8 In Major League Baseball (MLB), the ball has a mass of 150 g, and a fastball is traveling at about 100 mph (160 km/h). How many MLB fastballs need to strike a 1-m^2 target each second to provide the same average force as is provided by air at a pressure of 1 atm?

12   The Speed of Sound in a Gas

The speed of sound is intimately related to the resonant frequencies of many musical instruments – that is, to the frequency of the notes produced. The speed of sound is also important for room acoustics that contribute to how those notes are perceived. It is straightforward to measure the speed of sound in a gas under a variety of circumstances. Values for many different gases are readily available. For sound in air under normal conditions, the speed is about 340 m/s. It is more challenging to understand why the speed of sound is what it is. To do that, it is appropriate to start by considering gases other than air.

Going back to the vibrating string (Chapters 6 and 8), the basic relationship between the properties of the string and the speed of a disturbance on the string was found by considering the "springiness" of the string, which comes from the tension, and the "inertia," which comes from the mass of the string. For the string, it was convenient to use the linear density of the string – the mass per unit length – for the final result. The wave speed was ultimately proportional to the square root of the tension divided by the density. The same general approach will be used here for the speed of sound in a gas.

Measured Values

The inertial contribution to the speed of sound is expected to be proportional to the mass of a gas molecule or, more generally, to the density of the gas. For a gas, that can be well approximated using the ideal gas model. For a container with volume, V, at a specified temperature, T, and pressure, P, the number of gas molecules is $N = PV/k_BT$, and N does not depend on what gas is used. If the (average) mass of one molecule of the gas is m, the total mass of the gas is Nm, and therefore the mass density of the gas, ρ, is given by $\rho = Nm/V$. The density of a gas can be measured by weighing a beaker filled with the gas and subtracting the weight of the same beaker when it has been evacuated using a vacuum pump.

Table 12.1 Density and speed of sound at 0°C and 1 atm for a variety of gases.

Gas	Density (g/liter)	Speed (m/s)
He, helium	0.178	965
CH_4, methane	0.717	430
O_2, oxygen	1.429	316
Ar, argon	1.783	319
CO_2, carbon dioxide	1.977	259
SO_2, sulfur dioxide	2.927	213
SF_6, sulfur hexafluoride	6.16	150

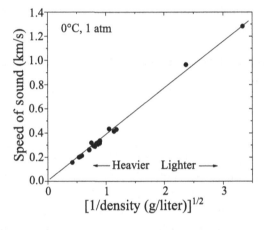

Figure 12.1 Speed of sound versus $1/\sqrt{\text{density}}$ for various gases at standard temperature and pressure.

Table 12.1 and Figure 12.1 show selected measured values of the speed of sound, obtained from various sources, for different gases at 1 atm pressure and 0°C, a condition known as **standard temperature and pressure**, or **STP** for short. If the argument that worked for the string also works for the speed of sound in a gas, then there should be a linear dependence with the inverse of the square root of the gas density $(1/\sqrt{\rho})$. If that is indeed the case, then all the data in Figure 12.1 should lie on a straight line that goes through the origin. While there is some deviation from such a line, much more than from experimental error, the trend seems to be at least somewhat valid. This deviation from the line will be addressed later.

The second piece required to determine the speed is to determine something equivalent to the "return force per unit displacement" – that is, the "springiness" of the gas. In order to get that, two additional ideas need to be considered: adiabatic processes and a more systematic look at how approximations are made and used for physics.

Adiabatic Processes

The gas law includes four variables: the pressure, temperature, volume, and number of molecules. As a gas undergoes changes, or **processes**, the values associated with these variables will change. In general, three of the values can be independently specified, and then the fourth will be determined by the gas law. For many situations of interest, the number of molecules is kept constant – no gas is added or removed. That will be assumed in what follows, so only three variables can change.

There are several special processes that often serve as models for real processes. In each case, some quantity is kept constant, thus reducing the number of values that change. Four special types of processes that are frequently encountered are:

- **Isothermal** (constant temperature)
- **Isobaric** (constant pressure)
- **Isovolumetric** (constant volume, or isochoric)
- **Adiabatic** (no heat transfer)

For each of these, it is straightforward to compare the state of the system (temperature, pressure, volume) before and after the process.

Example 12.1

A piston is used to slowly compress a volume of gas by a factor of 2. If the process is isothermal, what is the change in the pressure of the gas?

Using the gas law with constant temperature (and a constant number of molecules), $Nk_B T$ is constant, so the product of pressure and volume must also be constant. That is, $P_1 V_1 = P_2 V_2$, so if $V_2 = V_1/2$, then it must be the case that $P_2 = 2P_1$. Hence, for an isothermal process where the volume is decreased by a factor of 2, the pressure increases by a factor of 2.

Example 12.2

A piston is used to compress a volume of gas with provisions to keep the pressure constant. What is the result if the volume of the gas is reduced to one-half of its original volume?

This is an isobaric process since the pressure is constant. Use the ideal gas law and put all the constant terms on one side, with the remainder on the other. Then, $V_1/T_1 = V_2/T_2$; rearranging, this is equivalent to $T_2 = T_1 \cdot V_2/V_1$. If the volume is reduced to one-half, then $V_2 = V_1/2$, and hence the temperature must also be reduced by one-half. The "provisions" to keep the pressure constant in this case must have included some sort of refrigerator.

It is also important to distinguish between reversible and irreversible processes. A reversible process is one that can be undone by simply reversing what was done in the first place. For example, if a balloon is gently squeezed, the balloon returns to its original size when released. The process is reversible. On the other hand, imagine a container with a partition down the middle. On one side is water dyed blue, and on the other, water dyed red. If you remove the partition, the dyes will mix. If you now replace the partition, the colors will remain mixed and will not return to their separate sides of the box. Such a process is irreversible.

Another way to visualize the difference between reversible and irreversible processes is to make (or imagine) a video of the process in the forward direction. Now play that video in reverse. If what is observed in reverse simply would never happen in the forward direction, then the process is irreversible. In the example with the container and partition, the process in reverse would show all the blue dye spontaneously moving to one side while all the red dye moved to the other as the partition is put back into place. That just does not happen naturally; hence, the process is irreversible.

All real, everyday processes that we experience are irreversible, and some claim that this provides a way to define a forward direction for time. However, some processes are "close enough" to being reversible that it is okay to treat them as reversible. That is, some processes are appropriately modeled as being reversible.

Getting back to the processes listed earlier, **an adiabatic process is a process where there is no transfer of heat energy into or out of the system**. For example, if a gas is confined inside a thermally insulating container, it cannot transfer heat energy to the outside. There are no perfect insulators, so what really matters is the timescale for that energy transfer. If the timescale for energy transfer is "slow enough," or equivalently, if the process is "rapid enough," heat energy transfer will be negligible, and the system can be modeled using adiabatic processes.

For ideal gases, a gradual compression of a volume can be undone by a gradual decompression, and the process is reversible. But how rapidly can the compression be made and still have the compression be "gradual"? One has to ask, Gradual compared to what? Imagine a piston pushing on a gas to compress it. The gas provides a pressure on the piston due to the collisions with the rapidly moving molecules of the gas. Any piston that is moving very slowly compared to the molecules of that gas would be "slow" in this context, and the resulting compression would be gradual. Recalling that for air near room temperature and pressure, the molecules are traveling near 500 m/s (1,000 mph), changes can be quite rapid on a human timescale and still be "gradual" for the molecules in the gas.

Since sounds in the audible range all have wavelengths longer than a few centimeters and have frequencies of hundreds or even thousands of oscillations per second, there is reason to think that sound should be modeled as an adiabatic process.[1] There is insufficient time for heat transfer between the maximum and minimum of the wave. A few centimeters, or more, of air is a pretty good thermal insulator.[2] At the same time, examination of, for example, the motion of a speaker used to produce sound shows that the motion is quite slow compared to 500 m/s, so the process can also be expected to be gradual enough to be reversible. Hence, for sound, it seems reasonable to model the process as a "reversible adiabatic process." A reversible adiabatic process is sometimes referred to as being *isentropic* – a quantity called *entropy* is kept constant.

For ideal gases and reversible adiabatic processes, there is another relationship that can be used, in addition to the gas law. While the pressure, volume, and temperature will *all* change from the beginning to the end of such a process, they will also be related by the requirement that during the process, PV^γ remains constant. The value of the **adiabatic exponent**, γ, will depend on the particular gas and, to a lesser extent, on the circumstances.[3] As will be seen momentarily, it is expected that $1 \leq \gamma \leq 5/3$, and for air near room temperature, $\gamma = 1.4$. The adiabatic rule can be combined with the ideal gas law so that during the process, values at two different times are related by the following equations:

$$\frac{P_2}{P_1} = \left(\frac{V_1}{V_2}\right)^\gamma; \quad \frac{T_2}{T_1} = \left(\frac{V_1}{V_2}\right)^{\gamma-1} = \left(\frac{P_2}{P_1}\right)^{(\gamma-1)/\gamma}. \tag{12.1}$$

Example 12.3

A piston compresses air in an insulated container until the volume is reduced to one-half its original volume. If the air starts at normal room conditions (1 atm and 20°C), what is the final temperature and pressure?

$$\left(\frac{V_1}{V_2}\right) = \left(\frac{V_1}{V_1/2}\right) = 2 \rightarrow P_2 = (2)^{1.4} \, (1 \, \text{atm}) = 2.64 \, \text{atm}$$

$$\rightarrow T_2 = (2)^{(1.4-1)}(273 + 20) \, \text{K} = 387 \, \text{K} = 114°\text{C}.$$

---

[1] Isaac Newton initially made the assumption that the processes could be considered isothermal. He blamed the resulting 20 percent discrepancy between his theory and experiments on imperfections in the experiments. However, as experiments improved, the discrepancy persisted, thus showing that the problem was with his theory, not the experiments.

[2] Note that *rapid* is a comparative word, and to determine if the adiabatic model is appropriate, it is necessary to ask, Rapid compared to what? In this case, "rapid" is compared to the time it takes heat energy to go roughly a half wavelength through air. As the frequency gets extremely high, the wavelength gets extremely short, and although the motion is faster, it is not obvious that the process will remain rapid enough to justify the adiabatic model.

[3] The adiabatic exponent is sometimes also referred to as the *adiabatic constant, adiabatic index, isentropic exponent*, or *ratio of specific heats*, and in some disciplines, a Greek kappa, κ, is commonly used instead of a Greek gamma, γ.

Note that when the air cools back down to room temperature, which will eventually happen for real containers, the pressure will drop from 2.64 atm down to 2 atm during the same time frame. Working the problem in reverse, knowledge of the initial pressure (1 atm), the pressure immediately after compression (2.64 atm), and the final pressure after a long time (2 atm) is enough information to determine γ.

Hence, the speed of sound in a gas can be expected to depend on this adiabatic rule. While simple to write down, the equations are nonlinear due to the adiabatic exponent and are thus quite difficult to use without some approximation. Hence, a more thorough discussion of approximation techniques might be in order before proceeding.

The "Springiness" of a Gas

The gas laws for a reversible adiabatic process can now be used to find the springiness, an effective spring constant, of a gas. That can then be combined with the density to compute the speed of sound. To do this, some approximations are necessary.

Approximation Methods (Optional)

There are many simple problems in physics that are not easily solved, or even well understood, without making some additional approximations. Approximations are most often used when a solution is known for a similar situation and the known solution simply needs a small correction – that is, when changes from a known situation are, in some sense, "small." There is an art to making these approximations in a rigorous way. Many such approximations are based on mathematical results known as *series expansions*. The general theory behind these expansions is beyond the scope of this presentation; however, an example will prove useful.

A number of approximations are based on a relation known as the *binomial expansion*,

$$(1+z)^s = 1 + \frac{s}{1}z + \frac{s(s-1)}{1 \times 2}z^2 + \frac{s(s-1)(s-2)}{1 \times 2 \times 3}z^3 + \cdots, \tag{12.2}$$

where the number of terms on the right is finite if s is a positive integer (or zero) but is infinite for other values of s. While rather formidable-looking, the expansion is quite useful for approximation when the magnitude of z is small. It is important to remember that *small* is a relative term. In this case, *small* means "small compared to 1," so $(1+z)$ is close to 1. How much smaller than 1 will z need to be? That will depend on the accuracy of the result required. When z is small enough, then z^2 will be even smaller, and z^3 even smaller than that.[4] In such a case, keeping the first term or two involving z may be of sufficient accuracy, and the remaining terms, which are all much smaller, can be safely ignored.

To see how well such an approximation works, consider some numerical examples where only the first correction term is used. These, and similar examples, can all be easily checked using a standard electronic calculator.

Example 12.4

$$(1 + 0.03)^3 = 1 + 3 \times 0.03 + \ldots \approx 1.09,$$

and a more precise value directly from a calculator is 1.09272727, so the approximation using just the first correction term is accurate to 0.2 percent. Note that the symbol "$\approx$" should be read "is approximately equal to." Sometimes the symbol "$\cong$" is used with the same meaning.

Example 12.5

$$1/\sqrt{1.015} = (1 + 0.015)^{-0.5} \approx 1 + (-0.5)(0.015) = 0.99250,$$

whereas using the calculator directly gives 0.9925833, which differs in the fifth decimal place.

Example 12.6

$$(0.77)^{1.2} = (1 - 0.23)^{1.2} \approx 1 + (1.2)(-0.23) = 0.724,$$

whereas the calculator gives 0.7308, so the approximation is accurate to better than 1 percent.

For all these examples, a value near 1 was used, so it was easy to write it as 1 plus a small correction. Sometimes additional work is necessary to get results into this form. In each case, however, the goal is to get the problem into a form involving something known, or at least something easier to compute, and a correction. This process can also be done for something that is not near 1 by factoring out a common factor.

---

[4] For example, if $z = 0.01$, then $z^2 = 0.0001$, $z^3 = 0.000001$, and so forth.

Example 12.7

$$(25.862)^3 = (25 + 0.862)^3 = (25)^3(1 + 0.862/25)^3 = (25)^3(1 + 0.03448)^3$$
$$\approx 25 \times 25 \times 25 \times (1 + 3 \times 0.03448) = 17{,}241.25,$$

whereas the calculator gives 17,297.62, roughly 0.3 percent different. Note how the factor of 25 was factored out so that the approximation is still being used for "1 plus a small correction."

The "order" of an approximation corresponds to the exponent used for the small correction terms. The previous examples are all "to first order" since they involve terms with the small value to the first power (i.e., z) but no square (i.e., z^2) or higher powers. The next term for each of those, which will have the small number squared, would be referred to as the *second-order correction*, and so on.

For the previous examples, there is no reason why a calculator could not be used since the numerical values are known. There was really no need for such approximations. Approximation techniques are most powerful in cases where numerical values are not known, although there is some idea of what the range of the values might be. Approximations are also useful when extremely small changes are considered, where the number of digits available on the calculator is insufficient. Many problems that are very difficult, if not impossible, to solve without approximation become relatively easy to solve approximately. Such approximations are a basic tool of the physicist.

Spring Constant for a Gas (Optional)

Now return to the problem of the speed of sound in a gas and the adiabatic relationship. When there are multiple small quantities that depend on each other, the basic approach for making approximations is similar to those shown previously, although the problem certainly starts to look more complicated. The specific goal here is to determine the factors that contribute to the return force for a gas.

Consider a piston pushing on a gas inside a cylindrical chamber, as illustrated in Figure 12.2. If this piston moves fast enough, heat does not have time to transfer between the gas and the outside world, so the problem can be modeled as being adiabatic.[5] If the initial pressure

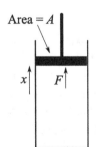

Figure 12.2 A piston pushing on an enclosed cylinder acts like a mass on a spring.

---

[5] Note that the friction is considered negligible, so no heat is being generated by friction.

and volume are P_0 and V_0, and the piston is only moved a little bit, then the final pressure and volume, P and V, will only be a little bit different, and they are conveniently written as the original value plus a small change: $P = P_0 + \Delta P$ and $V = V_0 + \Delta V$. Using the result for reversible adiabatic processes,

$$
\begin{aligned}
P_0 V_0^\gamma = PV^\gamma &= (P_0 + \Delta P)(V_0 + \Delta V)^\gamma = P_0 V_0^\gamma (1 + \Delta P/P_0)(1 + \Delta V/V_0)^\gamma \\
&\approx P_0 V_0^\gamma (1 + \Delta P/P_0)(1 + \gamma \Delta V/V_0) \\
&\approx P_0 V_0^\gamma + \Delta P V_0^\gamma + \gamma P_0 V_0^\gamma \Delta V/V_0,
\end{aligned}
\tag{12.3}
$$

where, in the last step, a term from the previous step involving the product $\Delta V \Delta P$, a product of two small quantities, and hence a second-order correction, is left out. This is appropriate since terms involving ΔV^2 and ΔP^2, which are also products of two small quantities, were already left out. It would not be correct to include some of these terms and not others. Equating the term on the left with the final result on the right, it then must be the case that, to first order,

$$
\Delta P = -\gamma \, P_0 \frac{\Delta V}{V}.
\tag{12.4}
$$

If the distance the piston moves is Δx and the area of the piston is A, then $\Delta V = A \Delta x$. Of course, the extra force required to make that change is $\Delta F = A \Delta P$. The result looks like a simple spring,

$$
\Delta F = -k \Delta x, \quad k = \gamma P_0 A^2 / V,
\tag{12.5}
$$

where, the effective spring constant, k, will depend on γ, a property of the gas, and P_0, the initial pressure. So long as the gas is not compressed so much as to invalidate the approximation – that is, so long as $\Delta V/V \ll 1$ – the return force from the gas in the cylinder acts like a simple Hooke's law spring.

While the mathematics may look daunting during the derivation of any approximate result, if done correctly, the payoff can often be great. The resulting relationship can be quite simple, especially if only first-order results are needed, and the final form may have a very simple interpretation. In the case of the piston pushing on the gas, the result after approximation says that it will act just like the piston is pushing on a spring. If the piston has a mass, M, the system will look like a mass on a spring. In fact, one method that is used to determine γ experimentally is to measure the oscillation frequency for such an arrangement. In addition, if the piston is taken to be the mass of air in the neck of a Helmholtz resonator, as seen previously (Chapter 5), then the equation for the resonant frequency of a Helmholtz resonator can be derived from this approximation.

The Adiabatic Exponent

The adiabatic exponent arises when considering what happens to energy during a reversible adiabatic process, and hence, it will show up in the speed of sound. For the molecules in a gas, one form of heat energy is the kinetic energy associated with the molecules' velocity. However, molecules can also rotate and vibrate, and such motions will also have energy associated with them. During an adiabatic process, the temperature of the gas will change, and hence the kinetic energy of the gas will change. For an adiabatic process, that change in energy is not due to an exchange of heat energy with the external environment; it is due to the mechanical process that was necessary to make the changes in pressure and volume (e.g., compressing or expanding the walls of a container). The details arise from an accounting of how that mechanical energy becomes heat energy, and vice versa. There is not the space to develop the full theory here. However, as mentioned previously, there is thermal energy associated with each degree of freedom, and so it should not be surprising that when the accounting is done, the number of degrees of freedom will be important – that is, how many different "bins" are available to store the energy. For the ideal gas model, the accounting yields

$$\gamma = \frac{\text{number of degrees of freedom} + 2}{\text{number of degrees of freedom}}, \tag{12.6}$$

where in this equation, the number of degrees of freedom is "per molecule."

The number of degrees of freedom is equal to the minimum number of coordinates necessary to specify the position and orientation of the atoms in the molecule. For a molecule in three dimensions, three variables are required (e.g., two horizontal and one vertical, such as x, y, and z) to describe where it is. If the molecule is spherical (e.g., a single atom), then that is all that is necessary – a rotated sphere looks unchanged. However, if the molecule is more complex, the molecule's orientation must also be specified. For a general shape, that will require three more coordinates (e.g., angles) since you can rotate about three different axes in three dimensions.

For the special case of a linear molecule – that is, atoms connected along a single line – rotation about the axis of the molecule makes no change, so only two coordinates are needed to specify the orientation. Any molecule made up of just two atoms, a diatomic molecule, will necessarily be linear since two points determine a line.

It is interesting to note that **the adiabatic exponent depends mostly on the symmetry (i.e., shape) of the molecule**. Table 12.2 shows measured results for several gases at 1-atmosphere pressure and near room temperature. The agreement is quite remarkable, given the simplicity of the ideal gas model. One possible exception is carbon dioxide, CO_2.

Table 12.2 Typical measured adiabatic constants for select gases near room temperature.

Type	Gas	γ
Single atom (spherical) $\gamma_{\text{ideal gas}} = 5/3 = 1.6667$	Ar	1.67
	Ne	1.64
	Xe	1.66
Linear $\gamma_{\text{ideal gas}} = 7/5 = 1.4000$	H_2	1.41
	N_2	1.40
	O_2	1.40
	CO	1.40
	CO_2	1.30
General $\gamma_{\text{ideal gas}} = 8/6 = 1.3333$	H_2O	1.33
	NH_3	1.31
	SO_2	1.29

For a gas that is a mixture of components, the predicted adiabatic constant is the weighted average of the component parts. Since air is mostly N_2 and O_2, it is not surprising that for air near room temperature, $\gamma_{\text{air}} = 1.40$.

Before going on, it is worth pointing out that a molecule will have bonds between atoms and an associated bond distance that can change, and if there are three or more atoms, there will also be bond angles that can change. This implies that even more coordinates are actually necessary to specify the molecule, and yet it appears that good results are obtained only if these extra motions are excluded. In fact, if the energy states of all the electrons orbiting the atoms, and the possible configurations for the nuclei of the atoms, are considered, there are many, many more coordinates required to define the molecule. If such a very large number of coordinates is required, then $\gamma \approx 1$ would be expected. Ultimately, the reason that result is not observed is due to quantum mechanics – not due to some fancy quantum mechanical calculations but simply due to the existence of quantum mechanics.

Quantum mechanics gives rise to discrete energy levels, so there is a minimum amount of energy required to make any change. At room temperature, the typical energy available per molecule is larger than the minimum required for making a change associated with position or rotation. However, most molecular vibrations require a minimum energy that is larger than what is available. The transitions associated with changing electron states, and even more so for nuclear states, require even more energy. So, at room temperature, only those degrees of freedom that can be expected to change at room temperature should be included, giving rise to an effective number of degrees of freedom. Indeed, for gases at high temperatures, where much more thermal energy (per molecule) is available, a value of γ somewhat closer to 1 should be expected – that is, until the molecules start to break apart into (spherical) atoms.

In the case of carbon dioxide, CO_2, nominally a linear molecule, there are two low-energy bending modes that can be somewhat active even at room temperature and even more so at higher temperatures, increasing the effective number of degrees of freedom and thus reducing the adiabatic exponent. This is the main reason the value for CO_2 seems a bit low.

Now, with these (approximate) results in hand, it is possible to compute the speed of sound. Notice that the results so far did not come from consideration of sound at all but, ultimately, started with the basic idea that a gas consists of separate molecules zipping around, making only occasional collisions with each other. From that came the ideal gas law and, although not derived in detail here, the law for reversible adiabatic processes. The adiabatic process, in turn, has an adiabatic exponent that is determined by the shape of the molecules.

Speed of Sound

Computing the speed of sound in a gas from scratch can be a formidable task, typically involving at least several equations and differential calculus. Instead of taking that approach, we use what is known for the mass on a spring (Chapter 5) and consideration of units.

The speed of sound, c, will have units of m/s. The speed will depend on the return force(s) per unit displacement and the mass. The mass will depend on the density of the gas, ρ. Using approximation methods (see earlier discussion), the return force per unit displacement depends on the product γP_0. The adiabatic constant γ has no units, although it must be present; ρ, the mass density, has units of kg/m^3; and the air pressure, P_0, has units of $N/m^2 = \left(kg \cdot m/s^2\right)/m^2 = kg/\left(m \cdot s^2\right)$. There is only one way to combine these that yields the correct units, and so it must be that

$$c = \text{const.} \times \sqrt{\frac{\gamma P_0}{\rho}}, \qquad (12.7)$$

where a unitless constant out front, which does not depend on the gas, may be present. Since γ, P_0, and ρ can all be determined independently, it is easy to compare this result to measurement. Table 12.3 shows values for several gases for $P_0 = 1$ atm, assuming the constant out front is just 1 (e.g., const. $= 1$), which is also what is obtained for a more rigorous derivation. The agreement between this theory and measured results is quite good, with differences of no more than a small percentage in the worst cases.

Table 12.3 Comparison between calculated speed of sound and experimental results for various gases at 1 atm and 0°C.

Gas	γ	ρ (kg/m^3)	c_{calc} (m/s)	c_{exp} (m/s)
He	1.67	0.178	973	965
Ar	1.67	1.783	308	319
N$_2$	1.40	1.251	336	334
HBr	1.40	3.50	201	200
SO$_2$	1.33	2.93	214	213
C$_2$H$_4$	1.33	1.26	327	317

Quantities are in Standard International (SI) units. Densities and measured speeds are from West (1980); the adiabatic constants are the values predicted by the ideal gas model.

Figure 12.3 The same data as in Figure 12.1 but grouped by molecular shape.

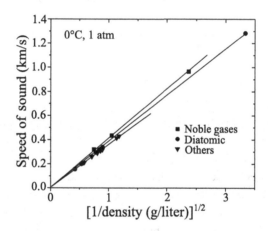

It is actually quite remarkable that such reasonable results come out of this derivation, given all of the assumptions and approximations that went into it. To start with, the ideal gas model was assumed, and that is based on an extremely simple picture of what gases are and how they work – that they are a collection of little balls bouncing around in a box. After that, several approximations were made. The final result can only be expected to be valid if all of the assumptions are correct. The comparison with measurement suggests that they are. In fact, the small deviation from straight-line behavior in Figure 12.1 can now also be explained as being due to variations in the adiabatic constant for differently shaped molecules. Figure 12.3 shows the same data but with separate lines for spherical (noble gases), linear (diatomic gases), and general molecular shapes. Considered separately, the deviation from straight-line behavior is significantly reduced.

The form of the equation for the speed of sound just derived (Equation [12.7]) suggests that the speed for a particular gas will change significantly if you change the pressure of the gas. That is misleading since **for a given gas, you cannot change the pressure without also changing the density** – the two properties are not independent. The density, the mass per unit volume, can be derived from the ideal gas equation, and that result can then be substituted into the equation for the speed of sound. The mass density is given by

$$PV = Nk_BT \rightarrow N/V = P/k_BT \rightarrow mN/V = \rho = mP/k_BT, \qquad (12.8)$$

where m is the mass of a single molecule or the average mass of the molecules for a mixed gas such as air. Substituting this expression for the density, ρ, the speed of sound is

$$c = \sqrt{\frac{\gamma k_B T}{m}}, \qquad (12.9)$$

which clearly has no pressure dependence at all but depends on the absolute temperature, the mass of the molecules, and because of the adiabatic constant, the shape of the molecules.

The dependence on the average mass of the molecules means that the speed of sound in air will depend somewhat on humidity. Water molecules have close to half the mass of both nitrogen and oxygen molecules, so as the humidity rises, the speed of sound increases. However, even on a humid day, the fraction of the air that is water is only a small percentage, so the increase in speed will also be no more than about 1 percent. For musical instruments where the frequency of the tone produced depends on the speed of sound in air (i.e., all the wind instruments), a 1 percent change due to humidity can have a significant effect on tuning.

The temperature dependence for the speed of sound near room temperature is often expressed using an approximation appropriate for small changes near room temperature. Such an expression can be obtained using the same approximation methods discussed earlier. For example, the speed of sound for temperatures expressed using the Celsius temperature scale will be

$$c = \sqrt{\frac{\gamma k_B (273.15 + T)}{m}} = \sqrt{\frac{\gamma k_B (273.15)(1 + T/273.15)}{m}}$$
$$= \sqrt{\frac{\gamma k_B (273.15)}{m}} \times \sqrt{1 + T/273.15}. \qquad (12.10)$$

The first square root gives the speed of sound at 0°C, and the second can be approximated, provided $T \ll 273.15$°C, so to first order,

$$c = c_{0°C}(1 + T/273.15)^{1/2} \approx c_{0°C}(1 + 0.00183 \times T). \qquad (12.11)$$

As an example, for nitrogen (N_2) gas with T in Celsius, this gives

$$c_{N_2} = (334 + 0.61 \times T)m/s, \qquad (12.12)$$

which agrees well with measured values near room temperature.

The temperature dependence of the speed of sound can be quite significant for musical instruments that rely on the speed of sound for tuning (i.e., all the wind instruments), especially when used outdoors. Remember that the difference in frequency from one note to the next in the chromatic scale is about 6 percent. The difference between a very hot day (30°C) and a cold day (0°C) can result in a change in the speed of sound by almost that much. Even the difference between a warm recital hall and a cool one can change the tuning by as much as 1 or 2 percent, which is certainly not negligible.

Now compare the results for the average speed of a molecule in a gas, derived in the previous chapter, and the speed of sound in that gas:

$$v_{rms} = \sqrt{\frac{3k_B T}{m}}; \quad c = \sqrt{\frac{\gamma k_B T}{m}}. \qquad (12.13)$$

The only difference is due to the difference between the number 3 and the value of γ. Since it is expected that $1 \leq \gamma \leq 5/3$, these two speeds will always be close for a wide range of temperatures and for many different gases. That these two speeds are closely related is not a coincidence. When energy is sent through air (e.g., as a sound wave), the speed at which the energy can be transferred from one molecule to the next is limited by the speed of travel over the distance between them – the molecules spend most of the time traveling and, in comparison, very little time colliding. The speed of travel is dominated by the thermal motion and not the collisions.

It is known that even at room temperature, many of the assumptions made here will break down when the pressure becomes extremely high, and hence some pressure dependence is inevitable. In particular, at very high pressures (e.g., many times 1 atm), the details of the interactions between molecules can start to play a much more significant role. Qualitatively, this is because the molecules are running into each other much more often – they spend proportionately more time colliding, so the details of the collision become more important. Also, when the pressure gets very large, the density of the gas becomes large – the molecules are more closely packed – so the size of the molecules can become a significant fraction of the total volume, although the ideal gas model ignores the size of the molecules.

Figure 12.4 illustrates the pressure dependence for dry air at room temperature. It can be seen that an increase in pressure by a factor of 10 results in a change in the speed of sound of less than 0.3 percent. Thus, although there is

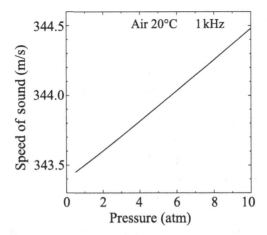

Figure 12.4 Measured variations in the speed of sound in dry air with pressure measured at 1 kHz and 20°C. The change is only about 0.3 percent for a factor-of-10 change in pressure. Data are taken from Zuckerwar (2002).

indeed some change, the pressure dependence is quite negligible for everyday purposes. While the extra effects at higher pressures can be taken into account with more advanced models, it is quite convenient to know that as long as the pressure stays near 1 atm, the ideal gas model seems to work very well, the adiabatic constant is constant enough, and the other assumptions remain valid.

To finish, a "cheat" used earlier must be acknowledged. For the ideal gas model, collisions between molecules were ignored, and only the interactions with the walls of a container were considered. The springiness of the gas was derived by considering an ideal gas in a closed system – the piston in a cylinder – that had walls. Sound, however, travels through air where there is no piston and there are no walls. For sound, the collisions with other molecules are necessary and play the roles of the walls and the piston. If you envision a small parcel of air, surrounded by similar parcels of air, then collisions between the molecules around the boundary of the parcels provide the "walls" and the (propagating) compression of the air. This is a crude description of what is done for a more correct derivation. The basic result is the same, however.

An experiment to try, by which to check all of this theory when used in practice, is to (1) measure the atmospheric pressure (e.g., with a manometer or similar technique), (2) measure the density of the atmosphere (e.g., weighing a full and evacuated beaker), (3) deduce the adiabatic constant of air from the ideal gas model (e.g., $\gamma = 1.4$), and combine these results to predict the speed of sound in air. Then use the theory of vibrations (Chapters 5 and 6) to predict the fundamental frequency for acoustic vibrations (e.g., for a simple flute, organ pipe, or Helmholtz resonator) and compare to measurement. None of these measurements includes sound, and the theory used several approximations. It is quite remarkable that the results will be close. That, however, is how physics is supposed to work.

Summary

Measured values of the speed of sound for various gases show a linear dependence on the inverse of the square root of the gas density, suggesting an analysis similar to that used for traveling waves on a string. To complete that analysis requires a determination of the return force per unit displacement – the springiness of the gas.

Gases can undergo changes in a variety of ways, referred to as *processes*. Four special cases were identified (isothermal, isobaric, isovolumetric, and adiabatic). The adiabatic process, involving no heat transfer, is appropriate for acoustic waves in air. Using a simple model and approximation techniques, the return force is found proportional to the adiabatic constant, γ, and the absolute pressure of the air (e.g., 1 atm).

Using an adiabatic constant based on the symmetry of the gas molecule, along with measured values for density and pressure, the speed of sound in various gases is accurately predicted.

ADDITIONAL READING

Fletcher, N. H. "Adiabatic Assumption for Wave Propagation." *American Journal of Physics* 42, no. 6 (1974): 487–489, and 44, no. 5 (1976): 486–487.

Mottmann, J. "Laboratory Experiment for the Ratio of Specific Heats of Air." *American Journal of Physics* 63, no. 3 (1995): 259–260.

West, R. *CRC Handbook of Chemistry and Physics*, 60th ed. CRC Press, 1980.

Worland, R. S., and D. D. Wilson, "The Speed of Sound in Air as a Function of Temperature." *Physics Teacher* 37 (1999): 53–57.

Zuckerwar, A. J. *Handbook of the Speed of Sound in Real Gases*. Academic Press, 2002.

PROBLEMS

12.1 Using a calculator, examine the approximation $\sin(\theta) \approx \theta$. How large can θ get, in radians, before the error is 1 percent? Ten percent? What are those values in degrees? (Recall that 1 radian = $180/\pi$ degrees).

12.2 Derive the result for the Helmholtz resonator frequency (see Chapter 5).

12.3 A moderately loud sound will cause the air pressure to change in time by about ±1 Pa. Estimate the amplitude of the periodic changes in the temperature of the air for the same sound.

12.4 At room temperature and at a constant pressure of 1 atm, estimate how much the speed of sound changes when the humidity changes from 0 to 100 percent.

12.5 A musician playing a wind instrument, and who has been sitting silent for a significant amount of time, will often silently blow through their instrument before they resume playing. Is there a good reason to do this based on the physics in this chapter?

12.6 How would changes in the speed of sound in air affect the tuning of a string instrument?

12.7 At the beginning of a game of pool, one billiard ball is used to break apart a number of closely spaced billiard balls. Is it appropriate to model that process as reversible?

12.8 For dry air, what is the percentage change in the speed of sound when the temperature changes from 0°C to 30°C?

12.9 Estimate the speed of sound on Mars using data from one or more of the Martian landers. Be sure to include your assumptions and cite your sources of data.

12.10 Use the results from the vibrating string to predict the lowest musical note expected from a flute (or fife) that is 50 cm long, for air at 1.0 atm with a density of 1.2 g/L. How well does the theory work in practice?

12.11 A plastic 2-L bottle filled with a gas at room temperature (20°C) and at an absolute pressure of 110 kPa is rapidly squashed. Immediately afterward, the pressure is 148 kPa. After a longer time (about half a minute in this case), the pressure has dropped to 132 kPa and remains steady at that value. What was the temperature of the gas immediately after the bottle was squashed? What is the adiabatic exponent for this gas?

13   Sounds We Hear

The sounds we hear come through the air, a gas. The basic gas laws were presented in the last couple of chapters, and the speed of sound was found by analogy to the one-dimensional vibrating string problem. There is still considerable detail that is missing. The purpose here is to fill in some of those loose ends – in particular, values found for quantitative measurements of the sounds we hear, particularly those of musical instruments compared to other sources of sound. Those results are often reported using decibels, a logarithmic scale. It is also important to consider the difference between measured sound intensities and "loudness," which is a perceived quantity.

Fluid Mechanics

Physicists refer to both gases and liquids with the more general term *fluids*. The major difference between the two is that a gas can be easily compressed or expanded, whereas for a liquid, such a compression or expansion is often quite difficult. In fact, in many cases, liquids are well approximated by treating them as completely incompressible. The practical theory describing all of these fluids is contained in a set of equations called the *Navier–Stokes equations*. Derived from Newton's laws, these equations are not for the faint at heart, and entire careers have been devoted to working with them. They appear in various forms, and one example is shown in Figure 13.1. These are nonlinear partial differential equations, and all of the equations must be satisfied simultaneously. Any further use of these equations is certainly well beyond the scope of this presentation.

For those well versed in mathematics and approximations, it is possible to show that, starting with a uniform gas at some pressure, sinusoidal ("plane-wave") traveling-wave solutions satisfy the Navier–Stokes equations to good accuracy, provided the pressure amplitude, p, of the wave is small compared to the overall pressure of the gas. That is the same criterion used when deriving the speed of sound (Chapter 12) – the changes should be small. It will be seen shortly

$$\frac{\partial \rho}{\partial t} + \frac{\partial (\rho u)}{\partial x} + \frac{\partial (\rho v)}{\partial y} + \frac{\partial (\rho w)}{\partial z} = 0,$$

$$\frac{\partial (\rho u)}{\partial t} + \frac{\partial (\rho u^2)}{\partial x} + \frac{\partial (\rho uv)}{\partial y} + \frac{\partial (\rho uw)}{\partial z} = -\frac{\partial p}{\partial x} + \eta \left[\frac{\partial^2 u}{\partial x^2} + \frac{\partial^2 u}{\partial y^2} + \frac{\partial^2 u}{\partial z^2} \right],$$

$$\frac{\partial (\rho v)}{\partial t} + \frac{\partial (\rho uv)}{\partial x} + \frac{\partial (\rho v^2)}{\partial y} + \frac{\partial (\rho vw)}{\partial z} = -\frac{\partial p}{\partial y} + \eta \left[\frac{\partial^2 v}{\partial x^2} + \frac{\partial^2 v}{\partial y^2} + \frac{\partial^2 v}{\partial z^2} \right],$$

$$\frac{\partial (\rho w)}{\partial t} + \frac{\partial (\rho uw)}{\partial x} + \frac{\partial (\rho vw)}{\partial y} + \frac{\partial (\rho w^2)}{\partial z} = -\frac{\partial p}{\partial z} + \eta \left[\frac{\partial^2 w}{\partial x^2} + \frac{\partial^2 w}{\partial y^2} + \frac{\partial^2 w}{\partial z^2} \right],$$

$\rho =$ air density, u, v, w are components of air velocity,
$p =$ pressure, $\eta =$ viscosity.

Figure 13.1 One form of the Navier–Stokes equations for fluid flow, clearly showing that fluid flow is much more complicated than just sound.

that, indeed, the **sounds normally experienced correspond to very small variations in pressure**, thus validating that criterion.

Other solutions, which are not sinusoidal traveling waves, are not at all rare. Turbulence in an airflow is an everyday occurrence. Whenever an everyday object moves through the air, some turbulence is created. Turbulence will show up as part of any discussion of air resistance, for example, around cars, and is crucial for the rapid combustion of gases, for example, in a gasoline engine. Acoustically, turbulence can produce undesirable background noise, for example, from a fan or a discontinuity in a heating duct. Turbulence is also at least partly responsible for the (nonlinear) sound-generating mechanism in the flute family, as a jet of air (another type of solution to these equations) impinges on an edge. Turbulence in liquids shows up every time a cup of coffee or cocoa is stirred.

Shock waves are another type of solution to the equations. Shock waves in air can ultimately result in a "sonic boom." A shock wave will be created when an object moves through air at a speed faster than the speed of sound. For example, the crack of a whip occurs when the speed near the tip of the whip exceeds the speed of sound. The same basic phenomenon occurs for the two-dimensional surface waves on water when a boat exceeds the surface–wave speed. In that case, a V-shaped "wake" appears on the water surface. Explosions and high-speed bullets can also produce shock waves in air.

Vortex rings are another example of nonsinusoidal behavior. So-called "smoke rings" are examples of vortex rings. These rings will travel much slower than the speed of sound and can travel for quite some distance before dissipating.

For the time being, however, the discussion will be restricted to sound waves in air – the approximate sinusoidal wave solutions to the very complicated Navier–Stokes equations. Anyone who is particularly interested in one or more of these other types of solutions might want to begin a course of study in fluid dynamics.

Sound Waves Are Longitudinal

As has been seen, all systems that exhibit oscillatory motion, such as the harmonic oscillator and vibrating string, have a return force. That is, when the system is displaced from its equilibrium, there is a force that tries to move the system back to equilibrium. Without such a force, there will be no oscillations.

There are two distinct types of forces that can be applied to any chunk of matter: **compression** (or extension) and **shear** (or bending). These are illustrated in Figure 13.2. A return force in response to compression will result in longitudinal waves, whereas a return force in response to shear yields transverse waves. The molecules in a fluid slide over one another very easily. Hence, if a shear force is applied to a fluid, there will be no return force. As an analogy, think about a stack of paper. If the pile is compressed, it pushes back and returns to its original height. However, if a shear is applied, the sheets of paper slide easily over one another, and there is no return force. For fluids, and for stacks of paper, waves can only be longitudinal.

A microscopic picture of sound (in a pipe) is illustrated in Figure 13.3. For that figure, the dots might represent dust particles suspended in the air. Each dust

Figure 13.2
Two distinct types of forces that can be applied to a chunk of matter or fluid are (a) compression and (b) shear.

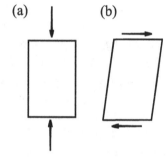

Figure 13.3 A moving piston on one end of an infinite pipe filled with air creates a longitudinal wave. The wave can be visualized by considering the motion of small dust particles suspended in the air. ★

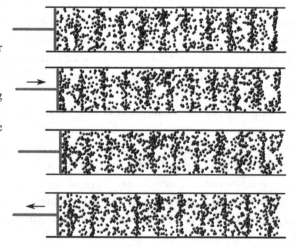

particle is undergoing sinusoidal motion about an average position. The particles themselves do not travel down the pipe, but the periodic compressions can travel. Close observation of any single particle shows a simple sinusoidal motion about a fixed average position.

The Amplitude of Sound

In the presence of a plane sinusoidal sound wave, the total air pressure is described by the combination of the background pressure (e.g., 1 atm = 101,000 Pa) and the changes to the pressure, p, due to the wave. For example, the pressure variations for a plane wave traveling in the x-direction in air at 1 atmosphere could be described by

$$p(x, t) = 101,000 \text{ Pa} + p' \sin(kx - \omega t). \tag{13.1}$$

The **acoustic pressure amplitude** of the sound corresponds to p'. For sound, the pressure is the simplest of the gas properties to measure – it is what most microphones sense – so descriptions of acoustic amplitudes tend to focus on the pressure. Hence, p' will sometimes simply be referred to as the *acoustic amplitude*. Of course, the other gas properties, such as density and temperature, will also be changing in time and position. Knowledge of p' is sufficient since the other properties can be calculated using the ideal and adiabatic gas laws (Chapters 11 and 12).

In the early 1880s, long before electronic instrumentation was common-place, the English scientist Lord Rayleigh (1842–1919)[1] devised a scheme to measure the amplitude of resonant sound waves in a pipe. His scheme was to suspend a small, lightweight, disk-shaped mirror in the pipe and measure the rotation of that disk in the presence of sound. The mirror's suspension provides a weak torsional return force, and the air flowing past the disk in either direction will tend to align the disk with the flow. Hence, the oscillatory longitudinal motion of the air will cause the mirror to rotate to a fixed angle. The rotation of the mirror is measured by shining a beam of light on the mirror and viewing the reflected beam. The expected rotation can be calculated based on prior work by the German physicist Rudolf König (1865–1927). Such a disk used in this way is referred to as a *Rayleigh disk*. Other direct measurements of sound amplitude around the same time included painstaking observation of suspended smoke particles as they moved under the influence of sound.

Measured values of p' for different sound levels illustrate just how amazing the sense of hearing is. The so-called **threshold for hearing** is the sound level, for a tone in the most sensitive frequency range for hearing, that has been adjusted so

---

[1] Also known as John William Strutt, Third Baron Rayleigh.

that in a very quiet room, most people can just barely hear the sound. An average value is about $p' = 20\,\mu\text{Pa}$. That corresponds to a change in the air pressure of 0.00000002 percent. **A moderately loud sound corresponds to $p' \approx 1$ Pa**, a convenient value to remember. There are two amazing facts about the latter value. First, it corresponds to a change in the air pressure of only 0.001 percent, still a very small change, but it is also 50,000 times larger than that for the threshold of hearing. Human ears are sensitive to extremely small (time-dependent) changes in air pressure over a *very* wide range of amplitudes.

One thing is clear. The amplitudes of the pressure changes due to everyday sounds are indeed very small compared to 1 atmosphere. This validates the approximations that are based on that assumption, for example, for the plane-wave solutions to the Navier–Stokes equations and during the computation of the speed of sound.

Sound in Decibels

To characterize values that occur over a wide range, it is often useful to compress the scale using logarithms. In addition, it seems that human perception of sound levels behaves logarithmically, at least approximately. Thus, it is natural to use **decibels** (dB) to describe sound amplitudes. Decibels are used for many other measures that cover a wide range and/or where the consequences are more or less logarithmic. Examples include the stellar magnitude ("brightness") scale for stars, the various scales used for earthquakes (e.g., the Richter scale), and so on.

Recall from Chapter 4 that decibels are based on the logarithm of a ratio. To describe one signal compared to another using dB,

$$\text{Comparison in dB} = 10 \, \log(\text{ratio of signal powers, intensities, or energies})$$
$$= 20 \, \log(\text{ratio of signal amplitudes}).$$

$$(13.2)$$

A signal described using dB is always based on a ratio. For sounds, it is often convenient to agree on a common reference sound level and express the amplitudes of sounds compared to that reference. A common value to use for sounds is an acoustic pressure amplitude of 20 μPa.[2] Such a description of the amplitude of sound compared to this reference, when expressed in dB, is referred to as the **sound pressure level** or **SPL** (usually pronounced as "es-pea-ell"). Then, if the acoustic pressure for a sound is p', the SPL(in dB) $= 20 \, \log(p'/20\,\mu\text{Pa})$. Note also that when referring to sound amplitudes using SPL (or any logarithmic scale), then when, for example, "the sound amplitude is doubled," that refers to

---

[2] Another agreed-upon scheme that yields almost the same values is based on a standard intensity of $10^{-12}\,\text{W/m}^2$ rather than a pressure amplitude. Some refer to those levels, in dB, as *sound intensity levels*, or *SILs*.

doubling p', which is inside the logarithm, not doubling the number of dB. Each time the amplitude, p', is doubled, the number of dB is increased by *adding* 6.02, not by multiplying by 2.

Example 13.1

When the amplitude of a sound that initially has an SPL of 45 dB is doubled, the new SPL is 51 dB, not 90 dB.

It is not hard to find tables of "typical values" for sounds in dB. Table 13.1 is one such table. Of course, in each case, the SPL will depend on many factors, such as the distance to the source, the environment surrounding the source, and so forth, so these values should be used in that context. Remembering that each addition of 10 dB corresponds to a change of power by a *multiple* of 10, the range between the threshold of hearing (0 dB) and a painfully loud sound (120 dB) corresponds to 12 factors of 10, or a change in acoustic power of 10^{12} – a million millions. Human ears are truly remarkable sensors to be able to cover such a range.

Table 13.1 Typical sound levels for a selection of sources.

Source	Sound Level (SPL dB)
Threshold of hearing (3 kHz)	0
Whisper	20
Quiet radio	40
Loud conversation (at 1 m)	65
Busy city street	70
Trumpet (max at 10 m)	75
1-mW sound source at 1 m	79
Bass drum (max, next to drum)	94
Riveter/jackhammer at 20 m	95
Large orchestra (max)	110
Crying baby (close up)	115
Short-term damage/pain	120
Nonlinear effects show up	135
Eardrum rupture	150

Intensity and Loudness

The connection between physical measurements of sound intensity and the human perception of "loudness" is complicated. **Loudness** is not a physical property that can be measured. Perhaps the most cited study, although certainly not the only study, to try to connect intensity to loudness dates is that by Fletcher and Munson at Bell Telephone Labs in the early 1930s. The question to be addressed is, Given a particular signal, can you predict how loud it will be perceived to be? Of course, with any issue involving perception, a group of individuals is tested and asked for their judgments. Results are based on average responses.

A first step is to establish a loudness scale to be used for comparison. Fletcher and Munson did this by defining the "loudness level" for 1,000-Hz pure tones to be equal to the intensity for those tones. The loudness level for other sounds is found by comparison to a 1,000-Hz pure tone – the observer adjusts the 1,000-Hz tone to sound "just as loud." The intensity of the matching 1,000-Hz tone becomes the "loudness level" for the sound being measured. This is a simple way to establish a measure, but care must be taken in its interpretation. If the loudness level of a sound increases by 10 dB, that does not mean that it now sounds 10 times louder. It means only that it sounds as much louder as a 1,000-Hz tone increased by 10 dB. To avoid confusion, later work sometimes refers to these equal loudness levels with the non–Standard International (SI) unit "phon" instead of dB, although this practice is not universal.

The next step is to compare other pure tones with the 1,000-Hz pure tone. This gives a measure of the frequency dependence of average hearing. Once again, "frequency dependence" refers to the response to the individual pure tones in the Fourier series of a complex sound rather than a time dependence in perception. A further next step is to consider how perception responds in the presence of complex sounds – that is, more than one pure tone at the same time, each of which may have a different loudness level. How loudness changes when one or both ears are used can also be considered. These last questions go a bit far afield from physics and so will not be considered here. Consult the Additional Reading sources for more information about those results. **The field of study that deals with how sounds are perceived is called *psychoacoustics*.**

Figure 13.4 shows representative contours of constant loudness for pure tones (sinusoidal waves) based on an average of responses (i.e., opinions) from multiple listeners. Such curves are often referred to as *Fletcher–Munson curves*.[3] The data show that at very low and very high frequencies, it takes much more intensity to achieve the same perception of loudness. For pure tones, human hearing is, on

---

[3] The original work by Fletcher and Munson has been updated by others several times since it was published, although the original names are still attached. More recently, the International Standards Organization (ISO) has established a standard set of such data, which is also updated from time to time, for use in engineering applications.

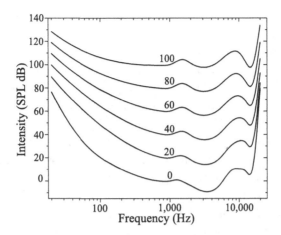

Figure 13.4 Typical Fletcher–Munson plot showing curves of constant loudness. Note that the loudness (in phons) is equal to the intensity (in dB) for a 1-kHz pure tone.

average, most sensitive for frequencies of a few kilohertz. Hence, it is not surprising that various alarms (such as smoke detectors, the backup beepers on trucks, etc.) will produce tones in that range. Acoustic waves significantly below 20 Hz might be felt, if they have enough intensity, but are not considered "audible" – they are not "heard." Such waves are referred to as *infrasound*. Acoustic waves above about 20 kHz are also not audible; they are "inaudible" and are referred to as *ultrasound*.[4]

Devices designed to measure sound intensities are often used to characterize the sound level in a room. When concerns about "loudness" are being addressed, those devices need to take into account the fact that human hearing is much less sensitive at low and high frequencies. To keep it simple, these devices use a weighted scale – lower and higher frequencies get less weight than those where the ear is most sensitive. There are two such scales in common use. One is referred to using units of dBA, and the other, dBC. It is a common practice in many disciplines to add a letter or two to "dB" in order to help the reader know what is being measured. Some of these letters are used in multiple contexts with different meanings. For example, one will see "dBc" used for radio signals to mean "decibels compared to the carrier," which is not at all related to the loudness of sounds. In each case, these extra letters need to be interpreted in context.

The **A weighting** scheme is based on a smooth curve that approximates the shape of the inverse of the 40-phon equal-loudness curve. The weight for a 1-kHz pure tone is adjusted to be 1. The **C weighting** scheme is much flatter but with cutoffs at lower and higher frequencies. The dBA scale serves as a crude measure related to loudness, and the dBC scale gives a measure of intensity, although confined to the normal range of human hearing. There are also B and D weightings that are

---

[4] The upper level of human hearing varies considerably from person to person and decreases significantly with age.

defined, but their use is rarer. "A" weighting is so common that often it is simply assumed, and the "A" is left off the units. The user is just supposed to know.

More experiments can be done to determine the scale factor between intensity and the perception of loudness. The listener is asked to adjust a sound so that it is, for example, perceived to be twice as loud as another. The intensities of those two sounds can be measured and compared. The result depends on the intensity level of the first sound, the types of sounds, and so forth, but a reasonable approximation, which works for the range of human hearing, is that **to make a sound twice as loud, the intensity should be increased by a factor of 10.** That is, a 10-dB change in intensity corresponds to (roughly) a factor of 2 in "loudness." The intensity range from 0 to 120 dB corresponds to 12 factors of 10. For loudness, this corresponds to roughly 12 factors of 2. Perception compresses the loudness to a range of a few thousand ($2^{12} = 4,096$) rather than a million millions. The non-SI unit "sone" is used to characterize perceived loudness, where 1 sone corresponds to a 1-kHz tone at 40 dB SPL (40 phon). Each increase (decrease) in the intensity by 10 dB doubles (halves) the number of sones.

Combining Signals

For the sake of those who are not experienced with calculations involving decibels, this chapter is concluded with a number of worked examples involving the comparison and combination of sound intensities from multiple sources. In each case, the intensities are expressed using dB. The same procedure applies if the units are dBA or other similar weighted units. If needed for reference, the properties of logarithms are summarized in Appendix A.

Example 13.2

A quiet room is measured to have a sound intensity of 15 dB. When a violin then plays, the intensity is 65 dB. How much larger is the acoustic amplitude in the presence of the violin compared to the quiet room?

The *acoustic amplitude* refers to the acoustic pressure amplitude, p'. The common reference (acoustic) pressure for these measurements, p_{ref}, does not need to be known. The calculation proceeds as follows:

$$\text{Change in dB} = (65 - 15)\,\text{dB} = 50\,\text{dB} = 20\,\log\left(p'_{\text{violin}}/p'_{\text{quiet}}\right)$$

$$50/20 = 2.5 = 20\,\log\left(p'_{\text{violin}}/p'_{\text{quiet}}\right)$$

$$10^{2.5} = 10^{\log\left(p'_{\text{violin}}/p'_{\text{quiet}}\right)} = p'_{\text{violin}}/p'_{\text{quiet}} \rightarrow p'_{\text{violin}} = 10^{2.5}\,p'_{\text{quiet}} = 316\,p'_{\text{quiet}},$$

so the acoustic (pressure) amplitude is 316 times larger in the presence of the violin.

Example 13.3

How many violins will sound twice as loud as one violin?

To make an estimate, assume that all the violins are playing the same music at the same volume, and they are all (roughly) the same distance away from the listener. It is also a good assumption that the sound energy emitted by one violin does not affect the energy emitted by another in a significant way. Thus, the total energy emitted by N violins is N times the energy emitted by 1 violin. Aside from directional effects, which are ignored here, that means that the intensity for N violins is N times that of 1 violin. Doubling from 1 to 2 violins increases the intensity by $10 \log(2) = 3$ dB, doubling again to 4 violins adds another 3 dB, doubling once more to 8 violins adds another 3 dB, and so on.

Using the approximate rule that an increase in intensity of 10 dB causes a sound to be perceived as twice as loud, it is clear that it will take roughly 10 violins to sound twice as loud as 1 violin.

Example 13.4

If a newborn baby emits sound at 85 dB and an audio system is playing rock music at 110 dB, what is the intensity for the combination, in dB?

Adding sounds expressed in decibels must be done with care. The values to be added are the intensities, which are inside the logarithm. To do that addition, the intensities must first be brought outside the logarithm. The solution proceeds as follows:

$$85 \text{ dB} = 10 \log(I_{baby}/I_{ref}) \rightarrow I_{baby}/I_{ref} = 10^{85/10} = 10^{8.5}$$
$$110 \text{ dB} = 10 \log(I_{rock}/I_{ref}) \rightarrow I_{rock}/I_{ref} = 10^{110/10} = 10^{11}$$

$$(I_{baby} + I_{rock})/I_{ref} = 10^{8.5} + 10^{11} = 1.00316 \times 10^{11}$$

$$(I_{baby} + I_{rock}) \text{ in dB} = 10 \log(1.00316 \times 10^{11}) = 110.0014 \text{ dB} = 110 \text{ dB}.$$

Note that knowing the value of the reference intensity is not necessary for this calculation. Within the accuracy of the input data, the result here is the same as the larger of the two values. That will occur when combining any two signals that are more than about 20 dB different. Can you still hear the baby over the rock music?

Example 13.5

If you are shouting at 82 dB next to a saxophone player who plays at 85 dB, what is the total intensity in dB?

The procedure here is the same as for the previous example; however, in this case, the two intensities are almost equal:

$$82 \text{ dB} = 10 \log(I_{shout}/I_{ref}) \rightarrow I_{shout}/I_{ref} = 10^{82/10} = 10^{8.2}$$
$$85 \text{ dB} = 10 \log(I_{sax}/I_{ref}) \rightarrow I_{sax}/I_{ref} = 10^{85/10} = 10^{8.5}$$

$$(I_{shout} + I_{sax})/I_{ref} = 10^{8.2} + 10^{8.5} \rightarrow$$

$$(I_{shout} + I_{sax}) \text{ in dB} = 10 \log(10^{8.2} + 10^{8.5}) = 86.8 \text{ dB}.$$

Here, the intensities of the two sounds are similar, so the result is not the same as the largest.

As with other skills, proficiency with these calculations takes some practice. Such practice is encouraged.

Summary

The equations that govern fluid motion are extremely complicated. Traveling sinusoidal sound waves are an approximate solution and are valid as long as the sound is not too loud. Other solutions include shock waves, turbulence, and vortex rings.

Sound waves in air – or more generally, fluids – will be longitudinally polarized.

Quantitative measurements of the typical pressure variations associated with sound show that the variations are very small compared to atmospheric pressure. The amplitudes of audible sounds encompass a very large range, leading to the use of decibels (dB), a logarithmic scale.

Loudness is a perceived quantity associated with intensity, and there is a significant frequency dependence to that perception. Other factors being equal, an increase in the sound intensity of 10 dB will be perceived as doubling the loudness.

When sound intensities expressed in dB are to be combined or compared, some care needs to be taken to avoid common errors.

ADDITIONAL READING

Fletcher, H., and W. A. Munson. "Loudness, Its Definition, Measurement and Calculation." *Journal of the Acoustic Society of America* 5 (1933): 82–108.

Minnix, R. B., and D. R. Carpenter, Jr. "A Variable-Volume Helmholtz Resonator." *Physics Teacher* 23 (1985): 49–51.

PROBLEMS

13.1 What is the ratio of the height of typical ocean surface waves to ocean depths found around the world? How does the experience of an ocean sailor depend on that ratio? How does this relate to sound intensities?

13.2 The lowest notes that can be produced with a contrabassoon have a frequency below 20 Hz. Can those notes be heard? Why or why not?

13.3 For most people, a 10-dB increase in the sound intensity sounds twice as loud. Using that value, an increase of how many dB would be four times as loud? How about half as loud? How much louder is 100 dB compared to 0 dB?

13.4 If one trumpet produces an SPL of 75 dB, what is the SPL (in dB) for seven trumpets?

13.5 In Chapter 12, the speed of sound was found using approximations based on the assumption that the acoustic pressure is very small compared to the pressure of the gas. For louder sounds, the simple model can be expected to break down. For sounds in air, if the acoustic pressure is 10 percent of one atmosphere, 0.1 atm, what is the SPL (in dB)?

13.6 Loudness is a perceived quantity, meaning, quite literally, it exists only in one's head. Intensity is a physical quantity that can be measured and is closely associated with loudness, at least over a limited range. What other such quantities are perceived and exist only in one's head, and what physical quantities are closely associated with them?

13.7 A pure tone at the frequency of A6 (1,760 Hz) is desired with a loudness equal to that of a 60-dB (SPL) pure tone at the frequency of the note A1 (five octaves lower). What intensity should be used for the higher-frequency tone? Refer to Figure 13.4.

14 Sound in Pipes

The behavior of the wind instruments, including woodwinds, brasses, organ pipes, and the human voice, can generally be modeled as one-dimensional (1-D) acoustic systems. That is, crudely speaking, they all involve sound propagation through pipes. For these examples, the pipes may be complicated – the dimensions along the length of the instrument may change, there may be holes, or there may even be connections to other pipes. For a musical instrument, there must also be openings in the acoustic system – the pipe(s) – to the outside world. Hence, consideration of the transmission of sound in a simplified version of such 1-D systems is in order.

There are many parallels between the transmission of acoustic energy through a pipe and the transmission of other types of waves in one dimension. Waves on a string under tension, the electrical signals carried in cables, microwaves transmitted through waveguides, and the optical signals carried in fiber optics are all examples. The theory behind all of these is often referred to as *transmission-line theory*. Of particular importance for wave transmission is a parameter known as the *impedance* of the system and how it changes from place to place. Sometimes a parameter called the *admittance* will be used instead. Mathematically, the impedance and admittance are inverses – if you have one, it is easy to determine the other.

In this chapter, a simple discussion of impedance and how it is important for musical instruments will be the main topic. Many of the ideas from previous chapters must be brought together in the process. Also, along the way, there will, of necessity, be a discussion of imaginary values and why a physicist might use them. There are several rather complicated mathematical expressions shown in the process. None of the more complicated equations should be memorized. To the extent possible, stand back and appreciate them in the moment, and know that they will still be here in the future should they be required.

Impedance

The steady-state result for a sinusoidally driven harmonic oscillator, shown previously (Chapter 5), had two frequency-dependent components: amplitude and phase. At any fixed frequency, the phase relationship is fixed, and the (steady-state) amplitude of the motion is proportional to the amplitude of the driving force. That is, there is a linear relationship between the steady-state response and the driving force. If the force is doubled, the response is doubled, and so on. Mathematically, the rather complicated expression for the steady-state harmonic oscillator (Equation [5.15]) can be written simply as

$$\text{Response} = \text{constant} \times \text{force}, \tag{14.1}$$

where the constant may depend on the frequency of the driving force, but it does not change in time. All of the complicated frequency dependence is now conveniently hidden inside the constant and is no longer visible. At the same time, the simple relationship between force and response, expressed in the frequency domain, is clear.

Impedance is a measure of how much a system, in some sense, impedes, resists, or otherwise suppresses the response. For a given force, a larger impedance results in less response than does a smaller impedance. The admittance is simply the inverse of the impedance and so has the opposite effect – smaller impedance implies a larger admittance. Thus, in the simple equation mentioned for the harmonic oscillator (Equation [14.1]), the constant shown is the admittance since the larger the constant, the larger the response for a given driving force.

In electronics, voltage, V, plays the role of a force; the response is the current, I; and often the impedance is simply a resistance, R. This rule is then $V = IR$, which is known as **Ohm's law**. A generalized version of Ohm's law can be used whenever there is a linear relationship between something that plays the role of a driving force and the corresponding response to that force. For mechanical systems, velocity, rather than current, is used to define the response.

Thinking in the frequency domain, the response to a general driving force can be written as the sum of the Fourier components. Each Fourier component obeys a similar Ohm's law equation. Writing the impedance with the symbol Z, as is commonly done in physics, and the velocity and force with v and F, respectively, the Fourier components at (angular) frequency ω will obey $v(\omega) = F(\omega)/Z(\omega)$. Due to the frequency dependence, the impedance may need to be described using a function (of frequency) and not just a single value.

Complex Impedances

In many cases, the response is out of phase with the driving force. That is, if the time dependence of the driving force is proportional to $\cos(\omega t)$, the time dependence of the response is proportional to $\cos(\omega t + \theta)$. The driven harmonic oscillator is an example that exhibits this property. In such cases, two values are needed to specify the relationship between the driving force and response – an amplitude and a phase.

When phase information needs to be included, scientists and engineers often do not specify the results using a description involving amplitude and phase. Instead, they resort to the use of complex numbers. The complex numbers arise when considering the square roots of negative numbers.

The square root of a negative value does not exist, at least not for measurements in our everyday lives. Such a value is thus considered imaginary. When a formula results in an imaginary value, it is sometimes the case that the formula is simply no longer applicable. After all, one can *never* measure something that is imaginary – there is a reason they are called imaginary. And yet, as will be seen, for impedances, those imaginary values are quite useful if used and interpreted correctly. They are an extra tool in the toolbox. The imaginary values still can never be measured, but they lead directly to quantities that can be measured.

Recall that the square root of a product is the same as the product of the square roots. That is, $\sqrt{a \cdot b} = \sqrt{a} \cdot \sqrt{b}$. Since a negative value can be written as -1 times a positive value, it is sufficient to worry only about the square root of -1. For example, $\sqrt{-289} = \sqrt{289} \cdot \sqrt{-1}$ and $\sqrt{289} = 17$ is easily computed using a calculator. To shorten the notation, the square root of -1 is replaced by a symbol. In physics, the lowercase i is often used, although in some physics and much of engineering, a lowercase j is used instead. Here, only i will be used. Hence, $\sqrt{-289} = 17i$.[1] A numerical value that includes multiplication by i is referred to as an **imaginary number**. To distinguish our usual numbers from imaginary numbers, our usual numbers are then referred to as the **real numbers**. A value expressed as the sum of a real number and an imaginary number is called a **complex number**. For example, $18 + 17i$ is a complex number with a "real part," 18, and an "imaginary part," 17. You cannot simply add real and imaginary parts together in a meaningful way because they are incompatible with one another – they come from different worlds. What is left is a number that has two separate parts.

A full discussion of the mathematics of complex numbers is beyond the scope here. Some information is included in Appendix A. Suffice it to say that the mathematics happens to do exactly what is required when considering the amplitudes and phase shifts for sinusoidal signals. It may seem an unlikely connection – after all, one can certainly measure amplitudes and phase shifts for real sinusoidal signals. There is nothing imaginary about those signals. When a measurable quantity is referred to as being imaginary, that means it can be represented using the mathematics of imaginary numbers, not that the quantity itself exits only in your imagination.

---

[1] Although recall that the result of the square root can be either positive or negative. The calculator will return only the positive result. Hence, the more correct answer here is $\pm 17i$.

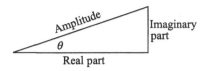

Figure 14.1 The relationship between descriptions using amplitude and phase and descriptions using complex numbers can be described geometrically.

The conversion from the realm of complex values to our real world of phase-shifted sinusoidal signals can be thought of geometrically. As shown in Figure 14.1, draw a right triangle where the length of one side is proportional to the real part, and the side perpendicular to that is proportional to the imaginary part. The length of the hypotenuse (the "diagonal") is proportional to the amplitude, and the angle, θ, is the phase angle. The mathematics of sinusoidal signals, complex numbers, and the geometry of right triangles are all interconnected in this way.

The discussion will return to complex impedances later on, in particular as applied to the discussion of woodwind finger holes.

Sound in Air-Filled Pipes

For sound in air, a longitudinal wave, the appropriate driving force to consider arises from the very small changes in air pressure associated with sound – the acoustic pressure. The response is the speed of the air; that is, the speed of the small bit of wind due to the sound as the air rapidly moves ever so slightly forward and backward. This motion of the air is referred to as the **particle velocity** – think of a small dust particle suspended in the air that moves to and fro with the wave. The ratio of the force, due to the pressure, to that response yields the **acoustic impedance**.

A careful derivation of the impedance associated with sound in air, even for simple 1-D motion, involves approximation methods, as discussed previously (Chapter 12), usually combined with additional mathematical methods beyond what will be assumed here. However, some basic results can be obtained using Newton's second law applied to a small volume of air in a pipe. Here, the pipe is considered to be narrow compared to the wavelength, so the waves propagate only along the length of the pipe.

Consider a sinusoidal, traveling sound wave moving in the $+x$-direction through a narrow, air-filled pipe of uniform cross-sectional area, A, a cylinder. At some fixed instant in time, look at a length, Δx, of the pipe between positions 1 and 2 (see Figure 14.2). The volume of air enclosed within that length is $A\Delta x$. On one side of that volume, the total pressure is P_1, and on the other, P_2. The net force on that volume, with positive values taken to correspond to a push toward

Figure 14.2 Newton's second law on a portion of the air in a pipe is used to find the impedance.

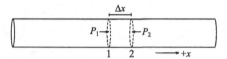

+x, is $(P_1 - P_2)A$. The mass inside that volume is given by the density times volume, $\rho A \Delta x$, and the average acceleration of that volume of air over a short time Δt is $(v_2 - v_1)/\Delta t$. The small changes in the density of air due to the sound wave are being neglected. Newton's second law, $F = ma$, for that portion of the air will be

$$(P_1 - P_2)A = (\rho A\, \Delta x)(v_2 - v_1)/\Delta t. \tag{14.2}$$

For a sine wave traveling at speed c, the wave moves a distance $\Delta x = c\Delta t$ in a time Δt, and hence a change in position can be achieved via a change in time. That is, a comparison of the wave measured at a single position for two different times can be equated to a comparison of the wave at a single time for two different positions.[2] Hence, the relation can be rearranged to give

$$\frac{P_2 - P_1}{v_2 - v_1} = \rho \frac{\Delta x}{\Delta t} = \rho c. \tag{14.3}$$

Since the right-hand side of this equation is a constant value, independent of the choice for positions 1 and 2, it must be true that for this traveling wave,

$$\frac{p}{v} = \rho c, \tag{14.4}$$

independent of position, where p is the acoustic pressure, and v is the associated average particle velocity.

The ratio $z = p/v$ is known as the **specific acoustic impedance** for the medium, in this case, air. That impedance depends only on the properties of the medium and not of the pipe. For flow through a pipe, it is useful to look at the total flow rather than the particle velocity. The flow measures how much material (e.g., air) is flowing at velocity v through a surface area, A, and that, of course, depends on the area A. That is, if a wind is blowing, the speed of the wind is the particle velocity. If the wind is blowing through an open window, how much air flows through the window per second also depends on the size of the window. A bigger window admits more air than a smaller window. The **acoustic impedance**, Z, is the ratio of the acoustic pressure to the acoustic *flow*. For a traveling wave in a pipe with a uniform cross section of area A, the impedance will then be given by

---

[2] For a fixed position, x, as time goes forward, the position relative to a wave traveling toward $+x$ will be the same as moving toward $-x$ at a fixed time.

$$Z = \frac{\rho c}{A}.$$ (14.5)

Note that a wider pipe will have a smaller impedance. That should be expected. It is easier to get flow through a fat hose than through a thin straw.

The specific acoustic impedance and the acoustic impedance should not be confused. They are related, of course, but different. The specific acoustic impedance is a property of the medium, the air, in the same way that density is a property of the air. Specific acoustic impedance will be in units of $kg/(m^2 \times s) = 1$ rayl. The acoustic impedance, without the word *specific*, corresponds to a particular situation, in this case, air in a pipe having cross-sectional area A, in the same way that the total *mass* of the air depends on the particular volume being considered. Acoustic impedance will be in units of $kg/(m^4 \cdot s) = Pa \cdot s/m^3 = 1$ acoustic ohm.[3]

Example 14.1

The specific acoustic impedance of air under normal room conditions is about 420 rayl.

The acoustic impedance for an air-filled pipe with a diameter of 3 in. (7.62 cm) under normal room conditions will be about 90,000 acoustic ohms.

The impedance for a pipe scales as $1/A$, and A scales as the diameter squared. Hence, the acoustic impedance of a 1-in.-diameter pipe will be nine times larger than that of a 3-in. pipe. Remember, it is harder to get air to flow through a smaller-diameter pipe, so **the impedance is larger for smaller diameters**.

The acoustic impedance just derived applies to a single sinusoidal *traveling* wave and is a characteristic of wave propagation in the pipe. Hence, it is often referred to as the **characteristic impedance**. In addition to the specific and characteristic impedance, the **input impedance** to a transmission line (e.g., a pipe) will also be important. The input impedance is the ratio of the pressure to the flow at the input to the pipe. For a semi-infinite pipe, where there are no reflections from the far end, the input impedance and characteristic impedance are equal. However, if the pipe is finite in length and there are reflections off the far end, then the pipe contains more than just a single traveling wave, and the situation becomes complicated.

The input impedance for a pipe is relevant when describing the behavior of wind instruments. For example, the nonlinear reed in the clarinet head joint is directly coupled to the input of a pipe, and the input impedance of the pipe

---

[3] The rayl and acoustic ohm both have mks and cgs versions that are not equivalent. When numerical values are important, it must be clear from the context which system of units is being used. Here, the meter-kilogram-second, or mks, unit is being used. The rayleigh (rayl) is named in honor of Lord Rayleigh.

characterizes the behavior of the pipe as seen at the input. That impedance as seen by the reed, the input impedance, is very important for tone production.

To compute the input impedance, the characteristic impedance can be used separately for traveling waves going in each direction and the total evaluated at the input to the pipe. An example will be seen later, but the details for reflection and transmission are considered first.

Reflection and Transmission

There will be reflection of a wave at any boundary where the impedance changes. **The larger the change in the impedance, the larger will be the amplitude of the reflected wave.** For sound in pipes, it is relatively straightforward to compute how much is reflected. As is done for other calculations, this is done in the frequency domain by considering just a single sinusoidal traveling wave since any other situation can always be considered to be a sum of such sinusoidal waves.

Consider a junction between two semi-infinite pipes, pipe 1 and pipe 2, having characteristic impedances Z_1 and Z_2, as shown in Figure 14.3. For a sinusoidal wave traveling in pipe 1 that arrives at the junction, some of the wave will be reflected back into pipe 1, and some will be transmitted into pipe 2. The relative amplitude of the reflected and transmitted waves, R and T, respectively, will be

$$R = \frac{Z_2 - Z_1}{Z_1 + Z_2} \quad \text{and} \quad T = \frac{2Z_2}{Z_1 + Z_2}. \tag{14.6}$$

A derivation of these **reflection and transmission coefficients**, R and T, is included at the end of this chapter for those interested.

If $Z_1 = Z_2$, then there is no reflection, and all of the wave energy is transmitted. In that special case, the pipes are **impedance matched**. When $Z_1 \neq Z_2$, some of the wave is reflected with either a positive or negative sign, depending on which impedance is larger.

The reflection and transmission coefficients are used to find the amplitude of the wave. Sometimes the energy (or intensity) is of interest. Since the energy in the wave is proportional to the square of the wave amplitude, the *fraction* of the

Figure 14.3
Configuration used to compute the reflection and transmission coefficients for acoustic waves in pipes.

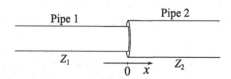

incident energy reflected will be $|R|^2$.[4] The fraction of the energy transmitted is not, however, simply $|T|^2$. That can be seen by considering the case where Z_2 is very, very much larger than Z_1. In that case, to a good approximation, $|R|^2 = 1$, so all the energy is reflected, but also $T = 2$. Clearly, if all the energy is reflected, none could have been transmitted. The problem is that the energy of the waves in the two different pipes cannot be compared solely based on the amplitudes – the properties of the pipes must also be included. That can be done, but it is much simpler to realize that the transmitted energy is the fraction of the energy that is *not* reflected, or $1 - |R|^2$. The extra complications and the need to put in the properties of the pipe can be avoided.

Example 14.2

Consider pipe 1 to be an air-filled pipe that is 3 in. (7.62 cm) in diameter, under normal room conditions (with $Z_1 = 90,000$ acoustic ohms), and pipe 2 to be a pipe with a diameter of 1 in. (2.54 cm). The cross-sectional area of pipe 2 will be $1/3^2 = 1/9$th that of pipe 1, so the impedance of pipe 2 is nine times larger – the smaller-diameter pipe impedes the sound more than the larger-diameter pipe. Then,

$$R = \frac{1-9}{1+9} = -0.8, \quad T = \frac{2 \times 9}{1+9} = 1.8, \quad |R|^2 = 0.64, \quad 1 - |R|^2 = 0.36,$$

and although the transmitted wave has close to twice the amplitude of the incident wave, it transmits only 36 percent of the incident energy. Note that sending sound in from pipe 2 toward pipe 1 – that is, reversing the roles of the two pipes – results in sound reflected back in pipe 2 with $R = +0.8$, and the fraction of energy transmitted into pipe 1 remains 0.36. Such an arrangement of pipes can never favor energy transmission in one direction over the other.

Pipe Resonances

The acoustic normal modes for simple pipes can be determined by analogy to the vibrating string problem. As was seen previously (Chapter 8), a traveling wave sent down a finite length of string will reflect from the end. If the wavelength and the string length match up satisfactorily, then a standing wave is produced – a resonance. The same behavior is observed for sound in pipes.

Cylindrical Pipes

The propagation of sinusoidal plane waves naturally leads to a consideration of plane-wave propagation within a cylindrical pipe. For the pipe, it is as if there

---

[4] The absolute value sign is needed here due to the fact that in some situations, imaginary numbers may be involved. The fraction of the power reflected can never be negative.

were a much larger plane wave, but only the waves within a cylindrical volume are of interest (Figure 14.4a). The waves inside the pipe are moving just as they would naturally move were the rest of the plane wave present. So, to a good approximation, the sound waves in a narrow cylindrical pipe are described mathematically using sinusoidal plane waves. When the end conditions – that is, reflections – are included, resonances occur that are harmonic, and the pipe becomes a useful component for a musical instrument.

Two types of end conditions were considered earlier for the string: the fixed end and the free end (Chapter 6). There are also two simple types of end conditions for a pipe: a closed end and an open end. An open end can be modeled as a connection to a pipe with a very large diameter, and a closed end to a pipe of extremely small diameter. In both cases, all of the acoustic energy is reflected.

A cylindrical pipe has a uniform cross section. By analogy with the vibrating string, the resonant frequencies for a cylindrical pipe of length L, when both ends are open or both ends are closed, can be expected to be given by

$$f = n\frac{c}{2L},\tag{14.7}$$

and when one end is open and the other closed,

$$f = (2n - 1)\frac{c}{4L},\tag{14.8}$$

where n is a positive integer, and c is the speed of sound.

Conical Pipes

The propagation of sound in a narrow conical bore can be considered based on propagating waves starting from a point source (Figure 14.4b). The sound propagating from a point source spreads out spherically. If a cone-shaped pipe is placed such that the apex of the cone (possibly extrapolated from the actual pipe dimensions) is at the source, then the natural propagation shows a wave that

(a)   (b)

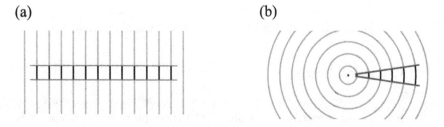

Figure 14.4 Illustrating propagation of (a) sinusoidal plane waves in a pipe and (b) spherical waves in a cone.

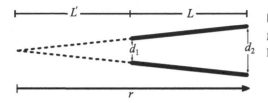

Figure 14.5 The geometry for a partial cone.

spreads out in the same way the cone spreads out. The wave inside the cone is doing what it would do if it were part of the full spherical wave.

For a section of a narrow conical bore, as shown in Figure 14.5, the appropriate mathematical description is of waves that arise from an isotropic point source combined with the spherically reflected waves sent back toward the source. For narrow cones, it is mathematically sufficient to consider "spherical waves," such as

$$p(r,t) = \frac{\sin(\omega t \pm kr + \varphi)}{kr}, \tag{14.9}$$

where r is the distance to the cone's apex (or extrapolated apex), and $k = 2\pi/\lambda$ as before.

Aside from the extra $1/r$ amplitude dependence, spherical waves in a narrow cone are mathematically similar to the plane waves in a cylindrical pipe. In particular, the possible wavelengths for solutions that are zero at both ends will be identical. Hence, a section of a cone of length L open at both ends can be expected to have the same resonant frequencies as will a cylindrical pipe of length L open at both ends. Interestingly, this result applies in the limit that the narrow end of the cone becomes complete – that is, the narrow opening becomes infinitesimally small.

The open–closed solution is more complicated. In that case, there is a maximum at one end and a minimum at the other. The location of the maximum is affected by the factor $1/r$, so a more careful calculation is required.

Figure 14.6 illustrates the resonant frequencies for a cone closed at the smaller end and open at the larger end. If $d_1/d_2 = 0$, then the cone is complete ($L' = 0$ in Figure 14.5), and the resonant frequencies match those of a cylindrical pipe of the same length, open (or closed) at both ends. At the other extreme, when $d_1/d_2 = 1$, which corresponds to $L' \to \infty$, the geometry, and hence the set of resonant frequencies, is identical to a cylinder of length L that is closed at one end and open on the other. The higher resonant frequencies are harmonic only for the two limiting cases. In general, this behavior illustrates that, when necessary, overtones can be tuned by adjusting the bore shape, much in the same way that xylophone bars can be undercut to adjust their overtones (Chapter 7).

A very practical illustration of the difference between the resonances of a conical bore and a cylindrical bore is to compare a standard clarinet with a

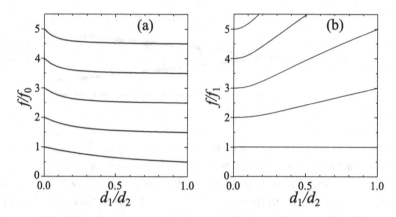

Figure 14.6 Resonant frequencies for a cone truncated at the smaller end and open on the larger end. In (a) the frequencies are compared to the fundamental frequency of a cylindrical pipe of the same length, in (b) the frequencies are compared to the fundamental frequency for the truncated cone.

Figure 14.7 A clarinet and a soprano saxophone are both single-reed woodwinds and are about the same length; however, they do not sound the same, and the clarinet, although shorter, goes to significantly lower pitches. This is largely due to the different shapes of the instruments. The clarinet is close to being a cylinder, whereas the saxophone is close to being a cone.

soprano saxophone (Figure 14.7). Both are single-reed instruments, and they are close to the same length, but they have a very different sound – that is, a very different set of overtones. This is particularly true when comparing the lower notes of the two instruments. In addition, the lowest note on the clarinet is almost an octave lower than that of the soprano sax. While the instruments are constructed of different material and the saxophone has much larger keys than

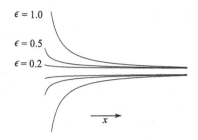

$\epsilon = 1.0$

$\epsilon = 0.5$

$\epsilon = 0.2$

x

Figure 14.8 Several Bessel horn shapes for $\epsilon > 0$.

the clarinet, the largest contributor to the difference in the sound is due to the fact that the clarinet is close to being a cylinder, closed at one end (i.e., $d_1/d_2 = 1$), and the saxophone is close to being a complete cone (i.e., $d_1/d_2 = 0$).

Several of the wind instruments used in a classical orchestra will have both cylindrical and conic (or nearly conic) sections, including flutes, trumpets, trombones, and others. The conical sections can be regarded as having a characteristic impedance that gradually varies with position. That is, the input impedance used to compute reflection and transmission coefficients will be different at one end of the pipe from the other.

Bessel Horns

Another interesting pipe is any one where the shape is defined by

$$r = a\,x^{-\epsilon}, \tag{14.10}$$

where a and ϵ are constants, and r is the radius of the pipe a distance x along the pipe. These are often referred to as **Bessel horns**. Several of these shapes are illustrated in Figure 14.8. These all approach an infinite radius when x approaches 0, so no real horn will match these exactly. However, finite-length sections of these shapes can be used as part of a model for an instrument with a flared bell (e.g., the brasses). For $\epsilon > 0$, the solutions for standing waves can be expressed using Bessel functions, "of order $\epsilon + 1/2$." Here it will be sufficient to know that solutions exist, and the details will not be pursued further. For the rest of the instrument, those Bessel solutions are matched at the boundaries to solutions for a cylinder and/or cone. The details of this mathematics will also not be pursued here. Note that $\epsilon = 0$ is a cylinder and $\epsilon = -1$ is a cone, so those shapes can be considered special cases of these Bessel horns.

Branches

When a pipe has branches, it is perhaps easier to consider the admittance, Y, rather than the impedance, Z, where $Y = 1/Z$. Recall that for a simple pipe, the acoustic impedance goes like 1/area, so the admittance is simply proportional to the area. For two pipes in parallel, what is important for the sound propagation is the total area. Hence, the input admittance at a branch will be the sum of the input admittances for the two branches. That result generalizes for all circumstances: whether or not from a simple pipe, the input admittance at the branch is the sum of the input admittances from each branch.

Woodwind finger holes can be treated as a short branch. The finger hole has a depth and hence acts like a very short length of pipe. That short pipe will have an open or closed end, depending on the note being played. The previous equations for the pipe impedance can be used to compute the extra admittance due to these finger holes. As discussed later in the chapter, branches are also important for models of the human voice.

Pipes with open ends, including those used to model finger holes, will also need to include an "end correction." This is usually included through the use of an effective length. The end correction arises from the fact that the real problem is three-dimensional, and the flow at the end does not immediately adapt to the (infinite) change in pipe radius. Hence, an open-ended pipe acts like it is a bit longer than its dimensions. For typical woodwind instruments, the end correction for a finger hole may result in an effective length that is significantly larger than the actual depth of the hole.

Pipes with Holes

The impedance for many situations can be derived, although usually with significant approximations made along the way. One such example, important for musical instruments, is a simple tube with an array of holes in it. For musical instruments, the size of the holes and the spacing between holes are typically small compared to the wavelength. The details on a length scale that is small compared to a wavelength are often not very important. For modeling purposes, the effect of a hole can often be lumped into an additional impedance, either at a single point or spread out over the spacing between the holes.

Using a model where the impedance is spread over the spacing between holes, an expression for the acoustic impedance for an infinite tube with uniform circular cross section and periodic circular holes in the side can be derived to be[5]

$$Z = \left(\frac{\rho c}{\pi a^2}\right)\left(\frac{2 + (b/a)^2 \cot(\omega t_e/c)\tan(\omega s/c)}{2 - (b/a)^2 \cot(\omega t_e/c)\cot(\omega s/c)}\right)^{1/2}, \qquad (14.11)$$

[5] For a derivation, including a discussion of the approximations involved, see Benade (1960).

where a is the tube's inner radius, b is the hole radius, c is the speed of sound, ω is the (angular) frequency, t_e is an effective depth of the holes,[6] and s is one-half the distance between the holes. The first term in parentheses is what is expected for a tube of radius a. The second term in parentheses results in a rather complicated frequency dependence. Note that the second set is raised to the power one-half; that is, it is a square root. Note as well that there is a negative sign in the denominator, and thus for some range of frequency, it appears that this formula might require one to find the square root of a negative value. Imaginary numbers will be showing up.

Example 14.3

The computed impedance for a pipe that is 3 in. (7.6 cm) in diameter with holes of 1 in. (2.54 cm) in diameter spaced 3 in. (7.6 cm) apart is shown in Figure 14.9. At higher frequencies, the values are all real numbers. At lower frequencies, they are imaginary numbers.

The dividing line – that is, the frequency below which the equation yields imaginary numbers – is also where the impedance becomes very large. That is referred to as the **critical frequency**. For this model, the critical frequency, f_c, is given roughly by

$$f_c \approx \frac{1}{2\pi} \frac{c}{a} \sqrt{\frac{b}{3s}}. \tag{14.12}$$

Note that at higher frequencies, the impedance is identical to that of a pipe of the same size but with no holes. It is as if the holes do not exist at higher frequencies.

What happens at the lower frequencies, where the impedance is imaginary? Surely, a measurement could be made at lower frequencies, so it is not an

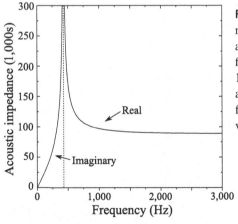

Figure 14.9 The magnitude of the acoustic impedance for 3-in. pipe with 1-in. holes spaced 3 in. apart. At lower frequencies, the values are imaginary.

---

[6] t_e is roughly $1.5 \times b$ larger than the actual depth.

impossible situation. However, there were assumptions that went into the theory that may now be suspect. In this case, the sound wave was taken to be a sinusoidal traveling wave within the pipe. Such a sine wave will carry energy into the pipe. Since the pipe is assumed infinite, that energy will never return and will be lost. The power necessary to sustain that wave comes from the product of the acoustic pressure and the flow rate. If the impedance is imaginary, then the product of the acoustic pressure and flow rate is also imaginary, so the energy being carried down the tube is imaginary. Hence, whatever real power that is put into the pipe is no longer traveling down the tube. The assumption that there is a traveling wave in the tube seems to be the problem.

At this point, it may be tempting to simply abandon the imaginary solution as being based on an invalid assumption, then start over for the lower frequencies. Here, however, the mathematics is actually revealing the real solution, if it is interpreted correctly.

Sinusoidal functions are part of a close-knit family of functions based on the exponential function. This was seen earlier for the vibrating bar (Chapter 7), where the connection to the hyperbolic sinusoidal functions was demonstrated. The functions can be related to each other through the use of imaginary numbers. The imaginary result for the impedance is not "wrong" but signifies that the actual behavior is not sinusoidal but from another function from this family. The details will not be pursued here, but in this case, the lower-frequency waves simply fall off exponentially with distance. To get this result, a more general form for the wave is assumed that involves imaginary values from the start.

Figure 14.10 shows the shape of the "wave" for solutions above and below the critical frequency. All of these, of course, oscillate sinusoidally with time, but below the critical frequency, they are no longer sinusoidal with distance, and they no longer propagate. Well below the critical frequency, the wave decays exponentially

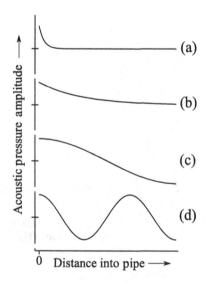

Figure 14.10 Snapshot of the pressure amplitude for a wave in a pipe with periodic holes. (a) Well below f_c, (b) below but near f_c, (c) above but near f_c, (d) well above f_c. Cases (c) and (d) are traveling sinusoidal waves, whereas in (a) and (b), the waves oscillate in time but do not travel. ★

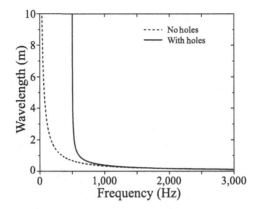

Figure 14.11 Wavelength as a function of frequency for acoustic waves in a pipe having periodic holes, resulting in a critical frequency of 500 Hz. The result is for the same pipe shown in Figure 14.9.

over a distance comparable to the spacing between the holes. That is, the wave does not propagate past the first hole it encounters. Near the critical frequency, the wave amplitude falls off over a distance encompassing several holes. Above the critical frequency, the wave propagates as if there were no holes, except that near the critical frequency, the wavelength is longer than expected (Figure 14.11).

Since a tube with periodic holes in it will let through the higher-frequency sounds, but lower-frequency sounds do not propagate, it can be used as a **high-pass filter**. The high frequencies pass through. In contrast, a **low-pass filter** can also be made with pipes. Appropriate combinations of the two types of filters can produce a **band-pass filter**, where a particular range of frequencies, referred to as a *frequency band*, will pass through, and both higher and lower frequencies are blocked. One can also produce a **band-reject filter**, where only the higher and lower frequencies are passed, and a specified frequency band is blocked. Low-pass and band-reject acoustic filters are used as sound mufflers. An automobile muffler is constructed based on these ideas. Some simple example designs are shown in Figure 14.12.

Woodwinds

A simple description of all woodwind instruments (flute, oboe, clarinet, etc.) would be that they are a pipe with two sections – a section without holes and a section with holes. The most basic woodwind instruments are the keyless flutes and fifes (Figure 14.13). To a first approximation, they are modeled with an all-pass filter (a simple pipe) connected to a high-pass filter (a pipe with holes). Throughout most of the playing range, the frequency of the sound is below the critical frequency of the high-pass section, and thus its impedance is imaginary. For those well versed in the use of complex arithmetic, it is straightforward to show that at a junction where the impedance changes from real to imaginary, the

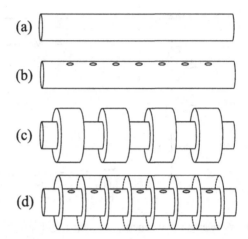

Figure 14.12 Pipes that can be used as acoustic filters: (a) all pass, (b) high pass, (c) low pass, and (d) band reject. Designs similar to (c) and (d) are used as sound mufflers. The design in (d) can be considered to be periodic Helmholtz resonators attached to the pipe.

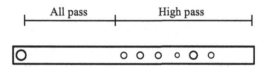

Figure 14.13 A simple flute or fife is a pipe with holes.

magnitude of the reflection coefficient will always be 1, as it also is at the end of the pipe. That is, the beginning of the high-pass section acts as an end to the pipe, and that will be near the first open hole.

When played at lower frequencies, the finger holes are covered systematically from left to right. That will effectively lengthen the all-pass section of the pipe, which lowers the resonant frequency. However, at the highest frequencies, above the critical frequency, the sound will travel through the full length of the instrument. To get the desired resonances, the fingerings are no longer systematic and, as a rule, are more difficult to learn. Figure 14.14 illustrates some computed resonant modes for a simple fife. For the lowest octave, well below the critical frequency, the first open finger hole acts like the end of the pipe. In the middle octave, some exponential behavior can be seen over the first few open holes. For the highest octave, the sound travels all the way to the end of the fife, even if there are several open finger holes.[7]

---

[7] Note that the frequency of the sound generated depends on the resonant modes and the (nonlinear) driving mechanism (Chapter 10). For the fife as normally played, the result is usually very close to the resonant frequency. Note also that traditional fife music generally uses the higher octaves and rarely uses the lowest octave.

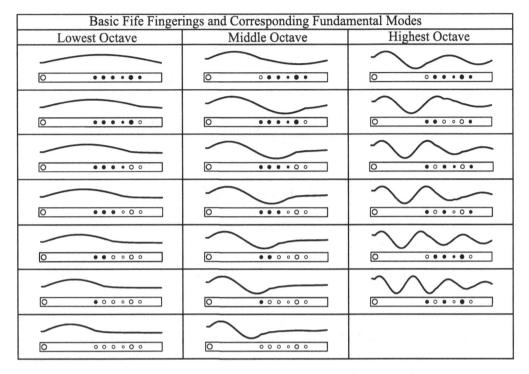

Basic Fife Fingerings and Corresponding Fundamental Modes		
Lowest Octave	Middle Octave	Highest Octave

Figure 14.14 Computed resonant acoustic modes for a simple fife. Open circles represent open holes, and filled circles represent holes covered by a finger. Note that for the lower octave, the wave is reflected near the first open hole, whereas in the highest octave, the sound propagates through the entire length of the tube despite the presence of many open holes.

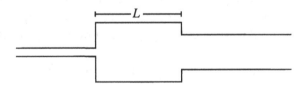

Figure 14.15 A schematic for the calculation of wave propagation through three connected pipes with different radii.

More Complicated Pipe Resonances

Now consider an arrangement of three pipes, as shown in Figure 14.15, with an acoustic traveling wave coming in from the left. The pipes on the left and right are considered semi-infinite, so reflections from their ends are not of concern, and the center pipe has a length L. This example serves as a simple system that illustrates how more complicated systems can be treated. The total wave in each pipe can be found by writing down the traveling waves, using the characteristic impedance to

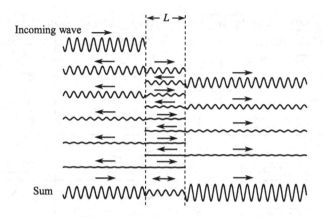

Figure 14.16 The incoming wave is reflected at the first boundary. Some of the energy is transmitted into the middle section. That wave bounces back and forth between the two boundaries. At each boundary, some wave energy is transmitted to the other pipes. The total wave is the sum of all of the waves. Pipe diameters, left to right, are 3, 6, and 2 units. ★

relate pressure and velocity, and matching at the boundaries.[8] It is perhaps more instructive to simply follow the wave through the system, keeping track of the reflection and transmission that occur as each boundary is encountered, as well as the extra travel distance required to reach each boundary.

Figure 14.16 illustrates the thought process for a particular three-pipe situation, frozen at a particular time. An incoming wave from the left strikes the first boundary. A portion of that wave is reflected, and a portion is transmitted. The portion that is transmitted strikes the second boundary and is reflected back, with some of the energy transmitted to the right. The wave in that middle section continues to bounce back and forth until all of its energy is lost due to transmission. The total wave in each pipe is found by adding up all of these component waves.

The total transmitted wave, on the right in Figure 14.16, is perhaps the easiest to calculate. Using R_{12} (T_{12}) to signify the reflection (transmission) coefficient for a wave in the first pipe when it encounters the second pipe, and so on for the other boundaries, and setting $x = 0$ at the first boundary, the wave on the far right will be

$$y_3(x,t) = T_{12}\,T_{23}\,\left(\cos(kx - \omega t) + (R_{23}R_{21})\cos\left(k(x + 2L) - \omega t\right)\right.$$
$$\left. + (R_{23}R_{21})^2\,\cos\left(k(x + 4L) - \omega t\right) + \ldots\right), \tag{14.13}$$

---

[8] Those well versed in the arithmetic and algebra of imaginary numbers will find that to be a particularly efficient method in this case. In the first step, the sinusoidal waves are rewritten using an identity relating sinusoidal functions to exponentials with imaginary arguments. All of the additional imaginary quantities will disappear in the final result.

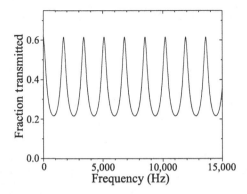

Figure 14.17 Amplitude of transmitted wave for the situation seen in Figure 14.15 with $L = 10$ cm and parameters for air.

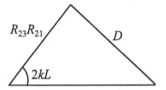

Figure 14.18 A geometric interpretation to help find the amplitude of the transmitted wave.

valid for $x > L$. Writing down such an expression is simply a matter of keeping track of the number of reflections and the extra distance traveled. With some effort,[9] it can be shown that the total amplitude of the transmitted wave on the right is

$$A_3 = \frac{T_{12}\, T_{23}}{\sqrt{1 - 2R_{23}R_{21}\, \cos(2kL) + (R_{23}R_{21})^2}}. \tag{14.14}$$

Figure 14.17 illustrates the computed amplitude of the transmitted sound for the particular arrangement of three pipes shown in Figure 14.15. The amplitudes of the waves in the other pipes will include similar maxima and minima.

When the product $R_{23}R_{21}$ is positive, this expression for the amplitude has a simple geometric interpretation related to the **law of cosines**, a generalization of the Pythagorean theorem. The denominator, for brevity denoted by D, is equal to the length of the third side of a triangle that has sides of length 1 and $R_{23}R_{24}$, respectively, with an angle between them of $2kL$ (radians). See Figure 14.18. When $2kL = 0$ the two known sides of the triangle lie atop each other, $D = 1 - R_{23}R_{21}$, and the denominator is a minimum, making the amplitude a maximum. When $2kL = \pm 2n\pi$, with n being any integer, the triangle is the same as when the angle is zero since each rotation of 2π corresponds to a complete rotation around a circle. Hence, the amplitude is a maximum whenever $L = n\lambda/2$. When the product $R_{23}R_{21}$ is negative,

---

[9] A most efficient method being the use of identities involving imaginary numbers that result in a simple geometric series that can be summed.

a similar interpretation is possible using $|R_{23}R_{21}|$ for the length and $2kL + \pi$ for the angle. That leads to maxima in the amplitude when $L = (n + 1/2)\lambda/2$. Note the similarity of these results to the resonance conditions found for the simple vibrating string with various end conditions (Chapter 6).

It is also possible to compute the input impedance to pipe 2. As mentioned, for musical instruments, knowing the input impedance is important since that describes what will be connected to the nonlinear driving mechanism (Chapter 10). To get the input impedance to pipe 2 for this example set of pipes, the pressure and flow are evaluated at $x = 0$ for a sinusoidal input at frequency ω. Since the pressure and flow are continuous across the boundary, this can be done using waves in pipe 1 or 2. Using the same procedure as previously, the total wave in pipe 2 that is traveling to the right will be

$$
\begin{aligned}
p_R = T_{12} \cos(kx - \omega t) + T_{12}R_{23}R_{21} \cos(2kL - kx - \omega t) \\
+ T_{12}(R_{23}R_{21})^2 \cos(4kL - kx - \omega t) + \dots,
\end{aligned}
\tag{14.15}
$$

and the wave going to the left will be

$$
\begin{aligned}
p_L = T_{12}R_{23} \cos(2kL - kx - \omega t) + T_{12}(R_{23})^2R_{21} \cos(4kL - kx - \omega t) \\
+ T_{12}(R_{23})^3(R_{21})^2 \cos(6kL - kx - \omega t) + \dots,
\end{aligned}
\tag{14.16}
$$

so the corresponding flows, taking positive to be to the right, are found using the characteristic impedance and will be $v_R = p_R/Z_2$ and $v_L = -p_L/Z_2$. The input impedance to pipe 2, as seen coming from pipe 1, will be the ratio of the total pressure to the total flow, or

$$
Z_{2,in} = \frac{p_R + p_L}{v_R + v_L} = Z_2 \frac{p_R + p_L}{p_R - p_L},
\tag{14.17}
$$

where the sums are evaluated for $x = 0$. Somewhat surprisingly, the infinite sums are not too difficult to evaluate. The resulting input impedance can be written

$$
Z_{2,in} = Z_2 \sqrt{\frac{1 + R_{23}^2 + 2R_{23} \cos 2kL}{1 + R_{23}^2 - 2R_{23} \cos 2kL}} \angle \theta,
\tag{14.18}
$$

where

$$
\theta = -\tan^{-1} \left(2 \sin 2kL \frac{R_{23}}{1 - R_{23}^2} \right).
\tag{14.19}
$$

Here, the notation $\angle \theta$ means that the flow will be out of phase with the pressure by a phase angle θ. That is, if the pressure has a time dependence of $\cos(\omega t)$, then the flow has a time dependence of $\cos(\omega t + \theta)$.

The more common and more compact way to write this result is in terms of the characteristic impedances, rather than reflection coefficients, and relies on the use of imaginary values,

$$Z_{2,in} = Z_2 \frac{Z_3 + iZ_2 \tan kL}{Z_2 + iZ_3 \tan kL},$$ (14.20)

where $k = 2\pi/\lambda$, as was used previously for traveling waves. That is, Equation (14.20) includes all the information that is in Equations (14.18) and (14.19) combined, but it is in a more compact form. In both cases, the periodicity of the trigonometric functions gives rise to periodic minima and maxima of the input impedance.

The concept of impedance is not limited to pipes. Any device (appropriate for use with sound) will have impedance and, as far as the rest of the system is concerned, is equivalent to an infinite length of pipe with the same input impedance. For example, a sound-absorbing material that, when placed near the end of a pipe, absorbs some of the incident sound energy and reflects the rest causes that end of the pipe to do just what an appropriately chosen infinite length of pipe would do. The point is that one does not need to actually have infinite pipes in order for them to be useful as part of an acoustic model.

The three-section model of Figure 14.15 can represent a pipe closed or open at the ends by taking the radius of the semi-infinite section to be very small or very large. For example, it can represent a simplified musical instrument if the first pipe represents a mouthpiece and the radius of the third is very small for a closed-ended tube or very large as a model for an open-ended tube.

This procedure can be extended to have many more sections, even hundreds of them, and to include conical and Bessel horn shapes, to extract results for more complicated pipe models. Systems with many more sections are probably best treated using computer modeling, as the three-section system was already complicated enough. The input impedances for a selection of pipe shapes, as a function of frequency, are shown by Olson (1967).

The Vocal Tract

Human speech can be modeled, at least in part, using pipes (see Figure 14.19). The simplest to analyze are the sustained sounds, such as the vowel sounds and sustained consonants such as m and n. The consonant sounds that are more percussive (such as b, k, and p) require a model that changes in time. In principle, a vocal model can be treated similarly to the three-part pipe system in Figure 14.15; however, the solution will clearly be much more complicated.

For the sustained sounds, the lungs provide a source of air that can set the vocal folds (or "vocal cords") into periodic motion. That motion sets the fundamental frequency for the sound and, like the reed instruments, contains a rich supply of overtones at harmonic frequencies. The amplitudes of those overtones are similar to those of the square wave.

The size and shape of the various "pipes" and openings are controlled by muscles, and each section is adjusted to provide the appropriate filter for the

Figure 14.19 A simple pipe model to describe the formation of some sounds made by the human voice.

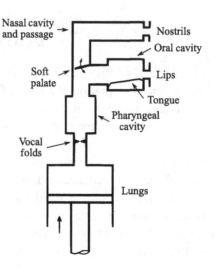

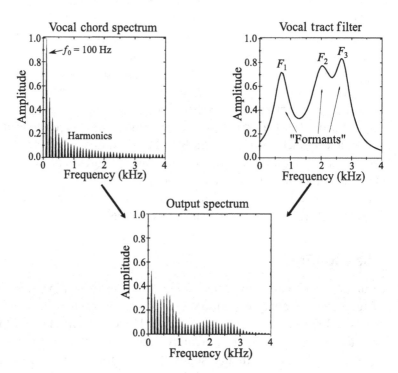

Figure 14.20 Sustained vocal sounds can be modeled as a sound rich in overtones, from the vocal cords, filtered by the vocal tract.

sound. Many of the sounds produced can be described using just three band-pass filters, resulting in three variable-frequency bands where the sound is concentrated. Those regions of the spectrum are referred to as **formants**. The process is illustrated in Figure 14.20.

Losses

So far, only lossless pipes have been considered. For resonant pipes, such as organ pipes, the energy lost during oscillation can be characterized using a quality factor, Q, as was discussed for the vibrating string (Chapter 6). The maximum Q is found to depend predominately on the diameter of the pipe and its length. When making a set of organ pipes (a "rank"), the length of the pipe primarily determines the frequency of the note, whereas the diameter influences the tone quality. If the pipes are to have a similar tone quality for the different notes – that is, to really act as a *set* of pipes – then the diameters need to be chosen appropriately. Such a choice is referred to as the **scale** of the pipes, not to be confused with the musical scale they may produce. A simple model from basic physics (Moloney and Hatten, 2001) shows that for a fixed-length pipe,

$$\frac{1}{Q} = C_{\text{wall}} \frac{1}{D\sqrt{f}} + C_{\text{rad}} f D^2 , \qquad (14.21)$$

where D is the diameter of the pipe, f is the frequency, and C_{wall} and C_{rad} are coefficients that depend on the pipe. The first term, which dominates at lower frequencies, is due to losses at the walls of the pipe, and the second is due to acoustic radiation losses. Based on that result, if the pipes are all to be operated near the maximum Q, then the diameter should be related to the length ($L \propto 1/f$) as

$$D \propto L^{5/6} , \qquad (14.22)$$

so if the length is doubled, then the diameter should increase by a factor of $2^{5/6} = 1.78$. Or put another way, while the length changes by a factor of 2 for each octave (12 half steps), the diameter should change by a factor of 2 for roughly 14 to 15 half steps. The actual scaling used for organ pipes includes a subjective component, as well as other considerations, and is often found to be in the vicinity of 16 half steps for a factor-of-2 change in length. The simple theory is surprisingly close.

Of course, a pipe used for music need not be operated at the frequency of maximum Q. If operated lower in frequency, the higher harmonics can become more prominent, and if operated higher in frequency, the higher harmonics can become suppressed. Thus, the timbre of, for example, organ pipes, will depend on the width-to-length ratio. This same effect also explains, in large part, why a flute and a piccolo playing the same note, in the same octave, do not sound the same.

Derivation of *R* and *T* (Optional)

The derivation of the reflection and transmission coefficients, R and T, is included here for those interested. Referring back to Figure 14.3, the position within the pipes is designated by the coordinate x, with $x = 0$ being at the junction. The characteristic impedance for traveling waves in these pipes is Z_1 and Z_2, respectively. To start, assume the only source of sound is a sinusoidal traveling wave sent from the left through pipe 1 toward pipe 2. To be explicit, $\cos(\omega t - kx)$ is used to represent the acoustic pressure amplitude of the incident wave with frequency ω and wavelength λ, where, as usual, $k = 2\pi/\lambda$. The acoustic flow is found using the impedance, noting that, by definition, the positive direction for the acoustic flow is the direction of wave travel.

Assuming there might be some reflection at the boundary, the waves in the two pipes are described by the following:

$$p_1 = \cos(\omega t - kx) + R\cos(\omega t + kx), \quad v_1 = \Big(\cos(\omega t - kx) - R\cos(\omega t + kx)\Big)/Z_1;$$
$$p_2 = T\cos(\omega t - k'x), \quad v_2 = T\cos(\omega t - k'x)/Z_2 ,$$

$$(14.23)$$

where R and T are constants to be determined. Notice that a minus sign appears within the expression for the flow in pipe 1 since the reflected wave's flow is positive when in the direction opposite to that of the incident wave.

Since the total wave in pipe 1 must match the total in pipe 2 at the boundary ($x = 0$) for all times, the time dependence (e.g., the frequency) has to be the same in both pipes; however, the spatial dependence away from the boundary (e.g., the wavelength) does not necessarily have to be the same.

The values of R and T can be determined by matching the waves at $x = 0$. Hence, $1 + R = T$ and $(1 - R)/Z_1 = T/Z_2$. With a little algebra, these are easily rearranged to give

$$R = \frac{Z_2 - Z_1}{Z_1 + Z_2} \quad \text{and} \quad T = \frac{2Z_2}{Z_1 + Z_2}, \qquad (14.24)$$

where R is called the *reflection coefficient*, and T is called the *transmission coefficient*.

Summary

Sound in pipes is important for many of the wind instruments. At any point in a pipe, the sound can be characterized by the ratio of the amplitude of the acoustic pressure divided by the size of the airflow, known as the *acoustic impedance*. In general, phase differences also need to be considered in addition to amplitudes. A common tool that can be used for both at once is the mathematics of complex numbers.

When two pipes are joined, acoustic energy can be reflected at the boundary. The larger the difference between the pipe's acoustic impedance, the larger the fraction of the energy that will be reflected. The fraction of the energy transmitted can be determined from the fraction reflected. The amplitudes of the reflected and transmitted waves are determined using the reflection and transmission coefficients, which in turn are computed from the acoustic impedances of the pipes. Multiple reflections can result in acoustic resonances and highly frequency-dependent impedances.

The input impedance can be used to characterize the response at the end of a pipe and is important for musical wind instruments since that is where the driving mechanism is attached.

A pipe with periodic holes is a model for understanding the effects of finger holes in woodwind instruments. A pipe with holes exhibits a critical frequency. Below the critical frequency, the impedance of the pipe can be described with an imaginary value, meaning that the sound does not propagate but is reflected. Well above the critical frequency, sound propagates the same as if there were no holes. Near the critical frequency, the wavelength differs from what would be expected in open air.

Branches in a pipe model can be used for more complicated models, including finger holes with a significant depth and models for the human vocal tract.

Energy losses in pipes depend on pipe dimensions. To maintain consistent sound, the width-to-frequency ratio of pipes used for music, such as organ pipes, scales differently than the length-to-frequency ratio. Likewise, similar instruments with different width-to-frequency ratios, such as the flute and piccolo, will sound different even when playing the same note.

ADDITIONAL READING

Ayers, R. D., L. J. Eliason, and D. Mahgerefteh. "The Conical Bore in Musical Acoustics." *American Journal of Physics* 53, no. 6 (1985): 528–537.

Benade, A. H. "On the Mathematical Theory of Woodwind Finger Holes." *Journal of the Acoustic Society of America* 32, no. 12 (1960): 1591.

Fletcher, N. H. "Scaling Rules for Organ Flue Pipe Ranks." *Acustica* 37, no. 3 (1977): 131–138.

Johnston, I. D. "Standing Waves in Air Columns: Will Computers Reshape Physics Courses?" *American Journal of Physics* 61, no. 11 (1993): 996–1004.

Moloney, M. J., and D. L. Hatten. "Acoustic Quality Factor and Energy Losses in Cylindrical Pipes." *American Journal of Physics* 69, no. 3 (2001): 311–314.

Olson, H. F. *Music, Physics, and Engineering*, 2nd ed. Dover, 1967. (See Chapter 4.)

Stinson, M. R. "The Propagation of Plane Sound Waves in Narrow and Wide Circular Tubes, and Generalization to Uniform Tubes of Arbitrary Cross-Sectional Shape." *Journal of the Acoustical Society of America* 89 (1991): 550–558.

PROBLEMS

14.1 What is the acoustic impedance of pipes of 1 and 2 cm in diameter?

14.2 Using the results from Problem 14.1, what are the reflection and transmission coefficients if a pipe of 1 cm in diameter is connected to a pipe of 2 cm in diameter? What fraction of the sound energy is transmitted across the boundary?

14.3 A pipe of 2 cm in diameter has equally spaced holes (see Figure 14.12b) 1.0 cm in diameter, 2 cm apart. What is the critical frequency? What musical note is that?

14.4 What are the magnitude and phase associated with the complex value $15 + 20i$?

14.5 At a particular frequency, the impedance at the input to a pipe is determined to be $(4 + 3i) \times 10^4$ acoustic ohms. What is the phase difference between the acoustic pressure and flow?

14.6 Use Equation (14.21), with $D = C_{\text{wall}} = C_{\text{rad}} = 1$, to plot $1/\sqrt{f}, f$, and $1/Q$ as a function of frequency, similar to what was done in Figure 6.17. Note that the frequency here will be in strange units due to the arbitrary choices for the diameter and the coefficients.

14.7 It was claimed that since

$$\frac{P_2 - P_1}{v_2 - v_1} = \rho c$$

for *any* two locations in a (uniform) pipe, it must be the case that $p/v = \rho c$ at each location in the pipe. Prove this claim. Remember that the uppercase P represents the total pressure, the lowercase p represents the acoustic pressure, and the lowercase Greek rho (ρ) is the air density.

14.8 Compare acoustics in a pipe with waves on a string. Does an open end of a pipe correspond to a fixed- or free-end condition?

14.9 Use the expression for a simple driven harmonic oscillator (Equation [5.15]) to derive the mechanical impedance of the oscillator at resonance. To get the oscillator's velocity from the expression for displacement, simply replace $A \times \cos(2\pi ft - \varphi)$ with $-2\pi Af \times \sin(2\pi ft - \varphi)$.

14.10 For a sinusoidal traveling wave, it is possible to get all the energy transmitted between two pipes of unequal diameter if a third pipe, one-quarter wavelength long (i.e., $kL = \pi/2$), with an appropriate diameter is inserted in between the two original pipes. If the first pipe has diameter d_1, and the second d_2, what is the appropriate diameter for the quarter-wavelength pipe?

14.11 Measure the pipe diameter, typical finger hole diameter, and typical hole spacing for a woodwind instrument that is mostly cylindrical (English horn, clarinet, flute, piccolo, etc.) and estimate the critical frequency for that instrument when none of the keys is pressed.

14.12 Show that complete reflection will result when a pipe with impedance Z_1 is connected to a pipe of impedance Z_2 if one of the impedances has a real value and the other an imaginary value. Hint: Convert the numerator and denominator of the reflection coefficient into polar format (i.e., using a magnitude and phase angle). See Appendix A, if necessary, for more information on complex values.

14.13 Use Equation (14.20) for the cases where $Z_2 \gg Z_3$ and $Z_3 \gg Z_2$ to determine the input impedance when the third pipe represents an open and a closed end.

15   Sound in Three Dimensions

The study of sounds in an enclosed space is often referred to as *architectural acoustics* or, more simply, *room acoustics*. This chapter and the next will present a brief introduction to the subject. In this chapter, the wave nature of the sound is temporarily put aside, to be discussed in more detail in Chapter 16. To start, simplified models for sources are considered. This is followed by models for sound propagation in a room that are motivated by the **ray approximation**.

Acoustic sources may be quite complicated. When characterizing the acoustic pressure amplitude at a distance from any source, there will generally be three regions to consider – the near, intermediate, and far fields. The **far field** is where the wavelength is small compared to the distance to the source. The solutions there correspond solely to sound waves that can propagate – that is, traveling waves. The near field is in the region closer to the source than about a wavelength, and the solution there includes both the propagating sound wave and acoustic energy, which may move around in the vicinity of the object but does not travel far. As the name suggests, the intermediate region is in between the near and far regions.

Many discussions of room acoustics will concentrate solely on the far field and the associated sound waves. That is where the sound is well described by simple propagating waves. For some musical situations, such a model may well be an oversimplification, but it is simpler to disregard the details near the sources for now for the sake of getting something done at an introductory level.

Isotropic and Point Sources

The simplest model for an acoustic source is that of the **isotropic source**. For an isotropic source, the sound spreads equally in all directions. Imagine a spherical shell of radius r that surrounds such a source, with the source at the center. If the total acoustic power emitted by the source is P, then all that power will intersect that shell, equally distributed in all directions. **Intensity is power per unit area**, and the area of a sphere of radius r is $4\pi r^2$, so the intensity, I, a distance r away is

$$I = \frac{P}{4\pi r^2}.$$ (15.1)

For sound waves (in the far field), the intensity, I, is related to the acoustic pressure, p, by $I = p^2/2\rho c$, where ρ is the density of, and c is the speed of sound in, the medium (e.g., air). Hence, the acoustic pressure, which is what drives the signal generation for many microphones, would be expected to be proportional to $1/r$.

Computing the intensity for other sources can become quite complicated. One approach is the use of "multipoles." Multipoles are idealized point sources – sources that are infinitesimally small – that cannot actually exist except on paper. They can be useful models for some real sources.

Monopoles

A simple isotropic source shrunk down to an infinitesimally small size is called a *point source*, or **monopole**. The strength of the monopole can be characterized by a value, Q, with dimensions of area × velocity (i.e., m³/s). Imagine a spherical balloon that expands and contracts in radius. The sound created – that is, how much air is pushed and by how much – will depend on the area of the surface and how fast the surface is moving.

The sound power emitted from such a monopole source with strength Q at frequency ω and the corresponding intensity a distance r away are given (in the far field) by

$$P = \frac{\omega^2 \rho Q^2}{8\pi c}; \quad I = \frac{\omega^2 \rho Q^2}{32\pi^2 c r^2}.$$ (15.2)

Such a source would create an infinite sound pressure at $r = 0$, and hence it cannot exist except on paper. However, a real source that is very small compared to the other dimensions in the problem may be well approximated by such a source. The important "other dimensions" here are the distance to the source and the wavelength of the sound.

More Complicated Sources

Other sources can be constructed using the sum of small, isotropic sources, such as monopoles. Some sources can be adequately modeled with a small number of such sources, in the same way that the monopole source can be used as a model for an isotropic source.

Dipole Sources

First consider two identical monopoles with strength Q a distance h apart. As the distance is decreased to zero, the two monopoles simply merge and act as one, with twice the strength. There is nothing new happening. However, now consider two monopoles, one with strength $+Q$ and the other with $-Q$. Here, the sign change corresponds to a 180° phase difference between the sources. When one source is emitting a positive-going sound (pressure) wave, the other is emitting a negative-going wave, and vice versa. Also, remember that *negative pressure* here refers to the acoustic pressure, which is the change from normal atmospheric pressure. More generally, one might include sources with other phase shifts, such as a 90° shift, although that will not be necessary here.

With $+Q$ and $-Q$ monopole sources a distance h apart, halfway in between and, more generally, in any direction equidistant from both sources, the intensity must be zero, no matter how small h gets. Hence, there is something different about this solution. Since there are now two monopoles, this is referred to as a **dipole**.

Figure 15.1a illustrates, schematically, a simple monopole with strength $+Q$. Figure 15.1b illustrates the dipole constructed with two monopoles a distance h apart.[1] Note that if the two monopoles are unequal in magnitude, then as the distance, h, becomes small, they become equivalent to the sum of a monopole and a dipole, which is not something new.

The power and intensity (in the far field) for such a dipole with a finite-sized, but small, spacing is

$$P = \frac{h^2}{3\lambda^2}\left(\frac{\omega^2 \rho Q^2}{8\pi c}\right); \quad I = \frac{\omega^4 \rho Q^2 h^2}{32\pi^2 c^3 r^2}\cos^2\theta. \tag{15.3}$$

The angle, θ, is the angle from the axis defined by the two monopoles. The product Qh is referred to as the *dipole strength*. When taking the limit to get a pure dipole, the product Qh is kept constant while the spacing is reduced. That is, the corresponding monopole strengths, $\pm Q$, are increased as h decreases, so the product is constant.

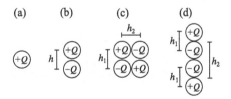

Figure 15.1 Monopoles (a) can be combined to approximate dipoles, shown in (b), and quadrupoles, shown in (c) and (d).

---

[1] Strictly speaking, it is not a pure dipole until *after* the separation has been reduced to zero.

Note that the dipole sound intensity depends on direction, in addition to the inverse square of the distance. The intensity is a maximum along the axis that connects the monopoles. The acoustic pressure along the axis changes sign at the center. That is, if two microphones are placed on the axis, one on each side of the dipole source, the signals from the microphones will be 180° out of phase with each other. A combination of a dipole source, which is positive on one side and negative on the other, and a monopole source, which is positive on both sides, can be used to generate a directional source that is strong on one side but produces little on the other side. The directional pattern is referred to as **cardioid**, in recognition of its heart-like shape.

The angular dependence for sound sources will often be displayed in a polar plot, where the radial component is given in dB and 0 dB corresponds to the maximum. Figure 15.2 illustrates such plots compared to more traditional linear polar plots, where the radius is simply proportional to intensity.

Quadrupole Sources

A **quadrupole** source can be constructed from two dipole sources that are near each other but pointing in opposite directions. If the dipoles point in the same direction, then when the separation is reduced to zero, a single dipole with twice the strength results, which is nothing new. There is more than one way to put two dipoles, pointing in opposite directions, next to each other to generate a quadrupole source. Figure 15.1c and d show two such arrangements – an axial and a lateral arrangement. It turns out that any other quadrupole source can be written as a combination of these two, with appropriate amplitudes and orientations, so understanding these two is sufficient for all. The axial arrangement is symmetric for rotations about the axis that connects the two sources. The lateral arrangement changes sign for 90° rotations about an axis through the center of and perpendicular to the plane of the four sources.

When the dimensions are very small compared to a wavelength and the distance to the observer, the far-field intensity for the quadrupoles is given by

$$I_A = \frac{1}{32\pi^2} \frac{\rho\omega^6}{c^5} \frac{Q^2 h_1 h_2}{r^2} \cos^4\theta, \tag{15.4}$$

$$I_L = \frac{1}{32\pi^2} \frac{\rho\omega^6}{c^5} \frac{Q^2 h_1^2 h_2^2}{r^2} \cos^2\theta \sin^2\theta \cos^2\varphi. \tag{15.5}$$

For the axial quadrupole, θ is the angle from the axial direction. For the lateral quadrupole, θ is from the axis that is perpendicular to the plane of the quadrupole, and φ is the angle around that direction. Polar plots, for results in one plane, for these sources are included in Figure 15.2.

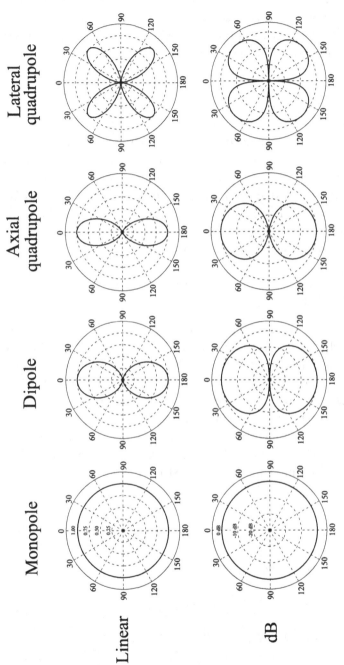

Figure 15.2 Polar plots for some model sources showing the far-field intensity using both a linear and a logarithmic (dB) scale.

The power emitted by quadrupole sources will go like the power expected for a monopole multiplied by $(h/\lambda)^4$. Higher and higher multipoles will involve higher and higher powers of the very small quantity (h/λ), and hence, since h is small, the power emitted becomes very small very quickly. Thus, most finite-sized sources can be modeled using just these first few multipoles, with the remaining contributing negligibly to the sound in the far field.

Example 15.1

A tuning fork can be approximated by a linear quadrupole source. The tines vibrate by as much as 1 mm and are about 1 cm apart. For a tuning fork at 440 Hz, the emitted acoustic power is then reduced from that of a monopole by a factor of somewhere between $(0.001/0.8)^4$ and $(0.01/0.8)^4$, or 10^{-12} to 10^{-8}. Very little power is emitted. In fact, it is very difficult to hear a tuning fork in the far field. To hear the fork, it is either held directly next to the ear or placed against a sounding board, and the sound heard is that from the vibrating surface and not directly from the fork.

Line Sources

A simple line source can be thought of as a string of monopoles along a line. In the simplest case, all the monopoles have the same strength. A line source consisting of a string of dipoles, as illustrated in Figure 15.3, is a simple model for a vibrating string. As the string moves up and down, for example, the air is compressed on one side while it is rarified on the other.

The sound from a vibrating string can also be approximated by the sound from a cylinder that is oscillating sinusoidally with a velocity amplitude, u, in a direction perpendicular to the axis of the cylinder. The power and intensity per unit length, L, derived for this model (Morse, 1948) in the far field are given by

$$\frac{P}{L} = \frac{\pi^2}{4}\frac{\rho\omega^3 a^4 u^2}{c^2}; \quad \frac{I}{L} = \frac{\pi}{4}\frac{\rho}{c^2}\frac{\omega^3 a^4 u^2}{r}\cos^2\varphi. \tag{15.6}$$

The angle, φ, is the angle around the cylinder, referenced to the plane of vibration.

Figure 15.3 A line of dipoles can be used as a simple model for a vibrating string.

Example 15.2

Estimate the intensity of the sound from the A string of a violin.

ρ = density of air $\approx 1\,\text{kg/m}^3$; $c = 340\,\text{m/s}$; $f = 440\,\text{Hz} \rightarrow \omega = 2{,}770\,\text{s}^{-1}$;
$a \approx 1\,\text{mm}$; $u \approx 1\,\text{m/s}$; $L = 0.3\,\text{m}$.

Putting in these values gives a radiated power of about $0.14\,\mu\text{W}$. At 1 m, this would be an intensity of about 11 nW/m^2, or a sound pressure level (SPL) of about 40 dB. That is just slightly louder than a quiet room. Such a calculation shows that a violin string does not produce very much sound. In fact, most of the sound that is heard comes from the vibrations of the wood of the violin. The vibrating string is coupled to the wood through the bridge, and the sound from the wood is what is heard.

On the other hand, an orchestral chime (or tubular bell) is more than 10 times larger in diameter. Since the power goes like the radius to the fourth power, the sound from the chime can be expected to be more than 40 dB larger than that from the violin string. Hence, the chime does not need to be coupled to any sort of board.

Plane Source in a Baffle

A "plane source in a baffle" is, among other applications, a model for a speaker in an enclosure. Here, the source is a solid disk (or piston). An example that uses a circular disk, of radius a, with a sinusoidal velocity of amplitude u, surrounded by a stationary wall, is illustrated in Figure 15.4. Intensities have also been derived for a number of other differently shaped sources.

The intensity for a circular baffle of radius a is given by Morse (1948) as

$$I = \pi^2 \frac{\rho c\, u^2 a^4}{\lambda^2\ r^2} \left[\frac{2\, J_1(\mu \sin \theta)}{\mu \sin \theta} \right]^2, \tag{15.7}$$

where

$$\mu = \frac{2\pi a}{\lambda} = \frac{a\omega}{c}. \tag{15.8}$$

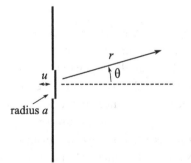

Figure 15.4 The geometry for a moving circular plane source in a baffle, a simple model for a speaker.

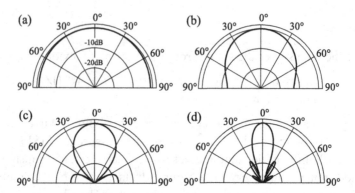

Figure 15.5 The radial-intensity pattern for a circular plane source in a baffle for $\mu = 1, 3, 5$, and 10 (a–d, respectively). At low frequencies (a), the pattern is almost isotropic. At higher frequencies (b–d), the pattern becomes increasingly directional.

The function $J_1(x)$ is a Bessel function (of the first kind, order 1). At low frequencies, where the wavelength is large, $\lambda \gg a$, the term in square brackets is very close to 1, and there is very little angular dependence. The source looks isotropic at low frequencies (at least when in front of the baffle). At higher frequencies, the angular dependence can be quite directional. The angular dependence is illustrated in Figure 15.5.

Note the dependence of the intensity on a^4/λ^2. The intensity is weaker for low frequencies (long wavelengths) than high, and the intensity rises very rapidly with increased radius. These different effects contribute to the development of speaker designs based on multiple speaker elements (in an enclosure) in order to treat the full range of frequency components in the sound more evenly.

The solution (Equation [15.7]) also hints at the explanation for how soundboards seem to amplify the sound. A vibrating system is coupled to a board that has much larger dimensions. The speed of sound in the board will be faster than it is in air, and hence the surface tends to move as a unit, at least as long as the frequency is not too high. Although the motion of the surface may be very small, the coupling to the air from such a large surface area can be quite significant.

Reciprocity

Consider two microphones wired in the opposite sense, spaced a distance, d, apart that act as point receivers sensitive to pressure. That is, if the microphones are sound sources, then if d is small, they will act as a dipole source. The principle of reciprocity says that when they are used as receivers, the relative sensitivity has the same dependence on distance and angle as would the intensity from an analogous source (e.g., as shown in Figure 15.2).

If the sound source is placed an equal distance to the two microphones, as illustrated in Figure 15.6, the microphone signals will cancel when combined,

Figure 15.6 Two microphones, wired in reverse, have a sensitivity pattern similar to the intensity pattern of a dipole source.

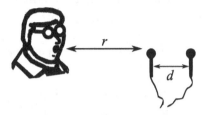

Figure 15.7 Two microphones, wired in reverse, will have an enhanced bass response when the source is along the line connecting the microphones.

resulting in no signal at all. This effect can be simulated using a stereo recording of music and almost any sound-editing program. Good recordings to use for this are digital studio recordings with a single lead singer (or instrumentalist). In practice, those recordings are usually done using a single microphone for the lead singer, and that single signal is then equally split between the left and right stereo channels – that is, the lead singer is placed exactly at center stage. The sounds from the accompaniment will be placed off-center, with unequal signals in the two channels. If such a recording is imported into an audio editor, one channel is flipped (multiplied by a minus sign), and then the two channels are combined into a single (monophonic) track, the signal from the lead singer will cancel, although much of the accompaniment will remain. You are then free to replace the lead singer with your own voice.

Proximity Effect

Consider the two microphones again – but now with a sound source along the line connecting the two microphones, a distance, r, away from the closest, as shown in Figure 15.7. If the source can be simply approximated as an isotropic source of traveling waves, then the acoustic pressure will go as the inverse of the distance. For a sinusoidal source, the combined signal from the two microphones will look like

$$S(t) = A\left(\frac{1}{r}\cos(kr - \omega t) - \frac{1}{r+d}\cos\left(k(r+d) - \omega t\right)\right), \tag{15.9}$$

where A is an overall scale factor. Using trigonometric identities (see Appendix A), the second cosine can be written

$$\cos(k(r+d) - \omega t) = \cos(kr - \omega t)\cos(kd) - \sin(kr - \omega t)\sin(kd). \qquad (15.10)$$

The following approximate expressions (see Appendix A) can be used when d is very small, in particular, $kd \ll 1$, and $d/r \ll 1$:

$$\cos(kd) \approx 1 - \frac{1}{2}(kd)^2 + \ldots; \quad \sin(kd) \approx kd; \quad \frac{1}{r+d} \approx \frac{1}{r}(1 - d/r). \qquad (15.11)$$

Using these approximations and keeping terms only to first order in d (that is, all terms involving d^n, with $n > 1$, are dropped), the signal is[2]

$$S(t) \approx A\frac{kd}{r}\left(\sin(kr - \omega t) + \frac{1}{kr}\cos(kr - \omega t)\right). \qquad (15.12)$$

When $kr \gg 1$ – that is, when the source is far away from the microphones – only the first term is significant, and the signal falls as $1/r$, as would be expected. However, if $kr \ll 1$, the second term dominates. Remembering that $k = \omega/c$, that means that low-frequency signals will get a boost when the source is close to these microphones.

While demonstrated here for two separate microphones, the same result occurs for any single microphone that acts like a dipole receiver or, more generally, any microphone with a directional pattern similar to the dipole shown in Figure 15.2. The result is known as the **proximity effect**. The extra bass boost helps give radio disc jockeys their deep, resonant sound and can be used to advantage in some other applications where extra bass response is desired.

The Inverse Square Law

As seen in the previous examples, if you are far enough away from a (finite-sized) source and move directly toward or away from the source, the intensity is proportional to the inverse of the distance squared. This is an example of an **inverse square law**. As a reminder, for acoustics, the inverse square law is only applicable for the propagating sound in the far field. It may not work very well when close to a source.

The inverse square law is useful for estimating changes in sound (or signal) levels with distance for environments where reflections are not significant. Reflections are considered separately later in the chapter.

Calculation of the intensity change with the change in distance is straightforward. There are two common types of calculations: those dealing directly with the intensity and those where the intensity is expressed in dB (SPL). Examples of each such calculation follow. As a first simple check on any of these calculations, remember

---

[2] Those who have learned calculus may recognize that what is being done here is the steps appropriate for "taking a derivative with respect to r."

that farther away from a source, sounds are quieter – the intensity is smaller. If the calculated result does not match this simple observation, the result is wrong.

Example 15.3

The sound intensity 2 m from a small speaker is measured to be $2.5\,\mu\text{W}/\text{m}^2$. Moving straight backward 3 m and remeasuring, what is the new sound intensity? Assume reflected signals are negligible.

Call the initial intensity I_1 and the final intensity I_2. Using the inverse square law, with distances in meters, write

$$I_1 = C/2^2, \quad I_2 = C/(2+3)^2,$$

where the constant C will be the same for each. A common mistake to avoid at this point is to forget to square the distances. Since the value for I_1 is known, C can be determined and used to compute I_2. Alternatively, save a computation by dividing one by the other to get rid of C, as follows:

$$\frac{I_2}{I_1} = \frac{2^2}{(2+3)^2} \rightarrow I_2 = \frac{4}{25}I_1 = 0.4\,\mu\text{W}/\text{m}^2.$$

Example 15.4

The sound level is measured with a sound meter to be 75.0 dB when it is 5 m from a trumpet. What reading would be expected 7 m from the trumpet? Assume the direction to the trumpet is unchanged and that reflections are negligible.

Here the intensity is being represented using dB (SPL). That is,

$$I_1 \text{ in dB} = 75\,\text{dB} = 10\log(I_1/I_{\text{ref}}),$$

where I_{ref}, which will not need to be known, is the 0-dB reference intensity being used.

Using

$$I_1 = C/5^2, \quad I_2 = C/7^2,$$

for an appropriate, but unknown, constant C, then dividing one by the other gives

$$\frac{I_2}{I_1} = \frac{5^2}{7^2} \rightarrow I_2 = \frac{5^2}{7^2}I_1 = \frac{25}{49}I_1.$$

(A common mistake at this point is to use $I_1 = 75$ dB and find $I_2 = 25 \times 75/49$ dB. This is incorrect. *The inverse square law cannot be applied directly to the values expressed in dB.*)

The intensity in dB at 7 m is given by

$$I_2 \text{ in dB} = 10 \log(I_2/I_{\text{ref}}) = 10 \log\left((25I_1/49)/I_{\text{ref}}\right)$$
$$= 10 \log(I_1/I_{\text{ref}}) + 10 \log(25/49)$$
$$= 75.0\,\text{dB} + 10 \log(25/49)$$
$$= 75.0\,\text{dB} - 2.92\,\text{dB} = 72.1\,\text{dB}$$

Note how the inverse square law is used *inside the logarithm*, and remember that terms multiplied inside the logarithm can be written as a sum of their separate logarithms. The ratio of the distances squared is all that is required to get the intensity *change* in dB. That change is added to, or subtracted from, the original value to get the final result. As a routine check, make sure that the farther source has the smaller intensity.

Echoes and Reverberation

The sound of an echo from across a large canyon often occurs as a single, delayed version of the original sound. Sound in a room will reflect off the walls and ceilings to produce many rapid echoes, referred to as **reverberation**. Hence, the measured acoustic intensity within a room may be more complicated than what is predicted by the inverse square law. In fact, simply by listening to a sound recording from a room, it is often possible to tell a lot about the room, or the "acoustic space." Often, a short-duration sound, such as a clap, is most useful for this purpose.

Special rooms engineered to minimize reflections, called **anechoic chambers** (*an-* referring to "no" and *-echoic* referring to "echoes"), can be used to characterize audio equipment in the absence of reflections.[3] The acoustic behavior of a real room can then be added to those results to predict what will happen in a real room.

Ray Approximation

The description of the propagation of waves in three dimensions can become quite complicated in general. If the wavelength is small enough, the ray approximation can be used. Here, "small enough" means that the wavelength is smaller than other significant dimensions in the problem. The ray approximation ignores the wave character and has the energy travel in straight lines away from the source. Hence, a sound traveling directly from a source to each receiver (e.g., listener) can be represented using a straight line.

The ray approximation provides a simple representation of wave propagation, which is often adequate for simple acoustic calculations. Hence, the extra complications arising from the wave nature of the sound can be postponed for the time being.

---

[3] Alternatively, an environment with no walls can be used – that is, outdoors over soft ground – although interference from other sources may be present. Note that the term *anechoic chamber* can also refer to rooms designed for other types of waves, in particular for radio waves, and the different types of rooms are not interchangeable.

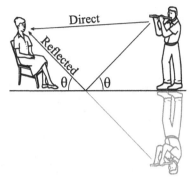

Figure 15.8 Sound can get to a listener both directly and after reflection. The listener experiences the combined signal.

When a wave strikes a surface that is flat enough – imperfections being small compared to a wavelength – then the wave can reflect off that surface. Within the ray approximation, the reflection will obey the **law of reflection**. That is, the incident and reflected waves make the same angle with the surface. For light, this describes the behavior of a mirror. For sound, when the ray approximation is valid, floors, walls, and other flat areas can act as acoustic mirrors. In the same way images can be seen in a mirror, acoustic images will result from these flat surfaces. This is illustrated in Figure 15.8, where a listener receives sound both directly from a source and also after reflection off the floor. The reflection off the floor acts as if it came from a second **image source**. The image source will be farther away, and hence the signal will be less intense and will be delayed in time when compared to the direct sound. The total sound received is the sum of the two.

Calculations based on the ray approximation and image sources can be used to simulate what might happen in an enclosed space. Figure 15.9 illustrates a source and listener in a rectangular room that acts like a house of mirrors. Figure 15.10 shows the received sound intensity for a very short impulse (e.g., a very short-duration clap) based on such a calculation. Since some energy may be lost with each reflection, a reflection coefficient – the fraction of the intensity that is reflected – is included, and for simplicity, it is assumed to be the same for all the surfaces. The time for the total sound intensity to decrease depends strongly on the reflection coefficient.

The apparent size of a room can be judged based on the time for the first reflection. A room where that first reflection occurs quickly is considered intimate – a quick reflection occurs when the walls are close by.

Reverberation Time

The reflections within a room are what define the space acoustically. It is relatively easy to distinguish a large empty room from a small intimate room just by listening to the sound of a clap. In a larger room, the sound takes longer to reflect back, and much more is reflected back if the surfaces are wood or brick compared to carpeting and drapes. Acoustically, perhaps the most important differences between rooms can be characterized using a single value, the **reverberation time** –

Figure 15.9 For sound, a simple rectangular room can act like a house of mirrors.

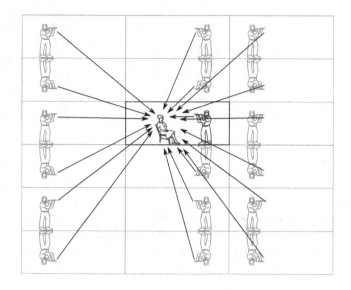

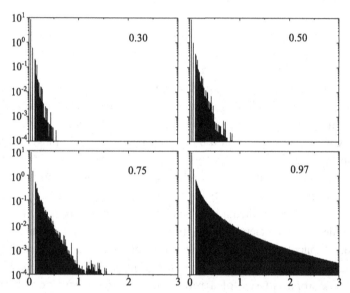

Figure 15.10 A computer simulation of the sound at the position of a listener after an initial short-duration source, such as a clap. In each case, an average reflection coefficient, shown in the upper-right corner, is assumed.

that is, how long a sound persists in a room. A room with a very short reverberation time is described as being acoustically **dead**, whereas a room with a long time is **live**.

Other descriptors of the sound of a room arise from the reflections. The sounds are said to have a smooth or rough **texture** if the sound decay, for example, as viewed on an oscilloscope, is smooth or looks jagged. A room is described as **warm** or **bright** if the lower-or higher-frequency sounds reverberate longer, respectively.

The reverberation time is usually taken to be the time for the acoustic intensity to decay by 60 dB. That is very close to the time it takes for the sound from a single firm clap of the hands to completely disappear in an otherwise quiet room.

As a first estimate, consider a one-dimensional model where sound bounces back and forth between two opposite walls that are a distance L apart. If the sound loses a fraction of its energy, ε, with each bounce, this is an exponential decrease (see Chapter 2), so after N bounces, the fraction of sound remaining, f, is

$$f = (1 - \varepsilon)^N. \tag{15.13}$$

Here, ε would be referred to as an *absorption coefficient*. Then the number of collisions necessary to reduce the intensity by 60 dB is found from

$$\begin{aligned} -60\,\text{dB} &= 10\log\left((1 - \varepsilon)^N\right) \\ -6 &= N\log(1 - \varepsilon). \end{aligned} \tag{15.14}$$

If $\varepsilon \ll 1$, then (see Appendix A)

$$\begin{aligned} \log_{10}(1 - \varepsilon) &\approx -\varepsilon/\ln(10) = -\varepsilon/2.3026, \\ N &\approx 2.3026 \times 6/\varepsilon = 13.8/\varepsilon. \end{aligned} \tag{15.15}$$

The time for the sound to travel a distance L at speed c is L/c, so the time, T_R, required for these N collisions is

$$T_R = 13.8\frac{L}{c\varepsilon}. \tag{15.16}$$

Since it is based on a decay of 60 dB, this reverberation time is often represented by the symbol T_{60}. Now this result needs to be generalized for a three-dimensional room.

To generalize, an appropriate representative length can be substituted for L. For a rectilinear room with dimensions w, l, and h, it might be tempting to use the average; however, that will unduly emphasize the longer distances. In fact, there are more collisions per second, a faster rate, for the shorter dimensions, so they will be more important. The collision *rate*, and hence the loss rate, will depend on speed/distance. That suggests the so-called *harmonic mean* might be used. The harmonic mean is the inverse of the average of the inverses. That is, replace the distance, L, using

$$\frac{1}{L} = \frac{1}{3}\left(\frac{1}{w} + \frac{1}{l} + \frac{1}{h}\right) \rightarrow L = 3\left(\frac{1}{w} + \frac{1}{l} + \frac{1}{h}\right)^{-1}. \tag{15.17}$$

Creating a common denominator to add the fractions yields

$$L = 3\left(\frac{lh}{wlh} + \frac{wh}{wlh} + \frac{wl}{wlh}\right)^{-1} = 6\frac{V}{A}, \tag{15.18}$$

Table 15.1 Typical absorption coefficients, ε, from various sources.

Material	Frequency		
	100 Hz	1,000 Hz	4,000 Hz
Concrete (smooth)	0.01	0.02	0.03
Glass	0.19	0.04	0.02
Plasterboard	0.20	0.08	0.02
Plywood	0.45	0.11	0.09
Carpet	0.10	0.32	0.60
Curtains	0.05	0.35	0.45
Acoustic tile	0.25	0.90	0.90

where V is the volume of the room, and A is the total area of the walls, floor, and ceiling. The estimate for the reverberation time is now

$$T_R = 82.8 \frac{V}{c\varepsilon A}. \tag{15.19}$$

This assumes that the losses are the same for all of the surfaces. To take into account different materials, define an **effective absorbing area**, where the total area is divided into pieces, each with an appropriate absorption coefficient, ε,

$$A_{\text{eff}} = \varepsilon_1 A_1 + \varepsilon_2 A_2 + \varepsilon_3 A_3 + \varepsilon_4 A_4 + \ldots. \tag{15.20}$$

Some typical absorption coefficients are shown in Table 15.1. Note that the effective area depends on frequency, meaning that lower-frequency components of a complex sound may persist for longer or shorter times than higher-frequency components. Then, the reverberation time, T_R, is estimated to be

$$T_R = 82.8 \frac{V}{cA_{\text{eff}}}. \tag{15.21}$$

This will be an overestimate since only collisions between opposite surfaces (opposite walls or ceiling to floor) are considered. There can also be ricochets around corners (e.g., from floor to wall to ceiling), which will shorten the average time between collisions. In addition, not all rooms are rectilinear, and rooms may contain other reflecting surfaces (pillars, furniture, etc.).

In the late 1800s, Wallace Sabine[4] made many measurements of reverberation times using a sound source and a stopwatch. Based on those measurements and subsequent analysis, he found that for a typical room,

---

[4] Wallace Clement Sabine (1868–1919) worked at Harvard University and is considered by many to be the "father of architectural acoustics." Pronounced "**Sey**-bahyn."

$$T_R = 55.3 \frac{V}{cA_{\text{eff}}},$$ (15.22)

which is very close to the cruder estimate found previously. Equation (15.22) is sometimes referred to as **Sabine's formula**. The number out front, $55.3 = 24 \ln(10)$, is to be considered typical and should not be taken too seriously. If dimensions are in meters, use $c = 340$ m/s, whereas if dimensions are in feet, use $c = 1,100$ ft/s.

Example 15.5

Estimate the reverberation time at 1 kHz for a room that is 5 m × 7 m with a ceiling height of 3 m, if the walls are made from painted brick, the floors are carpeted, and the ceiling has acoustic tiles.

The absorption from the brick walls will be very small compared to the carpeting and tiles, so it is not necessary to include it here. The effective area will then be

$$A_{\text{eff}} = 5 \times 7(0.35 + 0.90) = 44\,\text{m}^2.$$

The estimated reverberation time is then given by

$$T_R = 55.3 \frac{5 \times 7 \times 3}{340 \times 44}\,\text{s} = 0.4\,\text{s}.$$

The effective area, A_{eff}, is often expressed in units of "sabins" rather than ft^2 or, if in m^2, "metric sabins."[5] The effective absorbing area of various objects that may be present in a room (people, chairs, etc.) can be determined by their effect on the reverberation time. These measurements can be done, for example, by measuring the reverberation time for an empty room and then again with the object in the room. Such a measurement is most accurate if the reverberation time for the empty room is as long as possible.[6] The result will usually be expressed in sabins. Table 15.2 shows representative average absorption values (in sabins) for different objects.

Example 15.6

A room is constructed with a volume of 48 m^3 using hard walls, floor, and ceiling. The reverberation time of the empty room is measured to be 7.0 s. When three people enter the room, the reverberation time drops to 2.5 s. How many (metric) sabins does one average person contribute?

---

[5] Pronounced "cey-bins."
[6] A typical handball or racquetball court might be used for such measurements.

The effective area of the room is computed (i.e., estimated), and that result is used to find the effective area of the people:

$$T_{R1} = 7.0\text{s} = \frac{55.3}{340} \frac{48}{A_{room}} \rightarrow A_{room} = 0.90\,\text{m}^2.$$

$$\frac{T_{R1}}{T_{R2}} = \frac{7.0}{2.5} = \frac{A_{room} + A_{people}}{A_{room}} \rightarrow (2.8 - 1)A_{room} = A_{people}.$$

So

$$A_{people} = 1.8\,A_{room} = 1.6\,\text{m}^2.$$

The effective area per person is thus $(1.6/3)\,\text{m}^2 = 0.5\,\text{m}^2 = 0.5$ metric sabins. Note that, as an approximation, the surface area and volume of the open space in the room were considered to be unaltered when the people entered. Absorption of extended objects, such as auditorium seats, may instead be specified using an effective absorption coefficient, where the area is the (floor) area occupied by the objects, not the actual area of the object. There is considerable variation in all these average absorption values from one investigation to another.

The most desirable reverberation time, a perceived quantity, will depend on the volume and use of the room. Rooms used for music will have longer reverberation times (1 s or longer) than those used for speech (0.4–0.8 s). The reverberation tends to increase the total acoustic intensity in the room, which, if not too long, acts to reinforce the sound, and that is usually considered good for music. Larger performance halls will have a reverberation time of just under 2 s for midrange frequencies. On the other hand, sounds from speech (e.g., in lecture halls) change rapidly with time and will become muddy and unintelligible if the reverberation lasts too long. Recording studios also tend to use shorter reverberation times since it is easier to add any desired amount of reverberation to a recorded signal

Table 15.2 Average sound absorption, in sabins, for different objects. (Averaged from various sources.)

Object	Frequency		
	100 Hz	1,000 Hz	4,000 Hz
Adult person standing	1.6	9.7	11
Unupholstered seat	0.5	2.0	1.9
Upholstered seat	1.5	4.5	4.5
Adult person seated in upholstered seat	1.8	4.9	5.6

(a) (b)

Figure 15.11 A bare lecture hall during remodeling (a) and the corresponding hall after remodeling (b).

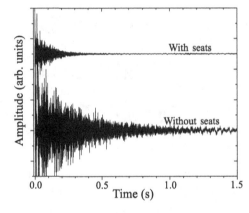

Figure 15.12
Recorded sound amplitudes from a clap in the lecture hall shown in Figure 15.11.

after the fact than it is to remove it. Rooms designed for multiple uses may include provisions to change the reverberation time, such as adjustable drapes.

Figure 15.11 shows a lecture hall both in its bare state (all chairs, ceiling tiles, etc., removed during remodeling) compared to the same room with the sound-absorbing seats, ceiling, and wall treatments restored. The corresponding recorded sound after a modestly loud clap is shown in Figure 15.12.[7] The difference in the reverberation time is quite significant, illustrating the importance of these acoustic treatments in room design.

Room Features of Note

Prior to the introduction of electronic amplification, larger halls used for the performance of music were often designed to maximize the sound energy directed

---

[7] This lecture hall is approximately 85,000 ft^3 (2,500 m^3).

Figure 15.13 A parabolic shape can redirect upward-traveling sound down to the audience.

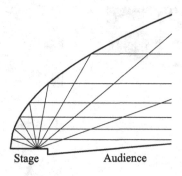

Stage   Audience

Figure 15.14 An example of a stage surrounded by near-parabolic architecture. (Courtesy of the Calumet Theatre, Calumet, MI.)

to the audience. One way to do this is to use a paraboloid-shaped surface, at least approximately, for the walls and ceiling near the stage. A paraboloid is obtained by rotating a parabola about its axis. The energy from a signal source located near a particular location, the focal point, will reflect straight out to the audience – the sound reflection is highly directional. This is illustrated in Figure 15.13, with an example shown in Figure 15.14. Satellite dish antennas use the same principle to send and, in reverse, receive microwave radio signals.

Surfaces that intersect at a right angle (90°) have the interesting property that they reflect signals back in the direction they came from. Refer to Figure 15.15. The effect can be demonstrated by looking into two mirrors that are at right angles (i.e., a corner). You will always see your nose at the intersection of the mirrors, no matter where you stand. In an acoustic environment, this can lead to hot spots, sounds reflected back to performers rather than the audience, and other undesirable behaviors, so such intersections are often avoided in performance halls. The same principle is used for stealth aircraft – right angles that would result in radar signals reflected back to the source are avoided. Right-angle reflectors can be used to advantage when strong directional reflections are desired.

Three reflectors that are mutually perpendicular, like intersecting walls and the ceiling of many rooms, make a corner reflector. If you look into the corner

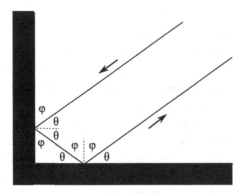

Figure 15.15
Reflectors that meet at right angles will send incoming sounds back toward the source.

formed by an arrangement of three mutually perpendicular mirrors, you will always see your nose very close to the corner. These corner reflectors, or **retroreflectors**, are quite effective at returning a signal to its source and may also be a feature to avoid (or otherwise treat) when designing a performance hall for music. There are certainly applications where corner reflectors are used to advantage. Many simple bicycle reflectors are constructed using multiple corner reflectors (for light) or a similar construction. Several retroreflector arrays have been placed on the moon to allow (laser) light from Earth to be reflected back to the source. A precise measurement of the light's travel time can then be used to determine the distance to the moon. Although not common, there are a few uses for acoustic retroreflectors as well, for example, in multisource environments where interference between sources is undesirable.

It is certainly possible to have a situation where the sound in a performance hall that travels directly from a source to an audience member, without reflection, is actually weaker than the combined reflected signals that arrive later. Interestingly, if the reflections are not too loud, a human listener will often correctly identify the location of the source of the sound. This ability is referred to as the **precedence effect**, where the (perceived) location is determined predominantly by the first sound heard rather than by the loudest.

Smaller rooms may also have noticeable resonances, for example, when the spacing between the walls is an integer number of half wavelengths. Sounds at resonant frequencies will be enhanced. Likewise, it is possible that some frequencies are suppressed in certain rooms. Even some objects, such as a vase that acts as a Helmholtz resonator, can affect the sound in a room. A full treatment of the sound in a room can be a very complicated endeavor.

The Cocktail Party Effect

A cocktail party serves as an example of a room with multiple sources competing for attention. Other examples, perhaps more important, include call centers, air traffic control centers, patrons in a restaurant, students working in small groups in a classroom, and other situations where each listener desires to hear a particular source of sound in the presence of many other sources.[8]

Using the party as a starting point, imagine that in a room, there are N subgroups of individuals having separate conversations. Within each group, the people are spaced an average distance d apart. Assuming only one person is talking in each group (the so-called "polite person approximation"), there are N sources. One question is, How large can you make N without having so much background noise that the talkers become unintelligible to members of their own group? Perhaps one of the more remarkable properties of this calculation is that it can be done, at least approximately.

Acoustic Energy Density

The discussion for isolated sources used intensity – power per unit area – to characterize the strength of an acoustic signal. With multiple signals that may be coming from many different directions, it is often useful to use **energy density** – energy per unit volume – instead. Imagine a simple acoustic signal with intensity I entering a (hypothetical) square box with sides of length L, as illustrated in Figure 15.16. The power at the entrance to the box – how much energy enters per second – is then IL^2. If the signal just gets to the box at $t = 0$, it will get to the far side of the box after a time $t = L/c$, at which point the box is filled with sound. The total energy that has entered the box is

$$E = I L^2 t = I L^3/c. \tag{15.23}$$

Hence, the energy per unit volume (i.e., inside the box), W, is

$$W = E/L^3 = I/c. \tag{15.24}$$

The energy density for an isotropic source a distance, d, away, emitting power, P, and in the absence of reflections, will then be

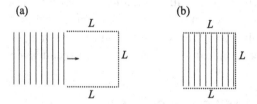

Figure 15.16 An illustration of sound entering a volume, a hypothetical box, used to find the sound energy density.

---

[8] The problem has also been extended to examine communications between animals in a group.

$$W_d = \frac{1}{4\pi d^2}\frac{P}{c}. \tag{15.25}$$

In a room with reflections, this is the energy that arrives before any reflections have occurred, and it would be referred to as the **direct energy**.

Reverberant Energy

Reverberant energy is the energy that arrives after one or more reflections. If the power emitted from a source is P, the total power that remains after the first reflection is $P(1 - a)$, where a is the average absorption coefficient for the room. The sound remains in the room, on the average, for some time Δt.

The reverberation time is the time for the intensity to drop by 60 dB, a somewhat arbitrary choice. To compute the rate of change, the appropriate time to use is the average time, or **lifetime**, that the sound is present. The lifetime, Δt, is the time to drop by $e^{-1} = 0.368$. That time is related to the reverberation time by

$$\Delta t = \frac{T_R}{6}\log_{10}(e) = 0.0724T_R = \frac{4V}{caS}. \tag{15.26}$$

The right side of this equation comes from Sabine's formula (Equation [15.22]). The effective area has been written as the actual area, S, multiplied by an average absorption coefficient, a.

The **reverberant energy density** in the room is then

$$W_R = P(1 - a)\Delta t / V = \frac{4P(1 - a)}{caS} = \frac{4P}{cR}. \tag{15.27}$$

Here, in this context, $R = Sa/(1 - a)$ contains all the information about the room and is sometimes referred to as the *room constant*.

Signal to Noise

Returning to the cocktail party, the energy density at the position of a listener in one of the groups will include the direct sound from the speaker in that group plus the reverberant energy from all the groups. This assumes that the other groups are all far enough away so that the direct signal from them is negligible. For simplicity, assume all the talkers emit the same power, P. The total energy density at the listener is then

$$W_{\text{tot}} = \left(\frac{P}{4\pi cd^2} + \frac{4P}{cR}\right) + (N - 1)\frac{4P}{cR}. \tag{15.28}$$

The first term is the desired signal, including both the direct and reverberant energy, and the second term is the reverberant energy density from all of the other conversations.

Whether or not the speaker is intelligible depends on the relative sizes of the desired signal and the background interference (the "noise"). There is no set value for what that should be, as it is a matter of perception, but it seems reasonable that the ratio, the *signal-to-noise ratio* (SNR), should be close to 1 or larger. Using SNR = 1, solve for N:

$$\left(\frac{P}{4\pi cd^2} + \frac{4P}{cR}\right)\Bigg/\left((N-1)\frac{4P}{cR}\right) = 1 \rightarrow N = 2 + \frac{R}{16\pi d^2}. \tag{15.29}$$

If N is larger than this value, it will become difficult to understand the speaker. Whether or not SNR = 1 is the best value to use here depends on circumstances and is a matter of ongoing research. It is a simple matter to use separate recordings of reverberant sounds ("crowd noise") and an orator, combined in different ratios using an audio editor, to test the reasonableness of the SNR = 1 assumption. Interestingly, the power, P, is not important, suggesting that as long as all the talkers speak at a low volume, they will be understood just as well as if they were all to speak very loudly. In practice, of course, once the reverberant sound becomes comparable to the direct sound, everyone raises their voice in an effort to be heard.

A room's **clarity** depends on the relative amount of indirect and direct sound. The more direct sound, the clearer the sounds will be. If there is more indirect sound, the room is said to have **fullness** – the room becomes filled with reverberant sound.

As a fine point for these calculations, if a talker is close to a wall, their image source can contribute almost equally, and the talker effectively has as much as twice the power. Standing in a corner of a room results in three image sources, so combined with the original source, there is effectively as much as four times the power. If you want to be heard over others, perhaps you should stand in the corner.

Calculations might be improved by considering the directional nature of human voices, male versus female voices, the effects of background music, or even a live band. Factors that may be more difficult to take into account include the visual portion of speech perception – the so-called ability to "lip-read."

Summary

The sound intensity from a finite-sized source, far from the source, will be traveling waves. The intensity of the traveling waves will be proportional to the inverse of the distance to the source, squared, and may be highly directional. Simple models, multipoles, based on point-sized collections of out-of-phase, ideal point-sized isotropic sources, can be used to approximate real sources. The radiation efficiency falls rapidly as the number of sources increases, so only the smaller models are necessary. Physically larger sources can be treated as the sum of a large number of in-phase sources, and they can be much more effective at radiating sound than any point-sized source.

The ray approximation is useful for the simple analysis of an acoustic space, such as an auditorium. One simple characteristic of such a space is the reverberation time. The reverberation time is proportional to the volume, and the inverse of the effective absorbing area, for the space. The ray approximation is also useful for understanding some directional reflections that may occur in a room.

Reciprocity allows sources and receivers to switch roles, and the directional patterns remain the same. Thus, for example, a two-source, one-receiver result can be used to understand a one-source, two-receiver setup.

In an environment with many distributed sources, it may be useful to consider the acoustic energy density rather than the intensity. One example is known as the *cocktail party effect*, where an estimate can be made of the number of conversations that can occur simultaneously in an enclosed space based on direct and indirect (reflected) acoustic energy.

ADDITIONAL READING

Beranek, Leo L. "Analysis of Sabine and Eyring Equations and Their Application to Concert Hall Audience and Chair Absorption." *Journal of the Acoustical Society of America* 120, no. 3(2006): 1399–1410.

Berg, R. E., and D. G. Stork. *The Physics of Sound*, 2nd ed. Prentice Hall, 1995. (See Chapter 8.)

Hardy, H. C. "Cocktail Party Acoustics." *Journal of the Acoustical Society of America* 31 (1959): 535.

MacLean, W. R. "On the Acoustics of Cocktail Parties." *Journal of the Acoustical Society of America* 31, no. 1 (1959): 79–80.

Morse, P. M. *Vibration and Sound*, 2nd ed. McGraw-Hill, 1948.

Morse, P. M., and R. H. Bolt. "Sound Waves in Rooms." *Reviews of Modern Physics* 16, no. 2 (1944): 69–150.

Pierce, A. D. *Acoustics: An Introduction to Its Physical Principles and Applications*, 3rd ed. Springer, 2019. (See Chapter 6.)

Wallach, H., E. B. Newman, and M. R. Rosenzweig. "The Precedence Effect in Sound Localization." *American Journal of Psychology* 62, no. 3 (1949): 315–336.

PROBLEMS

15.1 If a contrabassoon produces a sound at 20 Hz, what is the wavelength of the sound waves associated with the fundamental? How does that compare to the size of a typical room?

15.2 Provide a convincing argument that for an observer who is not too close, a group of three isotropic sources very close to each other is always equivalent to the sum of a monopole, dipole, and/or quadrupole source.

That is, show that "tripoles" are not necessary. This need not be a rigorous proof.

15.3 In an anechoic chamber, where reverberation is negligible, the sound intensity 2 m from a violin is measured to be I_{2m}.

a. Moving straight backward away from the violin, how far do you have to go for the intensity to be half this value?

b. How far back would you have to go for the violin to *sound* half as loud?

c. If the electrical signal from the microphone used to measure the intensity is proportional to the acoustic pressure *amplitude*, how far back do you have to go for the signal amplitude from the microphone to be reduced by half?

15.4 If a certain empty room has a reverberation time of T_R, what reverberation time would be expected for an empty room made from the same materials but with all dimensions scaled by a factor s?

15.5 A certain room has a reverberation time of T_R. The room has a highly reflective ceiling and floor, so most of the absorption occurs at the walls. If the ceiling is lowered to one-half of its original height, all other factors remaining the same, what is the new reverberation time?

15.6 Sometimes reverberation is characterized by the time for the sound to drop by 20 or 30 dB, designated by T_{20} and T_{30}, respectively, rather than 60 dB. Show that if the sound decays exponentially, $T_{60} = 2\,T_{30} = 3\,T_{20}$. Use this result to estimate T_{60} for the two situations shown in Figure 15.12.

15.7 When designing a large performance or lecture hall, aside from patron comfort, why might it be better to use upholstered seating rather than unupholstered seating?

15.8 Consider a directional source that combines isotropic and dipole characteristics that results in an intensity pattern that is described by

$$I = \frac{I_0}{4}(1 + \cos\theta)^2,$$

where I_0 is a constant, and θ is the angle away from the forward direction. Plot the directional pattern for this intensity using polar plots having linear and dB scales. See Figure 15.2 for examples of similar plots. Use $I_0 = 1$ when plotting the directional pattern.

15.9 Estimate the acoustic energy density during a loud concert. If the room has a reverberation time of 1 s and a volume of 8,000 m^3, how much acoustic power, in watts, is being produced? Include your assumptions.

15.10 Evaluate Sabine's formula for the reverberation time (Equation [15.22]) for a room that has perfectly absorbing walls, floor, and ceiling. Does it give a reasonable result? If not, why not?

16   Interference, Diffraction, and Diffusion

Reflection and reverberation of sound were treated in Chapter 15 using the ray approximation. While there was some mention of frequency dependence, that discussion put aside, for the moment, the wave nature of sound. In this chapter, interference and diffraction effects that are unique to waves are considered. In fact, the existence of these effects can be used as evidence of the wave nature not just of sound but also of light, radio waves, and even the "particle waves" of quantum mechanics. The chapter concludes with a discussion of acoustic diffusion, an important part of acoustic room treatments. A diffuser is an object designed for this purpose, and those designs often need to take into account the wave nature of sound.

Interference effects are frequency dependent and so are most often presented in the frequency domain, one traveling sine wave at a time. For interference and diffraction effects, it will be most convenient to express results not in terms of frequency but in terms of wavelength. Of course, if the wave speed is known, it is a simple matter to find frequency from wavelength.

Interference

Interference between waves can be observed when two (or more) traveling waves meet. Interference is most important when those waves have identical frequencies. However, for example, two violins playing the same note will not be identical enough. In practice, to truly be identical, the waves can be traced back to a single source. For example, the sounds from a single violin that get to a listener by more than one route, say, directly and via a reflection, would necessarily have identical frequencies. Likewise, two separate speakers driven by a common input signal would produce sounds with identical frequencies.

A more technical description of waves with identical frequencies would be that the two waves should remain **phase coherent**, meaning that their relative phase difference remains constant. The length of time two waves remain phase coherent

Figure 16.1

Illustrating sound traveling by two different routes from a source to a receiver.

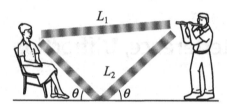

is known as the **coherence time**. To see the effects of interference, the coherence time must be long compared to the period of oscillation. If the wave speed is c and the coherence time is t_c, then the coherence length is ct_c. The **coherence length** should be long compared to a wavelength, and if the waves have traveled different distances, the coherence length must also be longer than the difference in the travel distances.

Consider the situation originally shown in Figure 15.8 but now with the wave properties included, as illustrated, frozen in time, in Figure 16.1. Here, only the part of a sinusoidal traveling wave that travels from the source to the receiver is included. The two portions of the wave shown start in phase at the source, but they travel different distances to the receiver. In this case, the two waves are close to one-half wavelength different at the receiver – one is positive when the other is negative. When the waves meet, the sum will be small. When the waves cancel where they meet, the result is referred to as **destructive interference**. If a different wavelength is used, it is possible that the waves can arrive in phase and add together. When the waves add to produce a larger amplitude, the result is called **constructive interference**.

Mathematics can be used to show what happens. The combined acoustic pressure for two sinusoidal traveling waves, labeled 1 and 2, that start identically and then meet after traveling different distances L_1 and L_2, respectively, will be

$$p_{tot} = p_1 \sin(kL_1 - \omega t) + p_2 \sin(kL_2 - \omega t), \tag{16.1}$$

where $k = 2\pi/\lambda$. The sound intensity, I, is proportional to the total acoustic pressure squared, averaged over time, or, using trigonometric identities (see Appendix A) and averaging over time,

$$I \propto \bar{p}_{tot}^2 = \left[p_1^2 + p_1^2 + 2p_1p_2 \cos\left(k(L_2 - L_1)\right) \right]. \tag{16.2}$$

If the two waves do not have identical wavelengths, as described earlier, the cross term involving the cosine will have a periodic time dependence and will average to zero over time. That cross term describes the interference effects – without it, the behavior is simply the sum of each wave's separate intensity. Note that only the *difference* in the travel distance, or "path length" $(L_2 - L_1)$, is important.

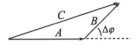

Figure 16.2 The result of interference between sound waves can be understood by considering a geometry problem.

While Equation (16.2) was not the result of a geometry problem, it can be understood by considering a triangle and the law of cosines, a generalization of the Pythagorean theorem (see Appendix A). Refer to Figure 16.2 and apply the law to get

$$C^2 = A^2 + B^2 + 2AB \cos(\Delta\varphi). \tag{16.3}$$

Comparing Equations (16.2) and (16.3), the "angle" describing the phase difference between the two waves is $\Delta\varphi = k(L_2 - L_1)$. When $\Delta\varphi = 0$, the waves add to the full extent possible, and $|p_{tot}| = |p_1 + p_2|$. When $\Delta\varphi = \pi$, the waves subtract to get the smallest amplitude possible, $|p_{tot}| = |p_2 - p_1|$. The cosine function is periodic, so adding any integer multiple of 2π (360°) to $\Delta\varphi$ does not change the result. Hence, the condition for the maxima and minima of the intensity will be

Maxima: $k(L_2 - L_1) = 2\pi n$,

Minima: $k(L_2 - L_1) = 2\pi\left(n + \dfrac{1}{2}\right), n = 0, \pm 1, \pm 2, \ldots.$ \hfill (16.4)

Rewriting this result in terms of half wavelengths,

Maxima: $(L_2 - L_1) = 2n\left(\dfrac{\lambda}{2}\right)$,

Minima: $(L_2 - L_1) = (2n + 1)\left(\dfrac{\lambda}{2}\right), n = 0, \pm 1, \pm 2, \ldots.$ \hfill (16.5)

In words, this most important result is:

The interference maxima occur when the path difference is an even number of half wavelengths, and the minima occur for an odd number of half wavelengths.

Example 16.1

A sinusoidal signal is played through a speaker and is heard by a listener. Sound travels directly to the listener and is also reflected off the floor. For the geometry shown in Figure 16.3, what is the lowest frequency that has maximal *destructive* interference?

Figure 16.3
Geometry for
interference in
Example 16.1.

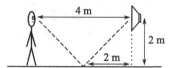

For maximal destructive interference, the path *difference* should be an odd number of half wavelengths. The lowest frequency corresponds to the largest wavelength, or $|L_2 - L_1| = \lambda/2$ (i.e., use $n = 1$ in Equation [16.5] to get the longest wavelength). Here,

$$L_1 = 4\,\text{m}, \quad L_2 = 2 \times \sqrt{2^2 + 2^2}\,\text{m} = 5.657\,\text{m},$$

where the Pythagorean theorem was used to find L_2 (the dashed path). Thus,

$$(5.657 - 4)\,\text{m} = \lambda/2 \rightarrow \lambda = 3.314\,\text{m},$$
$$c = \lambda f \rightarrow f = c/\lambda = (340\,\text{m/s})/(3.314\,\text{m}) = 103\,\text{Hz}.$$

Destructive interference will also be expected at three, five, seven, and so on times this frequency. A common error to avoid is the use of an odd number of wavelengths rather than an odd number of *half* wavelengths.

For multiple point sources, the wave pattern can be computed by adding the waves from each source. Image sources, such as those shown in Chapter 15, can often be used for reflections. Figure 16.4 illustrates a point source above a floor (or next to a wall). Figure 16.4a shows the wave pattern when the reflections are negligible. Figure 16.4b illustrates the wave pattern, computed using an image source placed an equal

(a)   (b)

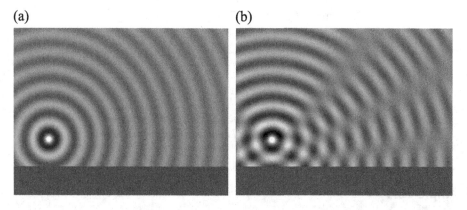

Figure 16.4 Waves from a point source above a floor when (a) the reflections are negligible and (b) there is perfect reflection. The waves in (b) can be computed by adding a second identical source, an image source, an equal distance below the floor. The pattern depends on the height of the source above the floor compared to the wavelength.

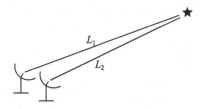

Figure 16.5 A path *difference* of just a few centimeters can have a large effect on the combined radio signals from a distant star.

distance below the floor, corresponding to perfect reflection. For the wavelength and geometry shown, Figure 16.4b has three lines of zero intensity going up to the right and similarly going to the left. Additional sources can be added, as necessary, to account for additional surfaces and multiple reflections, as was illustrated for the ray approximation in Figure 15.9. Of course, the wave pattern will change significantly with changes in frequency and the position of the source.

Wave interference, such as that shown in Figure 16.4, is often demonstrated using a ripple tank or wave tank. A ripple tank uses a pan of water, often illuminated from below, and various vibrators to create surface waves on the water. Edge effects, and others, can make these demonstrations somewhat problematic. Idealized computer simulations of these demonstrations are readily available that avoid these issues.

The idea behind interference can be used in reverse – that is, if the roles of source and receiver can be swapped. One application is in some radio telescopes. Radio waves from a single, distant source are received in multiple antennas (Figure 16.5) and then combined electrically. If the received signals are added, the result is a maximum if the path difference is an even number of half wavelengths, and a minimum if an odd number. One common radio wave used for radio telescopes is the "hydrogen 21-cm line," with a wavelength of 21 cm. Thus, a difference in the travel distance from a distant star or galaxy of just a few centimeters can make a big difference. While the distance to the star or galaxy cannot be determined to an accuracy of a few centimeters – not even close – the *difference* in the travel distances is what matters for the combined signal, and that is relatively easily determined.

Using two detectors for a single source can provide an accurate determination of the direction of the source with knowledge of the difference in path lengths. This is, in fact, an important result used by humans, who have two ears, to locate the source of sounds.

Diffraction

Diffraction is closely related to interference and refers to situations that can be thought of as interference from an infinite number of adjacent, identical (phase-coherent) sources. An important tool that can be used to understand

and calculate diffraction effects is known as *Huygens's principle*.[1] The principle says that a traveling wave, going forward, can be considered to be due to a continuous line of point sources along a "wave front." Imagine waves on water, frozen in time. If a line is drawn, for example, following the peak of a wave, or the trough between peaks, that line would describe a wave front. For three-dimensional problems, the wave front is described by a plane. The waves at points located on such a wave front will be in phase with each other.

Consider a plane wave that travels through an open window.[2] The signal across the opening of the window will be phase coherent. On the other side of the window, and as a first approximation, the window can be treated as a single point source. A better approximation would be to divide the window in half and treat each half as a point source with half the intensity. The process can be continued until the window is divided into an infinite number of infinitesimal windows, equivalent to a continuous array of sources. For numerical computations, the window simply needs to be divided into a very large number of smaller windows. This limiting process is illustrated in Figure 16.6.

The pattern depends strongly on the wavelength when the width of the window is comparable to a wavelength. This is illustrated in Figure 16.7. Figures 16.7d–f show the time-averaged intensity, using a gray scale based on the log of the intensity. When observed a distance from the window that is large compared to the size of the window, the intensity pattern is given by

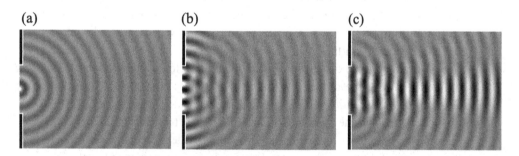

(a)   (b)   (c)

Figure 16.6 Approximations to the continuous sinusoidal traveling wave passing through a window using (a) 1, (b) 4, and (c) 1,024 identical point sources along a line. The wave is incident from the left.

---

[1] Also known as the *Huygens–Fresnel principle*.

[2] When used for diffraction of light, which has a much smaller wavelength than typical sounds, the size of windows that show significant effects are quite small, and the term *slit* is often used rather than *window*.

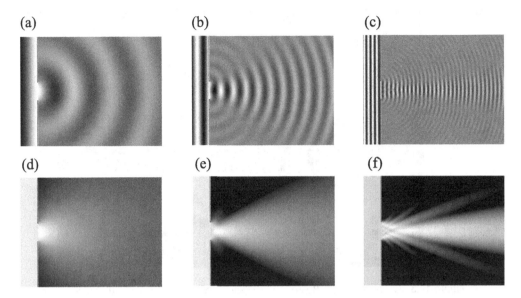

Figure 16.7 A sinusoidal plane wave is incident on a window from the left. Shown here are wave patterns for three different wavelengths: (a) about twice the size of the window, (b) comparable to the window, and (c) about half the size of the window. The figures in the lower row, (d) to (f), illustrate the corresponding time-averaged intensities.

$$I = I_0 \left(\frac{\sin \left(\pi d \sin(\theta)/\lambda \right)}{\pi d \sin(\theta)/\lambda} \right)^2, \tag{16.6}$$

where d is the width of the window, θ is the angle from what is expected from the ray approximation (i.e., straight), λ is wavelength, and I_0 is the maximum intensity. Most of the intensity will be within the range where $|\sin \theta| < \lambda/d$ or 1, whichever is smaller.

Diffusion

As mentioned in Chapter 15, a simple rectangular room can act as an acoustic house of mirrors. Alternatively, it can be thought of as an acoustic resonant cavity. In either case, the sound heard by a listener can be strongly dependent on the locations of both the listener and the source(s). **Diffusion** helps to fill the room with sound more uniformly. If a rectangular room is a house of mirrors, adding diffusion is like painting those mirrors white. The sound energy stays in the room, but it is spread more evenly in all directions. In many environments, particularly concert halls, people report that they prefer the diffused sound.

A room is said to have good **ensemble** if multiple performers can hear each other well, and a room has good **blend** if the sound is evenly spread from different

Figure 16.8 Symphony Hall, Boston, has a "shoebox" design incorporating decorations on the walls and ceiling that can provide diffusion. Removable curtains, which can provide additional absorption, can also be seen in this photo. (Courtesy of Len Levasseur.)

sources to all the listeners. Diffusers, along with other design considerations, are an important tool used to control these characteristics.

One way to cause diffusion is to include significant variations in otherwise flat walls. Ornamental decoration often serves that purpose. An example is Symphony Hall in Boston, shown in Figure 16.8, which is a "shoebox" design, well known for its quality acoustics. Notice, in particular, the decorative ceiling. Diffusion based on ornamentation is often ad hoc.

Other methods to aid in diffusion include the use of curved walls and the use of pillars and similar structures, thus distorting the "mirrors." This type of diffusion can be understood using the ray model discussed in Chapter 15.

There are also methods to design and engineer diffusion, which may be considered by some to be less artistic but that are easily defined and understood mathematically. Such designs include the variable-depth diffusers.

Variable-Depth Diffusers

A wall or ceiling with recesses, or indentations, can be used to provide diffusion. Some designs make use of periodic arrays of such recesses. Figure 16.9 illustrates three simple methods to control the reflection coefficient using a recess. A hard wall simply reflects (essentially) all of the sound. A deep recess filled with sound-absorbing material reflects very little. A recess that is a quarter wavelength will reflect most of the sound off the back of the recess, so when the sound returns to the entry point, the wave has traveled an extra half wavelength. The latter will

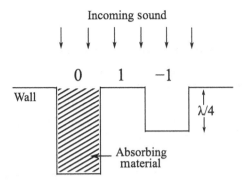

Figure 16.9 A simple scheme to produce reflection with (approximate) amplitudes of 0, 1, and –1.

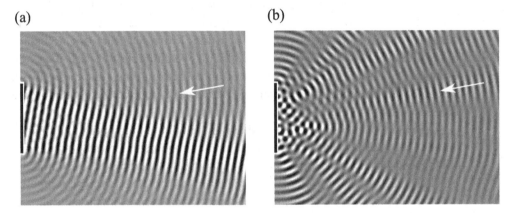

Figure 16.10 The resulting outgoing waves following the reflection of an incoming traveling wave, moving in the direction shown by the arrow, off a panel. A uniform panel, shown on the left, reflects with high directionality, like a mirror. On the right, the panel is divided into 16 segments, half of which are randomly chosen to reflect with a minus sign, yielding a large amount of scattering.

produce an effect similar to a negative reflection coefficient, at least for sound with a frequency near the design frequency. Other depths can be used to get any phase shift that is desired.

One simple diffuser design uses periodic recesses with reflection coefficients randomly chosen to be ± 1. An example of what happens is illustrated in Figure 16.10. When a uniform panel is used, the reflected waves from each part of the panel add constructively to create the reflected wave, and destructive interference is present in other directions, resulting in reduced intensity. The net result is a mirror-like reflection. By changing some sections to reflect with a minus sign – a 180° phase change – some of what was constructive interference is now destructive interference, and vice versa, and the simple mirror-like reflection is destroyed.

Another scheme, which is less dependent on frequency, uses absorbing ("0") and reflecting ("1") recesses. In that case, half the surface will be absorbing and half reflecting, so this scheme also functions as an absorber. In this case, the amplitudes of the reflected waves are randomly either 0 or 1, which also destroys what would have been the constructive and destructive interferences that give rise to the mirror-like reflection.

One way to generate a pseudo-random[3] surface is to use a **maximal-length sequence**. One such implementation, based on the number 7, is shown in Figure 16.11 for a diffuser based on 0 and 1. The first seven recesses are chosen randomly (say, with a flip of a coin, although all seven must not be the same), and then each additional recess is determined by the recesses 6 and 7 slots before it. If those recesses are the same, the next is a "0," and if different, a "1" (or vice versa). In logic operations, the operation "0 if the same, 1 if different" is known as the **exclusive or** (XOR) operation. More generally, the necessary operation used for these sequences is "0 if an even number of 1's, 1, if an odd number." Such an operation is referred to as determining the *parity*. For a design based on reflection coefficients that are ± 1, replace the 0's (absorbing recesses) with -1's (quarter-wavelength recesses).

A maximal-length sequence based on an integer n will repeat after $2^n - 1$ values. The recesses that can be used for the comparison, referred to as *taps*, will depend on the choice of n. For example, if $n = 9$, compare the 9th and 5th previous recesses, and for $n = 10$, use the 10th and 7th. There is always an even number of taps, and one of them will be the nth. Tables showing possible choices are readily available.

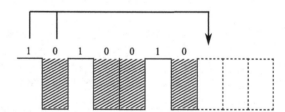

Figure 16.11 An illustration of one way to generate a diffuser/absorber with a maximal-length sequence based on the number 7. The next recess is determined by whether or not the sixth and seventh previous recesses are, or are not, the same.

---

[3] *Pseudo-random* refers to the fact that while the sequence of values is deterministic (i.e., not random), it has many of the mathematical properties that are associated with those of a random process.

Number-Theory Diffusers

Several variable-depth diffuser designs have been proposed that use results from number theory. These designs use multiple-depth recesses that each modify the phase of the reflected signal by a different amount. Once again, the goal is to scramble the relative phases to ruin the mirror-like reflection. Two designs, in particular, are those based on the **quadratic residue** and the **primitive root** sequences. Both rely on the "mod" ("modulo") function and **modular arithmetic**. Variable-depth diffusers based on these ideas are often referred to as **Schroeder diffusers**, after the German physicist Manfred Schroeder (1926–2009).

Modular arithmetic can be described by considering the face of an analog clock and is sometimes referred to as **clock arithmetic**. The clock face is treated as a number line, although with a limited number of values. To add two numbers, go clockwise, and to subtract, go counterclockwise. Figure 16.12 illustrates the clock face for "mod 12." This clock face is the same as a normal (12-hour) clock, except "12" is replaced with "0." To add 5 to 10 mod 12, start at 10 and move 5 places clockwise. The result is 3. Values mod m are always in the range from 0 to $m - 1$.

Many calculators have a "mod" or "rem" (remainder) button that can be used for these calculations. For positive values, the remainder returns the same value as the mod function. The remainder is what is left over when an integer is divided evenly into integer parts. For example, if 16 items are to be divided into equal-sized groups of 3, there will be 1 left over – the remainder is 1. Hence, 16 mod 3 = 1. If 25 items are put into groups of 11, there will be two groups of 11, with 3 left over, so 25 mod 11 = 3. If neither the mod nor the rem calculator button is available, the value can be found using a three-step process: division, subtracting the integer part of the result, then re-multiplication.

Example 16.2

To find the value of 59 mod 7 with a calculator without using a "mod" or "rem" button:

Divide 59 by 7: $59/7 = 8.42857$.
Subtract the value on the left of the decimal point to get 0.42857.
Multiply that result by 7: $7 \times 0.42857 = 3$.

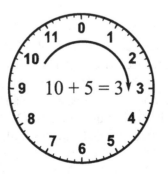

Figure 16.12
An illustration of the addition of 5 to 10 using mod 12 arithmetic.

Thus, 59 mod 7 = 3. The result will always be a nonnegative integer smaller than the "mod" value – in this case, the result should be in the range 0 to 6. If the result is not such an integer but is very close (e.g., 2.99999), then there was a rounding error in the calculation, and the result should be rounded to the nearest integer. If the result is not an integer (or very close to being an integer) or is outside of that range, then a mistake was made.

The two sequences often used to determine recess depths for variable-depth diffusers are the

$$\text{Quadratic residue sequence: } d_n = D(n^2 \bmod m), \text{ and the}$$
$$\text{Primitive root sequence: } d_n = D(J^n \bmod m),$$

(16.7)

where m is a prime number, and J is an appropriately chosen integer such that the primitive root sequence produces all values from 1 to $m - 1$. Not all combinations of J and m will work. Here, n is a sequence of integers, starting from any integer value; d_n represents the depth at the corresponding position n; and D is an overall scale factor with units of length (the number-theory sequences are the portion in parentheses). Both sequences will repeat.

Example 16.3

Write the quadratic and primitive root sequences for $m = 7$ and $J = 3$.
Starting with $n = 1$, the sequences are:

Quadratic residue:

$$d_1 = (1^2 \bmod 7) = 1, \ d_2 = (2^2 \bmod 7) = 4, \ d_3 = (3^2 \bmod 7) = 2,$$
$$d_4 = 2, \ d_5 = 4, \ d_6 = 1, \ d_7 = 0, \ d_8 = 1, \ d_9 = 4, \ \ldots$$

Primitive root:

$$d_1 = 3^1 \bmod 7 = 3, \ d_2 = 3^2 \bmod 7 = 2, \ d_3 = 3^3 \bmod 7 = 6,$$
$$d_4 = 3^4 \bmod 7 = 4, \ d_5 = 5, \ d_6 = 1, \ d_7 = 3, \ d_8 = 2, \ d_9 = 6, \ \ldots$$

Note that for $m = 7$, only $J = 3$ and $J = 5$ will work. The sequence for $J = 5$ is identical to the sequence for $J = 3$, except in reverse order.

The quadratic residue sequence will always be symmetric about zero. The primitive root sequences often occur in mirror-image pairs – that is, there are pairs of values for J that yield the same sequence, although in reverse order, as mentioned in Example 16.3. An example of an acoustic tile based on the primitive root sequence is illustrated in Figure 16.13. Two-dimensional versions of these diffusers have also been studied and are available commercially.

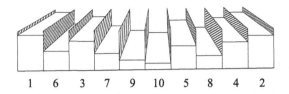

Figure 16.13 An example, shown in perspective, of a section of a diffuser "tile" based on a primitive root sequence with $m = 11$, $J = 6$.

In addition to its use for acoustic diffuser design, modular arithmetic is a very important part of algorithms used for the encryption of digital communications. The ability of modular arithmetic to produce a result that looks like random phase reflections for sounds is also of use in creating the random appearance of encrypted digital communications.

Active Diffusers

Arrays of microphones and speakers can be placed over a surface to produce **active diffusion** (or more generally, virtually any reflective property). The signals from the array of microphones are processed and then sent through amplifiers to the array of speakers to produce, or at least approximate, any desired "reflected signal." Noise canceling is a special-case use of this idea. Here, the word *active* is used to mean that external power must be supplied. In contrast, the previously described schemes – using ornamentation, curved walls, or an array of recesses – would be referred to as *passive* techniques.

Summary

When two or more waves having the same wavelength intersect, the waves can add or subtract, resulting in a significant effect on the observed intensity. For two waves that differ by a half wavelength, or any odd multiple of a half wavelength, the intensity will be a minimum. When the two waves differ by an integer number of wavelengths or, equivalently, an even number of half wavelengths, the intensity will be a maximum. When there is a relatively small number of sources, this is usually referred to as *interference*. *Constructive interference* refers to the case when the intensity is increased, and *destructive interference* to the case when the intensity is reduced.

The intensity from finite-sized sources, such as what results when waves pass through a slit or window, results from the interference of a continuum of closely spaced sources and is referred to as *diffraction*.

Both interference and diffraction are frequency (wavelength) dependent and can result in an intensity that is highly directional and may have a complicated intensity pattern.

In some cases, it is desirable to spread acoustic energy more evenly throughout a space. Diffusors are objects used to accomplish this. Some diffuser designs use randomness to accomplish this, whereas others may use mathematical properties associated with certain mathematical sequences from number theory. In each case, the conditions that result in the minima and maxima associated with interference and diffraction are, on average, removed.

ADDITIONAL READING

Cox, T. J., J. A. S. Angus, and P. D'Antonio. "Ternary and Quadriphase Sequence Diffusers." *Journal of the American Acoustical Society* 119, no. 1 (2006): 310–319.

Cox, T. J., and P. D'Antonio. *Acoustic Absorbers and Diffusers: Theory, Design and Application.* Spon Press, 2004.

Xiao, L., T. J. Cox, and M. R. Avis. "Active Diffusers: Some Prototypes and 2D Measurements." *Journal of Sound and Vibration* 285, no. 1–2 (2005): 321–339.

PROBLEMS

16.1 Estimate the frequency of the sound wave illustrated in Figure 16.1. Approximately what note is being played?

16.2 Generate the maximal-length sequence based on the number 4, where the third and fourth previous values are used for the XOR comparison. Start with four values (not all equal) determined by a flip of a coin and show that the sequence repeats after $2^4 - 1 = 15$ values.

16.3 A sound is generated by a speaker 3 m from a listener, as shown in Figure P16.3. What frequencies will have maximal destructive interference, due to reflections off the floor, at the listener?

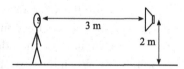

Figure P16.3

16.4 Provide a convincing argument that for integers k and n, $n > 1$,

$$k^n \bmod 7 = \left(k \times \left[k^{n-1} \bmod 7\right]\right) \bmod 7.$$

16.5 Write out one repeat for the primitive root sequence

$$d_n = 6^n \bmod 11.$$

16.6 Write out one repeat for the quadratic residue sequence

$$d_n = n^2 \bmod 11.$$

16.7 Can you find an integer value, m, where $(m \times 13) \bmod 7 = 1$? If so, then m is a multiplicative inverse of 13 for mod 7 arithmetic. Is there more than one such multiplicative inverse? Do all integer values have a multiplicative inverse?

16.8 If a plane-wave source of sound with a wavelength of 10 cm strikes a window that is 1 m square, the "beam of sound" that exits from the window will be 1 m square initially, and it will spread out with distance due to diffraction effects. Estimate the width of that "beam of sound" 10 m from the window.

16.9 How is the musical scale related to modular arithmetic? Can you compose a tune based on one of the sequences proposed for diffusers?

17 Faraday's Laws of Induction

Compared to the behavior of other everyday objects, the behavior of magnets is really quite strange. The odd behavior is easily observed simply by playing with a few small magnets. Those who have not done so recently are encouraged to temporarily put this text aside, find some magnets, and observe how they interact with each other. Sometimes the magnets attract, sometimes they repel, and they do so at a distance. That is, they need not be in contact for a force to be observed. There is some sort of mysterious invisible connection between magnets that crosses empty space.

The discovery of magnets goes back at least several thousand years. It is not clear exactly who first discovered them or when. The earliest known magnets are thought to have been lodestones, a naturally occurring magnetized piece of the mineral magnetite. Magnetite is a mineral form of an oxide of iron (specifically, Fe_3O_4), one common type of iron ore. Most magnetite is not naturally magnetized but will be attracted to a magnet. Compared to man-made magnets, especially the "rare-earth" magnets first developed in the 1970s, lodestones are relatively weak magnets. The difference between materials that are magnetic, meaning they are attracted to a magnet, and those that are magnetized, those that *are* a magnet, will certainly need to be addressed.

Magnets and, more generally, the "electromagnetic properties" associated with magnetism are essential for the operation of many everyday devices, including some used for music, such as speakers, some microphones, and electric guitar pickups. How magnetic properties are used to make such devices is the subject of this chapter. There is a lot of basic physics behind these uses. To start, some discussion of magnets and magnetism is in order.

Properties of Magnets

If a typical magnet is used to pick up small magnetic objects, such as paper clips or iron filings, certain regions of the magnet will exhibit the strongest attraction.

These locations are referred to as the **poles** of the magnet. Due to the way they usually become magnetized, most commercially produced magnets have two such poles.

If the shape of the magnet allows, a magnet with two poles can be hung by a thread in such a way that a line through the two poles is horizontal to the ground. If that magnet is far enough away from other magnetic materials, such as bookcases, desks, or building construction materials made of steel, then one of the magnet's poles will tend to be attracted toward Earth's north and the other toward the south. The magnet's pole that is pulled toward the north is labeled the *north pole* of the magnet, and the pole that is pulled to the south is called the *south pole*. For brevity, these are often labeled with an N or S as appropriate. This tendency to align north–south is, of course, what makes a magnetic compass useful for navigation.

There are only these two types of poles. That is, there is no west or east pole, for example. If there is a pole, it is either a north or a south pole.

Once the poles are labeled, observation of the interaction between two simple magnets (Figure 17.1) will show that **opposites attract** – that is, the north pole of one is attracted to the south pole of the other, and vice versa. In addition, **likes repel**. If the two north poles are brought together, there is a repulsive force. Similarly, the two south poles repel one another. The fact that a simple magnet interacts with Earth in such a way as to point in a specific direction suggests that Earth also acts as a magnet.[1]

It is not possible to make a magnet with just one pole, although it is possible to make one with more than two. When there are more than two, each can still be labeled with an N or S by checking for attraction to or repulsion from a small test magnet. In that case, the total strength[2] of all of the N poles will be exactly equal to the total strength of all the S poles. That is, any magnet will have just as much "northness" as "southness." Indeed, if you take a magnet

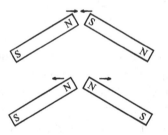

Figure 17.1 For magnets, such as the bar magnets shown, opposite types of poles attract, and the same type will repel.

[1] Note that *north magnetic pole* is the name given to Earth's magnetic pole that happens to be near the geographic north pole of Earth. Since the north pole of a magnet is attracted to that pole, the north magnetic pole is, magnetically, actually the south pole of Earth's magnet – that is, Earth's north magnetic pole should be labeled with an S.

[2] Unfortunately, it is not possible to rigorously define *strength* at this point, other than to simply appeal to one's intuition or to a simple task, such as determining how many paper clips it can hold.

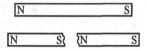

Figure 17.2 If the top magnet is broken in two, the result is two magnets, each with a north and south pole.

and break it in half in an attempt to separate the north and south poles, you end up with two magnets (Figure 17.2), each with just as much "northness" as "southness."

The Magnetic Field

When there is a force experienced through empty space, a phenomenon some-times referred to as *action at a distance*, it is common to describe the situation using fields. For magnets, the field is, understandably, called the **magnetic field**. The forces on a particular magnet are then described as being the result of the presence of the magnetic field caused by one or more other magnets. A convenience is that knowledge of the details of the source(s) for the field is unnecessary. Any set of sources that produce the same field will produce the same result for anything that is within that field.

The magnetic field at any point in space will have a magnitude and a direction. For a magnetic field, the direction is defined to be the direction that a magnetic compass needle would point if it were present. That is, the direction of the field is the same as the direction that the compass reports as north.

A uniform magnetic field is one where the strength and direction are the same everywhere. A magnet placed in a perfectly uniform field will experience no net force, but it will experience a torque – that is, a tendency to rotate. The field will pull the N side(s) of the magnet in one direction and the S side(s) in the opposite direction. If the field is uniform, the pulling and pushing forces will be equal in strength, and only a rotation remains. The familiar pull or push of one magnet in the presence of another arises only when the field is nonuniform. In general, it is not at all a simple matter to compute the net force on a magnet placed in a nonuniform field, and many simplifying assumptions are often necessary. Fortunately, such a computation will not be necessary here.

Magnetic fields are often described graphically using **magnetic field lines**, sometimes also referred to as *flux lines*. To create a magnetic field line that goes through a particular point in space, imagine putting a magnetic compass at that point and observing which direction the compass says is north. Now move a small distance in that direction and again look at the compass. Take each new step in the direction the compass says is north. It is an interesting property of magnetic

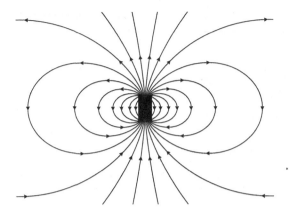

Figure 17.3 A two-dimensional view of the field lines for a uniformly magnetized cylindrical magnet. This magnet is magnetized parallel to the cylinder's axis, and north is at the top.

fields that if enough steps are taken (and with enough accuracy), the field line will return to where it started.[3] Thus, a magnetic field line will be a closed curve. The curve is labeled with arrows to indicate the direction of travel. To illustrate the entire field requires an appropriately chosen set of field lines.

An example that illustrates the magnetic field graphically using magnetic field lines is shown in Figure 17.3. Another interesting property of these field lines is that, at least for time-independent fields, the closed curves will always encircle at least one source of magnetic field – for permanent magnets, every field line will encircle at least a portion of the magnet. Note, however, that the portion of a field line that is inside a permanent magnet is often omitted from these figures.

The strength of the magnetic field is often referred to as the *magnetic induction*. In Standard International (SI) units, the field will be in tesla (T), named for the Serbian American inventor and engineer Nikola Tesla (1856–1943). Another popular unit is the gauss (G), where $1\ G\ =\ 10^{-4}\ T$, named for the German mathematician Carl Friedrich Gauss (1777–1855). In terms of field lines, the strength is proportional to the number of field lines that pass through a (small) window of area, A, divided by A. Hence, the strength is also sometimes called the *magnetic flux density*. Since a window can have different orientations, the strength is directional. In this case, the "direction of the area" of the window is taken to be perpendicular to the plane of the window. So, for example, the magnetic field strength in the vertical direction is found using a horizontal window.

The magnitude of the field corresponds to the maximum strength obtained when the orientation of the window is such that the field lines themselves are perpendicular to the plane of the window. Some examples illustrating typical magnitudes of magnetic field strengths are shown in Table 17.1.

---

[3] There usually will be a few special lines where you might have to travel infinitely far before you return. But for most lines, the distance is finite.

Table 17.1 Some representative magnetic field magnitudes.

	Strength	
	Tesla (T)	Gauss (G)
Smallest detectable	10^{-15}	10^{-11}
Of geophysical interest	10^{-9} ("1 gamma")	10^{-5}
Typical Earth's field at surface	0.5×10^{-4}	0.5
Close to a "refrigerator magnet"	0.03	$3 \times 10^{+2}$
Inside typical human magnetic resonance imaging (MRI) machine	2	$2 \times 10^{+4}$
Very high-field research magnet (continuous)	45	$45 \times 10^{+4}$
Highest man-made fields (short duration)	2,800	$2.8 \times 10^{+7}$
Some stars	Up to 10^{+11}	10^{+15}

In physics, the magnetic field is almost always signified using an uppercase B. The story told is that the choice to use B was made by the Scottish scientist James Clerk Maxwell (1831–1879), who systematically started labeling all quantities dealing with electricity and magnetism by starting with the letter A and continuing through the alphabet. The magnetic field was second on his list. Some of his choices have stuck, and some have not. The use of B for magnetic fields is one that stuck.

Magnetic Materials

Most materials are considered nonmagnetic. Glass, wood, copper, aluminum, brass, water, and ice are all examples of nonmagnetic materials. While there are many magnetic materials, most everyday examples will involve iron, nickel, or materials containing a significant amount of iron and/or nickel. However, the presence of those elements does not guarantee that the material will be magnetic. For example, steels are mostly iron, and although most steels are magnetic, most types of stainless steel are not. In fact, one way to distinguish stainless steel from other steels is to check for attraction to a magnet.

Magnets themselves must be made from magnetic materials, although not all magnetic materials can be made into usable magnets. However, all magnetic materials will act like a magnet when in the presence of a magnet. The magnet is

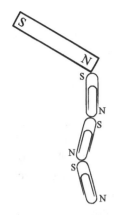

said to induce magnetism in the magnetic material, making it behave like a magnet. This effect can be simply demonstrated by picking up a steel paper clip with a strong magnet, then picking up a second paper clip using the first paper clip, and so on (Figure 17.4). Once the magnet is removed, the effect may last a short while, but after a short time, the paper clips will revert back to their unmagnetized state.

Figure 17.4 Paper clips are not normally magnetized, but in the presence of another magnet, magnetism is induced.

A magnetic material that is magnetized and stays magnetized without help from another magnet is referred to as a **permanent magnet**. That helps distinguish it from, for example, magnetic materials that have induced magnetism.

The magnetic properties of magnetic materials are temperature dependent. If the material gets hot enough, the material will become nonmagnetic. In the same way that ice is a low-temperature phase of water, being magnetic is a low-temperature phase of magnetic materials. *Hot* and *cold* are relative words, and so it is necessary to ask, hot compared to what? The answer depends on the particular material. For iron, the transition occurs at 770°C. At that temperature, the iron glows a reddish orange. On the other hand, the element gadolinium has a transition temperature of 20°C, very close to room temperature. Neodymium dihydride (NdH_2) needs to be below about –267°C, or just 6°C above absolute zero, to be magnetic. That is, for NdH_2, anything above –267°C is already too hot for it to be magnetic.

With rare exceptions, the magnetic phase transition is "second order." A first-order phase transition is one where you either have one phase or another. The familiar transitions of ice into water and of water into steam are examples of first-order phase transitions. For a second-order transition, there is, by some measure, a gradual change from one state to the other. As an analogy, a first-order transition is like an electric light with an on/off switch. It is either fully on or off. A second-order transition would be like a light that can be turned on and off gradually with a dimmer switch.

A magnetic material will contain atoms that can, somewhat crudely, be referred to as *magnetic atoms*. Such a magnetic atom has an electron configuration where a small subset of its electrons can work together to create an atomic-sized magnet. The strength of that atomic magnet is referred to as its *magnetic moment*. Whether such a moment exists will depend on the type of atom and on the environment surrounding the atom. There are very few types of atoms that can be used. Atoms that are good candidates for becoming a magnetic atom include some of the transition metals (Cr, Mn, Fe, Co, and Ni) and some of the so-called "rare-earth" elements (Eu, Gd, Tb, Dy, Ho, Er, Tm). Other atoms near these in the periodic table may also be magnetic in some circumstances.

Once there is a magnetic atom, its magnetic moment can interact with that of a neighboring magnetic atom. Based on observation of everyday magnets, it would be expected that the interaction between atoms would result in opposites attracting, so the magnetic moment of neighboring atoms would be pointed in opposite directions. However, the situation at the atomic level is more complicated. The direct magnetic interaction between magnetic atoms can be smaller than an indirect interaction involving other electrons in the system. That is, magnetic atom 1 interacts with the electrons in its vicinity, and magnetic atom 2 interacts with the electrons in its vicinity, and if those are the same electrons, there is an effective interaction between atom 1 and atom 2. This indirect interaction can sometimes lead to an effective "likes attract" magnetic interaction between the atoms. Thus, in general, the interaction between these neighboring atomic-sized magnets can be "opposites attract" or "likes attract" or maybe even something in between, depending on the balance between the different types of interactions.

The interaction between the atoms can give rise to an ordered state. This is illustrated schematically for two situations in Figure 17.5. When the atoms line up in the same direction, they combine their strength to make a strong magnet. When they line up in opposite directions, the state is still referred to as being magnetic, but outside of the material, the fields add to zero – that is, they cancel – and this will not result in a strong magnet.

When the atomic magnets line up parallel, as in Figure 17.5a, the ordering is called **ferromagnetic**. This is the type of ordering that can occur in iron. When the atomic magnets are anti-parallel, as in Figure 17.5b, it is called **antiferromagnetic**. Hematite, a common oxide of iron (Fe_2O_3), and chromium are examples of antiferromagnetic materials. In addition to these simple orderings, much more complicated ordered arrangements are possible. For example, the element dysprosium will be a "helical antiferromagnet" below 180 K, and it will then become a ferromagnet below 80 K. A helical antiferromagnet is a more complicated arrangement where the atomic-sized magnets order in a helical pattern in such a way that outside the material, their fields cancel. On the atomic scale, some such

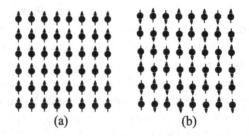

(a)   (b)

Figure 17.5 (a) Parallel and (b) anti-parallel ordered alignments at the atomic scale. The arrows point toward the north pole of each of the atom-sized magnets.

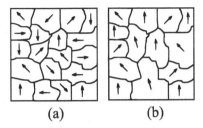

(a)   (b)

Figure 17.6 A two-dimensional schematic representation of magnetic domains that, in (a), tend to cancel each other out, although in (b), there is an alignment of the domains that can result in a strong magnetic field outside the material.

ordering is present in all magnetic materials when they are below their respective magnetic transition temperatures.[4]

The difference between a magnetic material and a magnetized material involves interactions over a longer length scale. In a ferromagnetic material, each region of the sample that has its atoms lined up with each other will act as a larger magnet, and that magnet can interact with other regions. The basic magnetic interaction over longer distances will be that opposites attract, so those regions tend to line up in opposite directions. When that happens, the fields external to the material cancel, and the material is not magnetized. These ordered regions are referred to as **magnetic domains**, and the boundary between domains is called a *domain wall*. To magnetize a material requires not only that it is below the transition temperature but also that the domains are lined up with each other. That is illustrated highly schematically in Figure 17.6. The size of these domains is temperature dependent. As the transition temperature is approached from below, the size of the domains shrinks to zero. Well below the transition temperature, the domains can become quite large, involving billions of atoms or even more. Impurities, or other material imperfections, may ultimately limit the size of the domains.

Domains can become aligned in a number of ways. Perhaps the simplest is to provide a large, externally applied magnetic field. The domains will all tend to point along the field lines of the applied field and, as a consequence, become aligned. In the process, some domains may grow at the expense of others. When this happens, the material has become magnetized and acts like a magnet. In order for the material to become a permanent magnet, the domains must stay aligned after the external field is removed. The ability of a magnetic material to keep its domains aligned is referred to as its *retentivity*. One factor that can help improve the retentivity is the presence of impurities or other imperfections. These imperfections act as obstacles that get in the way when the domains try to reorient

[4] The transition temperature is referred to as the *Curie temperature* for ferromagnetic ordering and as the *Néel temperature* for antiferromagnetic ordering.

their magnetism or change size. Other factors that can influence the orientation of domains include temperature changes and mechanical stress on and vibration of the material.

When domains change, they tend to do so rapidly. This leads to something known as the *Barkhausen effect*. The Barkhausen effect can result in extra electrical noise for devices that use magnetic materials, including some common electronic components. The discovery of this effect in the early 1900s is credited to the German scientist Heinrich Barkhausen (1881–1956). The effect is considered the first direct evidence for what was, at that time, the newly proposed magnetic domain theory.

Nonmagnetic Materials

Most materials encountered on a daily basis are considered nonmagnetic. This is not to say they do not have magnetic properties. The magnetic effects are simply much smaller than those for magnetic materials. There are two broad categories used to describe the continuum of this behavior: **paramagnetic** and **diamagnetic**.

When any nonmagnetic material is placed in a magnetic field, the electrons of the atoms in the material will adjust their orbits slightly. That change in the electron orbit influences the total magnetic field within the material. That is, the total field is the applied field plus the small change due to the electrons' interaction with the field. If the resulting field inside the material is increased, then the material is called *paramagnetic*; if it is decreased, it is *diamagnetic*.

Although the forces are weak, a paramagnetic material will be attracted to a magnet; a diamagnetic material will be repelled. Table 17.2 includes some representative examples of these materials. The size of the change in the field is usually

Table 17.2 Some examples of paramagnetic and diamagnetic materials. These are arranged from top to bottom with the strongest effects seen at the top. These are based on the response for 1 g of material.

Paramagnetic	Diamagnetic
Oxygen (liq.)	Bismuth
Iron oxide (FeO)	Silver
Uranium	Water
Platinum	Graphite
Aluminum	Copper

measured in parts per million (ppm) and, in many cases, is negligible compared to other effects. However, in very strong research magnets, the diamagnetic response from water is sufficient to magnetically levitate a small frog, and appropriately chosen rare-earth magnets can levitate a small piece of graphite. Due to its large paramagnetic effect, even a fairly modest magnet can deflect a stream of liquid oxygen or capture the liquid between the magnet's poles, if the poles are close enough together.

Electromagnets

Another way to make a magnet is to use electric power. Such a magnet is called an **electromagnet**. One advantage of an electromagnet is that its strength can be easily changed by varying the amount of electric power. That proves to be very useful for several music-related applications, most notably for speakers and earphones.

Without going into too much detail at this point, a basic description of the behavior of any electric circuit involves two things: voltage and current. For electric circuits, voltage differences play the role of a driving force, and current is the response to that force. The current is actually the flow of electrical charge, as will be discussed later, but at this point, it can be regarded as the flow of some sort of electric fluid through wires. In most cases, more voltage (difference) gives rise to more current. As an analogy, imagine a stream going down a hill. The current is represented by the amount of water per second that flows past a given point, and the voltage changes between two points are due to the steepness of the hill.

A battery can provide an **electromotive force**, or emf,[5] measured in volts (V), that gives rise to the electrical "push" for the current. If you take a length of wire and connect it across a battery, current will flow through the wire.[6] Electric currents are measured in amperes (A). What is interesting at this point is that if that current-carrying wire is brought near a magnetic compass, the compass needle can be made to deflect from north. This effect was thoroughly investigated in 1820 by the Danish scientist Hans Christian Ørsted (1777–1851), who is usually credited with its discovery. Ørsted used a voltaic pile as a voltage source for his measurements. The voltaic pile was invented in 1800 by the Italian scientist Alessandro Volta (1745–1827) and is an early form of battery.[7] To get a larger magnetic field, the wire can be coiled, as

---

[5] Also written *EMF* and read as separate letters: "ee-em-eff." To add to the confusion, *EMF* may also refer to "electromagnetic fields," which are one possible cause of electromotive forces.

[6] In practice, it is easy to exceed a real battery's capabilities by doing this. If the battery starts getting warm, the wire should be disconnected immediately and replaced with a much longer and/or much thinner wire.

[7] The word *battery* used for such an electrical device is often credited to the American inventor and statesman Benjamin Franklin (1706–1790), who, it is said, used the term in a different context prior to the invention of an electric battery.

Figure 17.7 A simple electromagnet constructed from a battery and a length of wire.

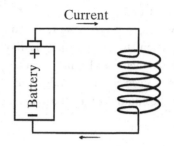

illustrated schematically in Figure 17.7, so that the contributions to the magnetic field from many sections of wire can combine. While any current-carrying wire will produce a magnetic field, an electromagnet is an arrangement specifically designed to concentrate the field.

The magnetic field produced by an electromagnet is indistinguishable from the field produced by a magnetic material. For a simple coil, as illustrated in Figure 17.7, there are north and south poles at the ends, along the axis of the electromagnet. Such a coil is attracted to, and repelled from, other magnets, whether they be magnets made from magnetic materials or other electromagnets, in the same way a permanent magnet would interact. The fact that magnetic fields can be produced using electricity, and not only from magnetic materials, was a first major clue to understanding that there is a very intimate connection between magnetism and electricity.

If a piece of unmagnetized magnetic material is placed in the field of an electromagnet, the field from the electromagnet will induce magnetism in the magnetic material, and the resulting magnetic field can be much larger than that of the electromagnet by itself. Hence, the wires of an electromagnet will often be wound around a magnetic material (e.g., iron or steel) to take advantage of this enhancement.

As mentioned, an electromagnet has an advantage over a permanent magnet in that the strength of the magnetic field can be adjusted by changing the applied electric power. If the battery is reversed, the field reverses – north becomes south, and south becomes north. If the battery voltage is doubled, the magnetic field doubles, and so on. If the electromagnet is supplied with a time-varying electrical signal, a time-varying magnetic field will be produced, and hence time-varying forces will be impressed upon neighboring magnets and magnetic materials.

Most audio speakers and earphones produce sound using a time-varying field from an electromagnet in the presence of another magnet. The audio signal is sent to the electromagnet, and the time-varying force that results is used to push on the air, causing time-dependent pressure variations – that is, sound. An example of a simple speaker showing the principle of operation is shown in Figure 17.8. Such a speaker is easily constructed. A commercial speaker will be engineered to produce sound more efficiently over a wider range of signal frequencies, so the

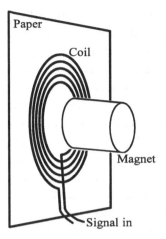

Figure 17.8 A simple speaker uses an electromagnet next to a permanent magnet. The coil of the electromagnet is mounted to a flexible membrane, such as paper, to push on the air more effectively.

geometry may vary considerably; however, the basic operating principle of the device is unchanged. With a careful inspection of a commercial speaker, the magnet, coil, and paper (or equivalent) can be identified.

Other devices that use the interaction between an electromagnet and a magnetic material include motors and solenoids. In this context, solenoids are simple electromechanical devices that convert electric power into a magnetically controlled push force or pull force. Such a force can, for example, be used to open and close the air valves on a pipe organ.

In Reverse?

Many of the effects seen for electricity and magnetism have the remarkable property that they also work, to some extent, in reverse. That is, if putting an electrical signal across a coil creates a magnetic field in the vicinity of that coil, then creating a magnetic field in the vicinity of a coil will create an electric signal. Indeed, it is easy to verify that this happens.

For example, a typical speaker produces sound when an electrical signal is applied to the coil, which then causes a time-dependent force, vibrating the coil, which creates pressure variations in the air. In reverse, pressure variations in the air due to sound cause a force on the speaker, which causes the coil to vibrate, which in turn gives rise to an electric signal. That is, a speaker can also be used as a microphone. This property can be easily verified by connecting a commercially available speaker to the input of an oscilloscope and then speaking directly into the speaker. In fact, some simple intercom systems and two-way radios will use their speakers for both sending and receiving sound. The quality is not particularly good, but they work. Some microphones, such as dynamic and ribbon microphones, discussed later in the chapter, use the same principle, although they have been optimized for use as a microphone.

A simple coil of wire and a magnet, along with a device to measure electrical signals, such as an oscilloscope or sensitive galvanometer, can be used to investigate this phenomenon further. It can be quickly ascertained that moving a magnet near a coil, or moving a coil near a magnet, will produce a signal and that only the relative motion seems to be important. What is also important is that a signal is observed only when there is motion. When there is relative motion, the coil will experience a time-dependent magnetic field from the magnet. Another feature, easily observed, is that the electrical signal reverses when the motion reverses. That is, if, say, a positive signal is observed when a magnet is moving toward a coil, a negative signal will be observed when the magnet is pulled away. The signal also reverses when the magnet is reversed. If a positive signal is observed when the north end of the magnet is brought near a coil, then a negative signal will be observed when the south end is brought near that same coil.

The rules of physics that can be used to quantitatively describe this type of electrical signal generation – exactly what happens when a coil experiences a changing magnetic field – are somewhat difficult to explain and use. Much of the initial understanding of the problem came from experiments by the English scientist Michael Faraday (1791–1867). In many ways, Faraday picked up where Ørsted left off. Among other important discoveries, Faraday is also credited with inventing the first electric motor. His law that governs electric signal generation is referred to as **Faraday's law of induction**. Some assumptions about the physical situation must be satisfied to use Faraday's law, although those assumptions are not always clearly stated along with the law. The more correct form is a generalization given later by Maxwell – one of the so-called "Maxwell equations."[8] Maxwell's version is significantly more abstract, relies on mathematics well beyond what is assumed here, and is often quite difficult to apply in practice. Here, the description will be more qualitative.

Qualitatively, the rule is simple: nature will, as best as it can, create electric currents to oppose *changes* in magnetic fields. Here, "as best it can" refers to the fact that there needs to be a path for the current, such as the presence of an appropriately positioned loop of wire or piece of metal, for this opposition to change to occur.

Consider a magnet with its north pole moving toward a simple coil of wire (Figure 17.9). Nature will try to oppose the change by creating a current in the coil. That current in the coil produces a magnetic field that, in turn, interacts with the magnet. The motion will be opposed if the coil's magnetic field has its north pole pointed toward the magnet since like poles repel. On the other hand, if the magnet is moving away from the coil, the current in the coil will create a south pole toward the magnet, trying to attract the magnet to oppose the change. That is, the force is always in a direction appropriate for reducing the magnet's motion.

---

[8] Maxwell's equations were many and were extremely complicated until they were condensed by Oliver Heaviside (1850–1925) into the form we usually see today. Heaviside was a largely self-taught British scientist and mathematician.

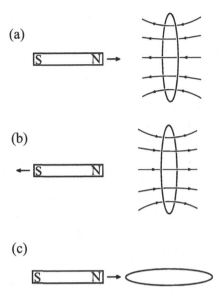

Figure 17.9 A magnet (a) approaching or (b) receding along the axis of a coil will induce a current, creating a magnetic field that opposes the motion. If the magnet approaches (c) or recedes (*not shown*) in the plane of the coil, there is no current.

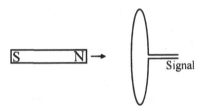

Figure 17.10 If the current path is broken, there will be no current, but the emf will still be present and can be measured with a voltmeter.

Since a coil produces a magnetic field with north and south poles along its own axis, no force on the magnet can be generated by the coil if a magnet is brought in from the side (Figure 17.9c). In that case, the coil cannot have a current that attracts or repels the magnet. Nature has no possibility to oppose the motion, and hence no current is created.

If the coil is broken at some point, the path for the current is broken, and nature cannot oppose the changing magnetic field. However, the emf, which would have created the current if the coil were whole, is still there and can be measured using a voltmeter (Figure 17.10). This is the origin of the signal observed when using a speaker in reverse.

The basic design of a dynamic microphone is the same as that of a speaker used in reverse, but it is optimized for reception. A ribbon microphone uses the same physics but with a single wide, flat, and very thin "wire" placed near magnets. The sound pressure moves the ribbon within the magnetic field, creating an electrical signal, as described by Faraday's law. While not as common, a speaker can be created that is essentially a ribbon microphone used in reverse – current through a single fat wire next to a magnet generates a force, causing the ribbon, and hence the air next to it, to move.

Quantitatively, the emf and/or induced current due to Faraday's law of induction will be proportionately larger for a stronger magnet or for a magnet (or coil) moving faster. Also, when comparing two coils, the coil that, when used as an electromagnet, creates the larger magnetic force on the magnet will have the larger effect when used in reverse.

The opposition to changing magnetic fields can be easily seen by observing the motion of a strong cylindrical magnet as it falls through a vertically oriented copper pipe. The copper pipe can be considered to be a stack of simple coils of copper wire. The magnet will act as if it is falling through a viscous fluid. In contrast, if the magnet is dropped through a plastic pipe, or if a nonmagnetized piece of material is dropped in a copper pipe, there is no noticeable opposition to the motion.

The Electric Guitar Pickup

Several of the concepts just described are combined in the typical magnetic electric guitar pickup. A simplified drawing of such a pickup is shown in Figure 17.11. A coil is wrapped around a piece of unmagnetized magnetic material. A magnet is used to induce magnetism in that material. The guitar string, which must also include magnetic material,[9] is in turn induced to be magnetic by the magnetic field in its vicinity that comes from the induced magnetism of the magnetic material. When the string is vibrating, the coil will sense the changing (induced) magnetic field from the string, and that provides a signal (an emf) that can then be amplified. Note that the magnetic field from the

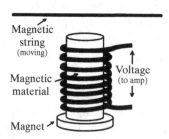

Figure 17.11 The basic construction of a typical electric guitar pickup. A real pickup will typically have a hundred or more turns of very fine wire.

---

[9] To verify these magnetic properties, note that a paper clip will stick to the guitar pickup, and a small magnet will stick to the electric guitar string. The unmagnetized paper clip will not stick to the string, however.

permanent magnet will be time independent, so it will not produce any signal in the coil. Only the time-changing magnetic field due to the moving string will create a signal. Since the string must be close to the coil to generate a strong signal, there will be a separate sensor for each string.

Such a pickup coil will also be susceptible to other sources of time-dependent magnetic fields, such as those produced by any alternating current (AC) power sources nearby that generate a hum on the output signal. To reduce that undesired signal, a second coil, of opposite polarity (e.g., with the coil wound in the opposite direction), can be placed nearby. The signal from a distant source will be almost the same for the two coils. However, the signal from the second coil will be opposite that of the first, so when combined, the extra noise is reduced. The nearby signal from the string will not be the same at both coils, so it will not cancel. This use of a pair of coils for each string to remove this undesired signal is referred to as a **humbucker** arrangement. A commercial set of such coils is shown in Figure 17.12.

The electric guitar pickup is sensitive to the motion near the pickup. Hence, pickups located at different locations, such as near the bridge or farther from the bridge, will emphasize different vibrational modes and thus can produce sounds that have a different timbre. In addition, the signals from multiple locations can be combined together to get in-between sounds. Hence, it is not uncommon to see two (or more) sets of coil pairs for each string of an electric guitar.

The **sustain effect**, sometimes used with electric guitars, can be implemented using two electromagnets and an amplifier. The signal from a pickup, due to a vibrating string, such as from the pickup illustrated in Figure 17.11, is amplified and sent to another similar, nearby coil that acts as an electromagnet interacting with the same string. As the string vibrates and moves closer to the pickup,

(a)   (b)

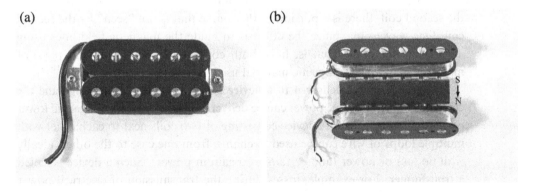

Figure 17.12 (a) A humbucker electric guitar pickup uses a pair of pickups wired so that time-dependent magnet fields from distant sources will cancel, but the signal from the nearby string will not. One large coil, with over 1,000 turns of very fine wire, is used for each set of magnetic posts, and one magnet, magnetized across its width, is used for all of the magnetic posts, as can be seen in (b),

current to the electromagnet is increased, providing an extra pull on the string. If the amplifier and electromagnet are strong enough, the string can continue to vibrate indefinitely. This is different from a sustain-effect box (or foot pedal) that acts as an automatic volume control, maintaining a constant output volume, but only as long as the input signal is present. Infinite sustain can also occur if the guitar is very close to its output speakers and the volume is turned up, allowing the strings to be driven by their own sound rather than an electromagnet. Wear earplugs when trying that experiment.

Transformers (Optional)

Since a magnetic field from a permanent magnet and that of an electromagnet are the same, it should not be surprising that a changing magnetic field from an electromagnet can also induce currents in coils and/or create electromotive forces. The changing field of the electromagnet looks similar to that of a moving magnet.

Imagine two coils next to each other, each consisting of a single loop of wire. As long as the axes of the coils are not perpendicular, a changing magnetic field created in one loop will cause a current in the other to oppose the change. If a time-dependent electric signal is put into one coil, a very similar time-dependent signal can be observed in the second. Note, however, that nature can only oppose *changes* in the magnetic field, so a constant signal will not be passed on. This, then, is one method that can be used to separate time-dependent from time-independent signals.

If two coils are close to each other, the coupling of time-dependent signals is stronger than if they are farther apart. The maximum effect occurs when all of the magnetic field produced by one coil, in some sense, passes through the other. That is, if any magnetic field lines from the source of the field do not pass through the second coil, there is a portion of the source that is not "seen" by the second coil. One way to maximize the effect is to guide the magnetic field lines using a magnetic material. That is, have both coils wound on a common piece of magnetic material. A magnetic material used in this way is called a **core**.[10]

The electric power delivered to a device is the product of the voltage and the current. That is, the same power can be delivered using a higher voltage and lower current, and vice versa. A device consisting of two coils next to each other with multiple loops of wire can be used to change from one case to the other, ideally with no loss of power (and certainly no gain in power). Such a device is called a **transformer**. For example, losses during the transmission of electrical power

---

[10] Due to their retentivity, magnetic materials may be able to retain some memory of previous signals. That property makes those with a larger retentivity usable for computer memory, and such memory, termed *core memory*, was common in early computers. While physically large and power hungry compared to semiconductor-based devices, magnetic core memory still may find some occasional and limited use in particularly harsh environments.

from a power plant are minimized if the voltage is very high and the current is low. Such high voltages are too dangerous and impractical to use in a home or business, so a transformer is used at the destination to lower the voltage. These transformers are usually seen on power poles near the point of use. The ability to change the voltage is the major advantage afforded by AC (or time-dependent) power compared to direct current (DC; or constant) power.[11] Likewise, a ribbon microphone produces a very small voltage but a more appreciable current. A low-power transformer can be used to increase the voltage to better match such a microphone to the amplifier that follows.

The discussion here has concentrated on magnetic properties that are strongly connected to electricity and electronics. Some basic knowledge of the most basic electronics – for example, knowledge of batteries and wires – was assumed. In Chapter 18, electricity and the connection between electricity and magnetism will be discussed further, and the latter will be shown to be quite strong.

Summary

Magnets have north and south poles. Unlike poles attract and like poles repel, even if the magnets are not in contact. That interaction at a distance is described using a magnetic field. Magnetic fields can be represented graphically using magnetic field lines. Magnetic materials near a magnet will have induced magnetism.

Materials that are magnetic often contain iron, nickel, or one of a few other elements. Some magnetic materials can retain significant magnetism even when no other magnet is present, creating a permanent magnet. Most materials are nonmagnetic and respond only very weakly to magnetic fields.

Electric currents also produce magnetic fields, indistinguishable from the field of a permanent magnet. The strength of the field will be proportional to the current supplied. A wire can be wound into a coil, or similar configuration, to concentrate the magnetic field, making an electromagnet. A time-dependent current can create a time-dependent field that can be used, for example, to make speakers or earphones.

The process works in reverse as well, in that a coil in the presence of an externally supplied magnetic field can produce a current, or if there is no complete circuit, an electromotive force (a voltage). In either case, the signal is only produced by changing magnetic fields. That process is called *Faraday induction*

---

[11] When electric power distribution was first being developed, Thomas Edison (1847–1931) championed "direct current" or DC power, which supplies a constant voltage. After buying Nikola Tesla's patents for AC power, George Westinghouse (1846–1914) pushed for AC power. Since time-dependent voltage levels can be easily and efficiently changed using transformers, there were considerable economic advantages to AC power distribution. Westinghouse eventually won that "war of the currents."

and is used for electric guitar pickups and some microphones. The direction of the induced current is such that the magnetic field created by the induced current opposes the externally applied changing field.

ADDITIONAL READING

Brice, R. *Music Engineering*, 2nd ed. Newnes, 2001.
Also see just about any introductory physics text that includes sections on electricity and magnetism. For example:
Knight, R. D. *Physics for Scientists and Engineers*, 3rd ed. Pearson, 2013. (See Chapters 32 and 33.)
Urone, P. P., and R. Hinrichs. *College Physics*. OpenStax, Rice University, 2020. (See Chapter 22.)

PROBLEMS

17.1 Considering the definition of what they are and what they show, under what circumstances can magnetic field lines cross?

17.2 If two electromagnetics are placed near each other and a constant current is sent through them, they are observed to attract each other. Explain what happens to the force when the current through both is reversed. How about if the current in only one of them is reversed? What about if a sinusoidal current is used?

17.3 A bar magnet falls vertically through a coil (Figure P17.3). If the magnet has its north end down, what is the direction of the magnetic field due to the current induced in the coil (a) just before the magnet reaches the coil and (b) after the magnet has just passed through the coil?

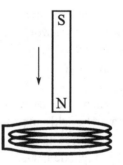

Figure P17.3

17.4 For the electric guitar sustain effect, the string can continue to vibrate indefinitely, but only after the string has been plucked. Why does the string need to be plucked to get the signal started? What are two different ways that can be used to stop the string vibration?

17.5 The typical acoustic hum heard when standing near a power transformer is predominantly at twice the frequency used for the power (e.g., at 100 Hz for 50-Hz power, 120 Hz for 60-Hz power, etc.). Why would this sound be at twice the power frequency? Why does this occur for power transformers much more than for, say, a transformer used within a microphone? Note that what is being considered here is the sound created directly by the transformer.

18   RC Time Constants

When using comparative words such as *fast* and *slow*, it is important to ask, Fast compared to what? Slow compared to what? In everyday language, such terms are often based on human experience – a qualitative comparison to everyday human activities. Physics, and science in general, will separate quantitative descriptions of nature from human experience, so such comparisons are problematic. However, for many physical phenomena, there is a natural dividing line between fast and slow that is based on the physical situation. In this chapter, one such dividing line is discussed: the **RC time constant**. The RC time constant is important for the behavior of electronic circuits and serves as an example for all time constants.

RC time constants show up in electronic circuits, including those that create signals, as well as circuits used to modify (e.g., filter) signals created by other means. The latter is an important part of signal processing. To start, some discussion of electricity, electric fields, electric circuits, capacitors (the "C" in "RC"), and resistors (the "R") seems appropriate. Capacitors, as part of an RC circuit, are also used as sensors and detectors, including for condenser and electret microphones. As was the case in Chapter 17, it will take a while to get to these applications.

Electric Fields

Perhaps the earliest direct experience that humans had with electricity was that due to **static electricity**. Such an experience is easily re-created in a number of ways that often involve rubbing two dissimilar materials together – for example, by rubbing a plastic rod with a piece of cloth (e.g., wool or silk) or by walking across carpeting in stocking feet. After the static electricity is created, its effects can be observed – for example, the plastic rod can pick up small bits of paper or cause an empty soda can to roll, and after someone walks across a carpeted floor, a spark may jump between their fingertips and a doorknob when the two are close enough to one another. Mechanical devices have been developed to produce this "static" electricity. Two examples are the van de Graaff generator, which uses motors and belts, and a Wimshurst generator involving two counter-rotating spinning disks.

After enough experimentation, these electric effects can be explained by the presence of electric charges, of which there are two types: *positive* and *negative*. Unlike the two magnetic poles (north and south), these charges can occur separately. The most common sources for the charges, responsible for effects seen in everyday life, are the protons and electrons of atoms, discussed previously (see Chapter 11).[1] As was the case for the magnetic poles, **like charges repel** and **unlike charges attract**, and those forces act at a distance.

Everyday objects contain many, many atoms, and hence there is a tremendous amount of charge present. Since the forces due to positive charges are opposite those of negative charges, when there are equal numbers present (in the same volume), there will be no net force – even if the number of charges is very large. The electric effects observed are due to a charge imbalance. When discussing charges and electric fields, only that imbalance, often referred to as the **net charge**, is considered. When specifying the quantity of charge "on an object," Q, what is specified is the net charge.[2]

The Standard International (SI) unit of charge is the coulomb (C),[3] equal to 1 A·s. A *net* charge of 1 C is a huge quantity. A highly charged van de Graaff generator, which can produce a long spark, will have a net charge of about 1 to 10 μC. In contrast, the charge of a proton is $q_e = 1.60217662 \times 10^{-19}$ C, and the charge of an electron has the same magnitude but is negative.[4] All observed particles have a net charge that is an integer multiple (including zero) of this **elementary charge**.[5]

As was the case for magnetism, forces that act at a distance can be described using a "field." **Electric fields** describe the force that would be present on a positive charge. The force on a negative charge would be in the opposite direction. In diagrams, **electric field lines** are used to show the direction of that force, and the density of the lines indicates the magnitude of the force – the denser the lines, the stronger the force. The field lines either start and stop at a charge or continue to an infinite distance. They do not simply stop, or appear, in midair.

---

[1] Other charged particles encountered in everyday life on Earth's surface include those due to cosmic rays and those due to some forms of radioactive decay. There are a surprisingly high number of these.

[2] Note that Q here is unrelated to the quality factor discussed for vibrations, even though the same symbol is used.

[3] Named for the French scientist and engineer Charles-Augustin de Coulomb (1736–1806).

[4] The choice of which is positive and which is negative is a result of history – the choice could have been the reverse.

[5] Some particles, protons and neutrons being examples, are made up of elementary particles called *quarks*. Quarks come in several varieties that have charges of ±1/3 and ±2/3 of the elementary charge; however, quarks are observed only in combination, never separately, and the only combinations observed have a total charge that is an integer multiple of the elementary charge.

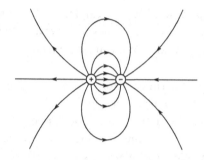

Figure 18.1 Electric field lines in the presence of a positive and negative charge show the direction of the force that would be present for a positive charge.

Thus, if there is a (static) electric field, there are charges somewhere, even if those charges may not be apparent or even nearby. An example showing a two-dimensional representation of electric field lines between two opposite charges is shown in Figure 18.1.

An electric current, such as those used in an electromagnet (Chapter 17), involves charges that are moving – they are acted on by a force, that is, an electric field, and create a magnetic field. Time-dependent magnetic fields can create currents, so those time-dependent magnetic fields are also a source of electric fields. Electric and magnetic fields thus are coupled. In fact, those fields are best thought of as one combined field – the **electromagnetic field** – and not as two separate phenomena. As a simple thought experiment, consider a beam of charged particles (such as electrons in a cathode ray tube[6]). Such a beam is an electric current and will produce a magnetic field. Now imagine an observer who is moving along with the beam. To that observer, the charges are not moving, and hence, to that observer, the current is zero, so no magnetic field would be expected. However, any effects due to the moving charges on other objects will still be present. Hence, whether an electric field or magnetic field (or some combination) is measured depends on the observer – they are two different views of one field. The unification of the two fields – electric and magnetic – was firmly established in the late 1800s by the Scottish scientist James Clerk Maxwell (1831–1879) and others.

Voltage, measured in volts (V), is the product of the electric field strength and the distance over which the field is applied. Voltages are conveniently measured with a voltmeter. One result is that electric fields are often specified in units of volts per meter (V/m) or similar units – a value for the magnitude of an electric field, when multiplied by a distance (e.g., in meters), results in a voltage. For this usage, it is only the distance along the direction of the electric field that is important.

---

[6] If you are unfamiliar with these, it is a simple matter to look them up using an Internet search.

Conductors and Insulators

Materials can be classified as electrical **conductors** or electrical **insulators**. A very simple device to make this classification is shown in Figure 18.2. If the light illuminates, the material is a conductor; otherwise, it is an insulator. Within a conductor, there are significant numbers of charges that can move easily within the material (usually, these are electrons), and in insulators, there is an insufficient number of mobile charges.

The ability of a material to conduct is not really an all-or-nothing proposition but a continuum. The ability of a material to conduct is characterized by a material property known as its **electrical conductivity**. Electrical conductivities can range over many orders of magnitude, as illustrated in Figure 18.3. The best conductors are metals. The ability of a particular piece of material to conduct also depends on its size and shape. Longer and thinner pieces conduct less well than shorter and fatter pieces, in the same way that it is harder to get flow of a fluid through a long, thin pipe than through a short, fat pipe.

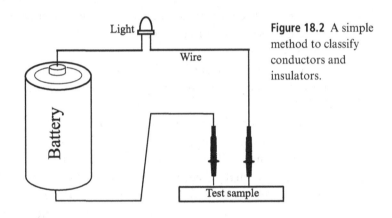

Figure 18.2 A simple method to classify conductors and insulators.

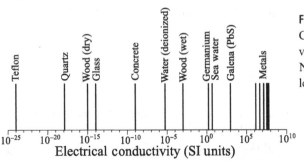

Figure 18.3 Conductivities for a variety of materials. Notice the logarithmic scale.

When a conductor has a net charge on it, the like charges will repel and will spread out as much as possible, which means they will all end up on the surface. Once the charges in a metal reach equilibrium – that is, they stop moving – then it must be that the electric field within the conductor is zero. If the field were not zero, there would be a force on the charges, and they would move. If the electric field is zero, then the *changes* in voltage – the field multiplied by distance – inside the material will also be zero. Thus, the interior of a conductor will be self-shielded from external electric fields, and a wire, a long, thin conductor, can usually be expected to have no significant voltage change from one end to the other.

Electrostatic Discharge (Optional)

The sparks in air that occur for static electricity will occur when the electric forces on atoms become large enough to separate electrons from their host atoms, creating free electrons and positively charged ions. **Ions** are atoms that have an imbalance between the number of electrons and the number of protons. Once the charges have been separated, they can move independently. If the forces are large enough, those moving free charges can collide with other atoms, causing many more charges to be separated. In the case of sparks from static electricity, those moving charges will quickly discharge the original source of the electric field, shutting off the process. When separated charges recombine, they can emit light – a spark.

Sparks are one form of **electric breakdown**. Electric breakdown can occur in any substance that normally does not have mobile charges. The details of the mechanism may differ somewhat for different circumstances. The minimum electric field strength necessary for breakdown is known as the **dielectric strength**. Table 18.1 shows some representative values. Note that creating electric breakdown in air is relatively easy compared to many solid materials.

The voltages used to power everyday battery-powered devices range from about 1 to 15 V. For devices that plug into the wall, 100 to 200 V may be required. To make a long spark requires much more. A good van de Graaff generator can produce upward of 100,000 volts.

Example 18.1

What voltage is necessary to make a 1-cm spark between a finger and a doorknob?

From Table 18.1, the dielectric strength for air ranges from about 1 to 3 kV/mm, which depends on conditions (humidity, temperature, etc.). Hence, a 1-cm (10-mm) spark will need 1–3 kV/mm × 10 mm = 10,000 to 30,000 volts.

Table 18.1 Dielectric strengths for a selection of insulators.

Material	Dielectric Strength	
	MV/m = kV/mm	kV/in. = V/mil
Mica	50–200	1,300–5,000
Teflon	20–60	500–1,500
Paper (dry)	15–50	400–1,200
Neoprene rubber	15–30	380–770
Window glass	10–14	250–350
Air	1–3	20–75

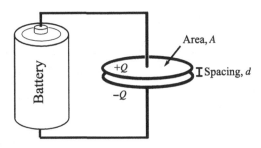

Figure 18.4 A simple capacitor is charged using a battery.

A sustained voltage higher than a few tens of volts can be hazardous to humans, and possibly lethal, depending on how and where it is applied. Fortunately, static discharge between a finger and a doorknob, even though the voltage is very high, is of very short duration and involves very little power.

Capacitors

Capacitance is a measure of the ability to store net charges. Whenever an electronic circuit produces electric fields, there is capacitance. Electronic devices specifically designed to create confined electric fields are **capacitors**. Capacitors are constructed by placing two conductors near each other, with an insulator in between. The electric field created is generally contained within the device.

The simplest capacitor design, which is sufficient to understand the behavior of all capacitors, is the **ideal parallel-plate capacitor**. Two conducting plates of area, A, are placed parallel and spaced a relatively small distance, d, apart. In this case, "small" is in comparison to the other dimensions of the plates. The plates can be charged using a voltage source, such as a battery, as shown in Figure 18.4. The net positive charge on one plate is attracted to the net negative charge on the other; however, since they are separated by an insulator, they cannot combine. In between the plates is an electric field. If a positive charge were to be placed between the plates, it would be repelled from the positively charged plate and attracted to the negatively charged plate.

Figures 18.5a and b show calculated electric field lines for parallel circular disks. In Figure 18.5a, the charge is assumed to be uniformly distributed, and in Figure 18.5b, the disks are assumed to be conductors, so the voltage measured at any point on the disk will be the same. Note that the field between the plates is very similar, and the only differences are near the edges. As an approximation, the simplified electric field shown in Figure 18.5c, which ignores the edge effects entirely, can be used as a model. If the voltage source provides a voltage V, then the magnitude of the electric field is $E = V/d$.

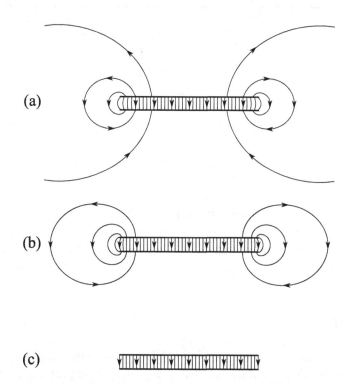

Figure 18.5 Electric field lines between circular plates: (a) for uniformly charged plates, (b) for conducting plates, and (c) a simple model that ignores edge effects.

Example 18.2

A 12-V battery is used to charge a parallel-plate capacitor having a plate spacing of 1 mm. What is the magnitude of the electric field between the plates?

$$E = 12\text{V}/1\text{mm} = 12\text{V}/10^{-3}\ \text{m} = 12{,}000\ \text{V/m} = 12\ \text{kV/m}.$$

Capacitance, C, relates the voltage on the plates, V, to the net charge on the plates, Q by

$$Q = CV. \tag{18.1}$$

Thus, capacitance has units of A·s/V. In honor of the English scientist Michael Faraday (1791–1867), a pioneer in electromagnetism, 1 A·s/V = 1 farad (F). One farad is a very large capacitance. For the ideal parallel-plate capacitor, ignoring edge effects,

$$C = \varepsilon \, \varepsilon_0 \frac{A}{d}, \tag{18.2}$$

where $\varepsilon_0 = 8.8541878$ pF/m is the "permittivity of free space," a constant of nature; ε is the relative permittivity of the insulating material used between the plates and has no units; A is the area of the plates; and d is the spacing of the plates. Representative values for relative permittivity are shown in Table 18.2.

Example 18.3

What is the capacitance of a parallel-plate capacitor with circular plates 30 cm in diameter and 1 mm apart if polypropylene is used for the insulator?

$$A = 2\pi \, r^2 = 2\pi(0.30 \text{ m}/2)^2 = 0.1414 \text{ m}^2,$$
$$C = (2.4 \cdot 8.542 \cdot 0.1414/10^{-3}) \text{ pF} = 2.9 \text{nF}.$$

A larger plate area allows like charges, which repel each other, to spread out more, and a smaller spacing allows the unlike charges, which attract, to be closer.

Table 18.2 Relative permittivity for a variety of insulating materials.

Material	ε
Vacuum	1
Air (dry)	1.0006
Liquid nitrogen (77 K)	1.45
Mineral oil	2.1
Teflon	2.0
Polypropylene	2.4
Paper (dry)	3.5
Silicon dioxide (SiO_2)	3.9
Salt (NaCl)	5.9
Silicon	11.8

Thus, a larger area and smaller spacing require less effort (e.g., smaller voltage) to store the same amount of charge.

A capacitor is a device constructed so as to have a particular capacitance. If the capacitor plates are too close together, the electric field can exceed the dielectric strength of the insulating material between the plates, and a spark will result. This is one reason why commercially available capacitors will always have a voltage rating, indicating the maximum that should be used with the device, and why capacitors designed for higher voltages tend to be larger in size.

Real capacitors, used in electronic circuits, are often alternating layers of conductors and insulators, rolled into a cylinder, or other more compact geometries, rather than parallel plates.

Resistors

An electric current is created in a conductor when a voltage (V) is applied, for example, when a conductor is connected across a battery or other power source. The magnitude of the current is usually signified using I.[7] For most conductors, the current is proportional to the applied voltage. This is referred to as **Ohm's law** and is written as

$$V = IR, \tag{18.3}$$

where the proportionality constant, R, is the **resistance**. The SI unit for resistance is the ohm, symbolized by an uppercase Greek omega, Ω.[8] More generally, resistance is one form of electrical impedance, and there is a direct analogy between this electrical law and the previous discussion of impedance for sounds in pipes (see Chapter 14). When using Ohm's law, be sure to distinguish the variable representing the voltage, V, from the abbreviation for the unit of voltage, V.

A **resistor** is a device that is constructed so as to have a particular resistance. Resistors turn electrical energy into heat. The rate of creation of heat energy, the power, is given by

$$P = I^2 R = V^2 / R, \tag{18.4}$$

and a resistor's power rating must be sufficient to handle the heat generated. As a result, resistors designed for higher-power applications tend to be larger in size.

---

[7] Historically, electric current was originally referred to as *current intensity*, or in French, the language used for many early papers important for the study of electric current, *intensité du courant*. Hence the use of I to refer to the intensity.

[8] Named for the German physicist Georg Ohm (1789–1854). An uppercase O would be confused with zero.

Circuit Diagrams

Electronic circuits are often described using **schematic diagrams**. In schematic diagrams, each component is represented symbolically. The symbols used for the components discussed here are shown in Table 18.3. An electric circuit will have paths where, for any starting point, there is at least one path that begins and ends at that point but that does not retrace itself. If there is a section of the circuit that must be retraced, then there is an "open circuit," and no (net) current will flow in that section of the circuit.

RC Time Constants

Consider the circuit shown in Figure 18.6a, which shows a capacitor, C, and resistor, R, connected together. If the capacitor has a (net) charge, $\pm Q$, on the corresponding plates (as shown), then there is a voltage across the capacitor, $V_C = Q/C$. When the switch is closed, the voltage across the resistor must equal

Table 18.3 Schematic diagrams for some of the basic electronic components.

Component	Diagram	Description
Wires		Allows flow of current between devices, ideally with no change in voltage.
Switches		When "open," there is no connection, and no current can flow. When "closed," the wires are connected, and current flows as if through a wire.
Batteries (power sources)		Provides an electromotive force to push current through the circuit. Characterized by a voltage, V, which increases from "−" to "+."
Resistor		Converts electrical energy to heat. The voltage decreases in the direction of the current, with a magnitude given by Ohm's law.
Capacitor		Stores electric charges. The voltage across the capacitor is proportional to the net charge stored on each side.

Figure 18.6
Examples of simple
RC circuits.

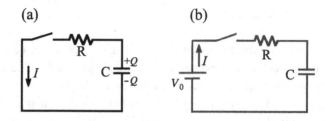

V_C since both sides of the devices are connected with wires. The current will be in the direction shown (from positive to negative), and $I = V_C/R$. That current is the flow of charge, and as long as that current is flowing, the net charge on the capacitor is being reduced. That process will continue until the net charge on the capacitor goes to zero.

Consider that same circuit (Figure 18.6a) for a short amount of time, Δt, after the switch is closed. That time is chosen to be short enough so that the fractional change in the net charge on the capacitor is small. As an approximation, the current (the rate of change of the charge) can be taken to be constant during that short time; there has not been enough time for a significant change. Then the (small) change of the net charge on the capacitor, ΔQ, during that time is

$$\Delta Q = -I\Delta t = -\frac{V_C}{R}\Delta t = -\frac{Q}{RC}\Delta t, \tag{18.5}$$

where the minus sign reflects the fact that the magnitude of the charge is decreasing. Now divide both sides by Q to get

$$\frac{\Delta Q}{Q} = -\frac{\Delta t}{RC}. \tag{18.6}$$

On the left side is the fractional change in the charge. The equation says that the *fractional* change in the charge will be the same for any (short) fixed time interval, Δt, and the size of that change depends on the product RC. For each change in time, Δt, the capacitor loses a fixed *fraction* of the electric charge. Equation (18.6) clearly describes an exponential decrease (see Chapter 2) with a timescale determined by RC. The larger the product RC, the smaller is the fractional loss of charge during the fixed time Δt.

For electronic circuits of this type, the characteristic time, often described using a lowercase Greek tau (τ), is referred to as the **time constant**, and here, $\tau = RC$. By the way, since the left side of Equation (18.6) has no units – they cancel – the right side also must have no units. Since Δt is a time (e.g., in seconds), then it must be the case that the product, RC, also has units of time. For SI units, it is indeed straightforward to show that

$$1\,\Omega \times 1\,\text{F} = 1\,\text{s}. \tag{18.7}$$

Example 18.4

What is the time constant for the circuit of Figure 18.6a if $R = 100$ kΩ and $C = 1$ μF?

$$\tau = RC = 100 \text{ k}\Omega \times 1 \text{ μF} = 10^5 \text{ } \Omega \times 10^{-6} \text{ F} = 0.1 \text{ s}.$$

Note also that Δt will be "small enough" when it is much smaller than RC. In the limiting case that Δt becomes infinitesimally small, the effects of the approximation used to get Equation (18.6) also become infinitesimally small. In that limit, the exponential changes of the charge on, and voltage across, the capacitor, as functions of time, can be shown to be

$$Q(t) = Q_0 \, e^{-t/\tau} = Q_0 \, \exp(-t/\tau),$$

$$V_C(t) = V_C(0) \, e^{-t/\tau} = V_C(0) \, \exp(-t/\tau), \qquad (18.8)$$

where Q_0 is the charge at $t = 0$, $V_C(0)$ is the voltage at $t = 0$, and e is Euler's number (2.7182818 …).[9] The **exponential function** ("exp") on the right is an alternate way to represent powers of e. The result is shown graphically by the solid line in Figure 18.7. For each time interval equal to one time constant, the fractional charge that *remains* is $e^{-1} = 0.36788$, about one-third, and the fraction *lost* is $1 - e^{-1} = 0.62312$, or about two-thirds.[10]

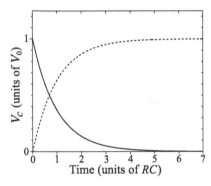

Figure 18.7
Exponential discharging (*solid*) and charging (*dashed*) of a capacitor in a simple RC circuit.

[9] The process of taking limits and solving for the result is usually taught in calculus courses. Knowledge of calculus is not assumed here. The constant is named for the extremely prolific Swiss-born mathematician/scientist Leonid Euler (1707–1783). Included in his works are studies on the propagation of sound and studies related to the mathematics of the musical scale.

[10] The procedure is the same as that used for half-lives in Chapter 11, except instead of using one-half remaining after each half-life, use exp(–1) for how much remains after each "lifetime." In some other applications, τ would be referred to as the "characteristic time."

Example 18.5

For the circuit in Fig. 18.6a, what fraction of the net charge is lost after three time constants?

After a duration equal to the time constant, $\exp(-1)$ of the original charge remains, so the amount remaining after three time constants is $Q(t) = (0.36788)^3 Q_0 = 0.04979 Q_0$. The fractional amount lost is then $(1 - 0.04979) = 0.9502$; that is, about 95 percent of the net charge was lost after three time constants.

Now consider the circuit in Figure 18.6b, where a battery has been added. Once the switch is closed, the current through the resistor will depend on the *difference* between the voltage from the battery and the voltage from the charge on the capacitor. If the battery has the larger voltage, the current will be in the direction shown by the arrow, which will cause more charge to build up on the capacitor. The charge will build up until the voltage on the capacitor equals the voltage from the battery. On the other hand, if the voltage on the capacitor is initially larger than that on the battery, then the current flows in the opposite direction until the voltages are equal. Compared to Figure 18.6a, the battery has simply added an offset – instead of discharging to zero, the capacitor will charge or discharge to the battery voltage. In fact, Figure 18.6a can be considered a special case of Figure 18.6b in which a 0-V battery is being used.

The offset supplied by the battery affects the endpoint, but it does not affect the rate of change. Hence, the charge on the capacitor will be as in Equation (18.8), but with an offset,

$$Q(t) = Q_1 \, e^{-t/\tau} + Q_2. \tag{18.9}$$

Dividing both sides by the capacitance,

$$V_C(t) = V_1 \, e^{-t/\tau} + V_2. \tag{18.10}$$

Evaluating at $t = 0$ and $t \to \infty$ to determine the constants,

$$V_C(t) = \left(\frac{Q_0}{C} - V_0\right) e^{-t/\tau} + V_0, \tag{18.11}$$

where Q_0 is the net charge on the capacitor at $t = 0$. In particular, if the capacitor is initially uncharged, then

$$V_C(t) = V_0\left(1 - e^{-t/\tau}\right)$$
$$Q_C(t) = Q_0\left(1 - e^{-t/\tau}\right), \tag{18.12}$$

where $Q_0 = CV_0$. This result is shown in Figure 18.7 (*dashed line*).

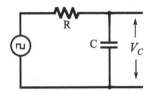

Figure 18.8 An RC combination with a time-dependent source.

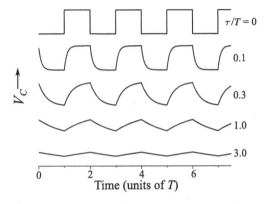

Figure 18.9 Voltage across the capacitor of Figure 18.8 for different RC time constants.

Now consider the circuit shown in Figure 18.8, where the battery has been replaced with a square-wave voltage source. The changes in the voltage across the capacitor will occur over a timescale RC. As the frequency of the square wave is increased, there is insufficient time for the capacitor to keep up with the changes in the square-wave input. If the square wave switches after each time T, so the frequency is $1/(2T)$, then the voltage across the capacitor, V_C, will change as a function of time, as shown in Figure 18.9. If RC is very small compared to T, then V_C is very nearly a square wave. If RC is larger than T, then V_C is diminished in amplitude and becomes very nearly a triangle wave. The square wave is said to have been "filtered" by the circuit.

If a sinusoidal input is considered, that is, thinking in the frequency domain (see Chapter 4), V_C will also be sinusoidal, although with possible changes in amplitude and phase that depend on frequency. As the frequency is increased more and more, V_C cannot keep up with the changes. Since the total signal is across both the resistor and capacitor, the voltage across the resistor, V_R, makes up the difference. See Figure 18.10, which shows the frequency dependence of these voltages and where $f_0 = 1/(2\pi RC)$. Since the power in the signal goes like the amplitude squared, the power is reduced by a factor of 2 when $f/f_0 = 1$, which is where the amplitude is reduced by the square root of 1/2. If the voltage across the capacitor (resistor) is considered an "output voltage," then the RC combination is a low-pass (high-pass) filter, with a "cutoff frequency" of f_0.

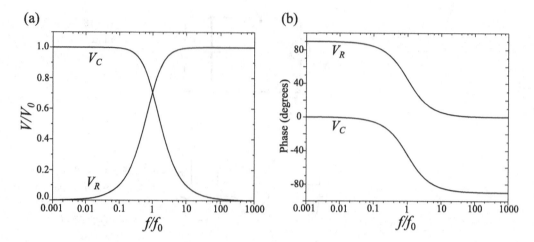

Figure 18.10 The relative (a) voltage and (b) phase across the resistor and capacitor for the circuit in Figure 18.8 but with a sinusoidal source at frequency f.

Example 18.6

A low-pass filter is to be used to block signals above 20 kHz. For a simple RC filter, what capacitance should be used along with a 1,000-Ω resistor?

$f_0 = 20,000 \text{ Hz} = 1/(2\pi RC)$, so $C = 1\,\text{F}/(2\pi \cdot 1,000 \cdot 20,000) = 8 \times 10^{-9}\text{F} = 8\text{ nF}$.

Condenser and Electret Microphones

At one time, capacitors were also referred to as *condensers*. That name is still used for microphones that are based on a time-variable capacitor. The so-called **condenser microphones** will have one or both plates of the capacitor designed to be flexible enough so that small, time-dependent air pressure changes, for example, due to sound, will cause a time-dependent plate spacing, which in turn results in a time-dependent capacitance. If the charge on the capacitance is kept constant, then a time-dependent voltage across the capacitor – an electrical version of the signal – will be produced. In practice, a long RC time constant is used. Here "long" is long compared to the time-dependent changes due to the sound – the charge simply does not have time to change and stays constant to a good approximation. The circuitry is illustrated in Figure 18.11.

Condenser microphones require a voltage source to charge the capacitance. This may be supplied by a battery in the microphone or using **phantom power**. The voltage used is usually a few tens of volts, with 48 V being a common value. Phantom power is often supplied through the same microphone cable that is used for the electrical signals. Since the phantom power is constant in time, it is easily separated from the time-dependent audio signals.

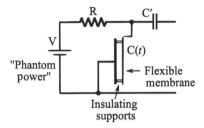

Figure 18.11 A condenser microphone is constructed from a capacitor sensitive to changes in pressure due to sound.

Electret microphones are a newer technology and use "plates" made of highly insulating material into which charge has been injected. The insulating properties are sufficient to retain the charge for decades or longer. Hence, the plates are, essentially, permanently charged. The signal from an electret microphone element is small, so such microphones usually have a built-in transistor amplifier. That amplifier will require some power, usually a few volts, to function. That supply should not be confused with phantom power.

Electret microphones are very inexpensive and are quite common in less demanding consumer applications, such as cell phones, computer microphones, headset and lapel microphones, and similar. They are not generally used for high-fidelity, low-noise applications.

Microphone Sensitivity (Optional)

The signals from microphones are relatively small. The size of those signals depends on the microphone, of course, and can be estimated based on data from the microphone manufacturer. The specifications are often given in somewhat problematic units: dBV/Pa. It is the purpose of this section to show how to interpret those units in order to make such an estimate.

Decibels are used in many circumstances. Remember that in each case, decibels deal with a ratio. As was previously seen (Chapter 13), sound levels are often expressed in dB SPL, where the ratio is that of the intensity of the sound divided by a specific reference intensity. In other cases, the ratio is between the intensity of two sounds. For example, when discussing how much louder a trumpet is compared to a violin, the answer (expressed in dB) involves the ratio of the intensity of the trumpet to the intensity of the violin. For some levels, an extra letter is added to "dB" to indicate what ratio is being used. Previously, "dBA" and "dBC" were seen, where the extra letter referred to a frequency-dependent reference. In the case of microphone signals, the letter V is added to indicate that the reference is 1 volt. Note that in this case, "volts" is an amplitude, so a factor of 20 is used for the dB calculation. Another common unit seen in electronics is dBm, which describes power levels compared to 1 mW.

Examples 18.7

$$A \text{ 1-mV signal is } 20 \log_{10}(1 \times 10^{-3}\text{V}/1\text{V}) = -60 \text{ dBV}.$$
$$A \text{ 3-mV signal is } 20 \log_{10}(3 \times 10^{-3}\text{V}/1\text{V}) = -50.5 \text{ dBV}.$$

For microphones, the output voltage depends on the sound intensity. To specify the sensitivity of the microphone, the reference sound intensity must also be specified. It is common practice to use a sound pressure amplitude of 1 Pa (94 dB SPL). A problem arises for the computation of the size of the signal for a different pressure amplitude. The temptation is to view dBV/Pa as a simple rate, then multiply that value using the number of pascals. That does not work. It is not even close. The units given are a result of history and can be misleading. In this context, it is best to read dBV/Pa not as "dBV per Pa" but as a shorthand for "dBV when 1 pascal is used." The calculation is done by computing the size of the signal, in volts, for 1 Pa, then multiplying that voltage, not the number of dBV, by the number of pascals.

Example 18.8

A manufacturer specifies that a microphone has a sensitivity of -55 dBV/Pa. What is the amplitude of the signal, in volts, for sounds with an amplitude of 3 Pa?
  First compute the signal amplitude, S, for 1 Pa:

$$-55 = 20 \log_{10}(S/1\text{V}) \rightarrow -2.75 = \log_{10}(S/1\text{V})$$
$$\rightarrow 10^{-2.75} = S/1\text{V} \rightarrow S = 10^{-2.75} \text{ V} = 1.77 \text{ mV}$$

Now multiply by the desired sound amplitude:

$$(1.77 \text{ mV for 1 Pa}) \times (3 \text{ Pa}) = 5.31 \text{ mV}.$$

A microphone's sensitivity also depends on frequency. It is common practice for a manufacturer to specify the sensitivity for sounds at 1 kHz, then to provide a graphic showing the changes, in dB, for other frequencies. Figure 18.12 is an example of such a graphic. Now, to compute the expected signal, start with the value at 1 kHz and 1 Pa, adjust for the new frequency, then proceed as in Example 18.8.

Example 18.9

A manufacturer specifies that a microphone has a sensitivity of -55 dBV/Pa at 1 kHz, with the frequency dependence as shown in Figure 18.12. What is the amplitude of the signal, in volts, for sounds with an amplitude of 3 Pa at 100 Hz?
  From the graphic, the signal, S, at 100 Hz is 4 dB less than at 1 kHz. So

$$(-55 - 4) = 20 \log_{10}(S/1\text{V}) \rightarrow -2.95 = \log_{10}(S/1\text{V}) \rightarrow 10^{-2.95} = S/1\text{V} \rightarrow$$

$$S = 10^{-2.95} V = 1.12 \text{ mV}.$$

Now multiply by the desired sound amplitude:

$$(1.12 \text{ mV for 1 Pa}) \times (3 \text{ Pa}) = 3.36 \text{ mV}.$$

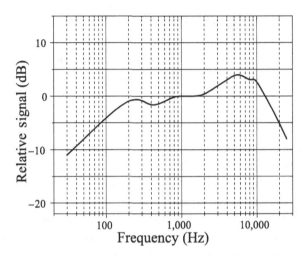

Figure 18.12
Microphone sensitivity used for Example 18.9.

Microphone signals may also exhibit a phase shift relative to the sound source. As was seen previously (Chapter 9), sometimes those phase shifts can be important. Unfortunately, not all microphone manufacturers routinely supply phase-shift information.

Summary

Electric interactions at a distance are due to an imbalance of electric charge. Electric charge comes in discrete units and can be positive or negative. Normal everyday objects contain an enormous amount of charge, although usually with equal amounts of positive and negative, so electric effects external to the object are small. When there is a charge imbalance, a net charge, those effects are significant. The force between charges is attractive for unlike charges and repulsive for like charges.

The interaction of charges with each other is present even if the charges are not in contact and, like the magnetic field used for magnets, can be treated using an electric field. Electrical conductors contain charges that are mobile, usually electrons, that can respond to an applied electric field, creating a current. Electrical insulators do not have these mobile charges; however, if a large enough electric field is applied to an insulator, some charges can break free to produce a spark or similar breakdown, creating a practical limit to the insulating properties.

Capacitors are electronic devices consisting of two conductors, called *plates*, separated by an insulator. An equal and opposite net charge can be maintained on the conductors through the application of electric power, such as from a battery or other voltage source, causing an electric field to be present in the insulator. The amount of net charge on each plate is proportional to the applied voltage. The proportionality constant is called *capacitance*.

A resistor is an electrical conductor that has some resistance – a friction-like effect for electric currents. The current through an ideal resistor obeys Ohm's law. That is, the current is proportional to the applied voltage. The proportionality constant is the conductivity. Resistance is often specified and is the inverse of the conductivity.

The net charge on the plates of a capacitor can be changed by connecting a resistor across the capacitor, possibly including a voltage source. That change will occur exponentially with time, with a time constant equal to the product of the resistance and the capacitance. The time constant is roughly the time for the change to be two-thirds complete. The electrical circuits involved can be described using schematic diagrams.

The RC time constant is a specific example of a time constant. The changes of an RC circuit in the presence of sound are responsible for the signals from some types of microphones. Microphone sensitivity is often expressed using units that may be confusing.

ADDITIONAL READING

Froehlich, B. "A Demonstration Condenser Microphone." *Physics Teacher* 50, no. 8 (2012): 508.

Suits, B. H. *Electronics for Physicists: An Introduction.* Springer, 2020. (See Chapters 1 and 3, or similar sections in any other introductory electronics text.)

Syed, M. Q., and I. Lovatt. "Seeing Is Believing: Demonstrating the RC Time Constant Visually." *Physics Teacher* 58, no. 6 (2020): 402–405.

PROBLEMS

18.1 A van de Graaff generator creates a 4-in. (10-cm) spark through air. What voltage was present just before the spark?

18.2 Estimate the voltage between a cloud and the ground just before a cloud-to-ground lightning strike. Be sure to clearly state your assumptions.

18.3 A capacitor is constructed based on specifications that had all dimensions given in centimeters; however, the dimensions were mistakenly read as inches. How will the capacitance of the resultant capacitor compare to the capacitance that was intended? (Note: 1 in. = 2.54 cm.)

18.4 If a 0.1-μF capacitor is to be used to obtain a time constant of 1 ms, what resistance should be used?

18.5 Consider the circuit shown schematically in Figure P18.5, where the capacitor is initially uncharged. The switch is closed at $t = 0$. What is the voltage across the capacitor (a) after one time constant and (b) after 0.1 s?

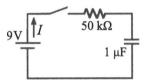

Figure P18.5

18.6 A high-pass RC filter is to block signals below 120 Hz. If $C = 1$ μF, what value of R should be used?

18.7 Starting with the coefficients of the Fourier series representation of the square wave (Equation [4.3]), show that if each term is reduced in proportion to the frequency of that term, the magnitudes of the coefficients resemble those of the triangle wave (Equation [4.4]). Compare to Figure 18.9.

18.8 In Figure 18.5, the electric field is shown for two charged plates, one of which has an excess positive charge and the other with an excess negative charge. Which plate (upper or lower) has the excess positive charge? Be sure to explain how you can tell.

18.9 Given their nature, under what circumstances, if any, can electric field lines cross?

18.10 What capacitance will have $V_C = 1$ V for a net charge equal to the elementary charge? If a parallel-plate capacitor with that capacitance is to be constructed with circular plates with a radius equal to their spacing, with vacuum as the insulator, what is the spacing of the plates? How does that compare to the size of an atom?

18.11 A certain microphone has a sensitivity of –52 dBV/Pa at 1 kHz, with a frequency dependence as shown in Figure 18.12. What signal amplitude, in volts, would be expected for a sound with a pressure amplitude of 5 Pa at 10 kHz?

19   Physics and Recording Technology

The ability to record and play back a musical performance has had an enormous impact on the availability and the economics of music. In its earliest form, standard musical notation can be used, in a sense, to record and play back a performance in the same way that a stenographer might record and read back a dictation. Musical notation dates back several thousand years. However, the recording technologies of interest here are those used for the direct recording of the sound vibrations made during a particular performance. That technology is relatively recent in human history, dating back to the latter half of the nineteenth century. Since the sounds themselves will disappear shortly after they are produced, such recording technologies all involve a transformation of the time-dependent pressure variations of the sound into another form that can be preserved, perhaps indefinitely. Of course, to play back the music, the stored information must be turned back into time-dependent pressure variations (sound), which are then an "acoustic representation" of the original signal. The more accurately this process can be accomplished, the more successful is the technology.

This exploration of how physics is involved in recording technologies will be somewhat historical in nature. Modern storage increasingly relies on digital electronic devices. Of interest here is the physics necessary for the storage and retrieval of that signal. Unfortunately, electronic storage devices have either too little interesting physics and/or entirely too much interesting physics to be discussed here. To delve into the operation of semiconductor physics and how that makes transistors and integrated circuits is very interesting but is too much, and without that, there is little physics left. While some discussion of digital signals will follow, that discussion is done without the necessity that the signal be stored electronically.

In general, there are two important considerations for any sound-storage technology: the medium and the encoding. The *medium* refers to the physical device or object that is storing the information, whereas the *encoding* describes how the sound information is actually stored. General categories of storage media include **physical storage** (wax cylinders, vinyl records), **magnetic storage** (magnetic tapes, computer disks/drives), **optical storage** (sound on film tracks, optical discs, CDs, DVDs), and of course, **electronic storage** (memory circuitry, solid-state computer drives, thumb drives, etc.).

Two distinct types of encoding are **analog** and **digital**. Analog storage is where one (or more) properties of the medium are varied continuously to represent the signal, at least within a certain range of amplitudes. Sound itself is an analog signal, where the air pressure varies continuously with time, so a one-to-one correspondence between the sound and the stored analog signal is straightforward. A digital signal is one where the amplitude of the signal at any given time is represented numerically. A stored digital signal is simply a sequence of stored numbers that represent the size of the signal at different times. Those numbers are often encoded within the medium using a binary (on/off, 0/1, plus/minus) coding scheme. If sound is to be recorded and played back using digital techniques, then it is necessary to add an initial conversion from the analog sound signal to the digital format to be stored, then the reverse process when the sound is played back.

Physical Media

The earliest known recorded sounds come from the mid-1800s and were traces created by a vibrating stylus (needle) on a smoke film on glass or paper. The stylus itself was attached to a larger diaphragm that was set into vibration by a sound – the variations in air pressure produced a time-dependent force on the diaphragm, which resulted in vibrations of the diaphragm and hence also of the stylus. The glass plate or paper was then moved under the stylus at a constant speed. As the stylus vibrated, it physically altered, in this case, scratched, the smoky film. The resulting "image" of the sound is akin to what one would expect from an oscilloscope trace. At the time, there was no method to play back those recorded sounds. Some of those earliest recorded sounds have since been re-created, more than a hundred years later, using computer imaging and analysis. The quality of those recordings leaves much to be desired, but it was a start.

While many worked on variations of the moving stylus technique, the first commercial success is generally considered to have resulted from Thomas Edison's **phonograph** of 1877, which included a simple means for playback. As was done previously, he used a stylus connected to a diaphragm. In the presence of sound pressure on the diaphragm, the stylus moved in an up-and-down motion, making indentations in a thin piece of moving tin foil.[1] A similar stylus could be used in reverse for playback. For playback, the foil would cause the

---

[1] If one were to attempt to replicate this equipment, it is important to note that at that time, "tin foil" was actually made from tin. In modern usage, *tin foil* often refers to aluminum foil.

Figure 19.1 A magnified and contrast-enhanced photo of the grooves on a 33 1/3 rpm vinyl record. The average spacing between these tracks is about 0.13 mm.

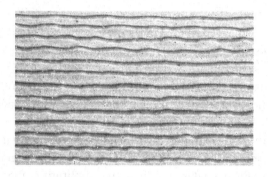

stylus to move, which would then move its diaphragm, re-creating the sound. The foil was flimsy and would not last long. Later developments, largely by others, replaced the flimsy foil with a more durable hard wax and then, later, with plastics.

The earliest phonograph recordings were generally in the shape of a cylinder, with the recorded information stored along a helical path around the outside. The disc-shaped record, with the information encoded using a side-to-side motion of the stylus, rather than up and down, and in a spiral on the face of the disc from outside to inside, appeared early as one of several competing technologies. The disc-shaped records, and the equipment to play them, were initially developed largely by the Gramophone Company and were referred to as **gramophone records**, as opposed to the cylindrical phonograph records. Early commercial versions of gramophone records were made from shellac or celluloid until vinyl eventually took over in the late 1940s. Discs were much easier to mass produce, store, and ship than were the cylinders and so became the market favorite. The invention and development of electronic amplifiers in the early 1920s also led to more efficient recording and playback machinery. The sound no longer needed to connect directly to the stylus for recording or playback, so the stylus began to be treated as part of a sensor system rather than as part of the mechanical sound-production system. A close-up of the grooves in a vinyl record is shown in Figure 19.1.

Physical grooves, as a means to store and play back sound, were the predominant commercial technology throughout most of the twentieth century. There were, of course, many different record formats developed and sold as manufacturers fought to make theirs play longer, better, cheaper, and/or with the most desirable performers. One of the interesting developments was the advent of stereo. The sound is recorded simultaneously as two tracks, one for the left side and one for the right. The new technology had become known as **stereophonic sound** (*stereo* for short), and although it did not previously need a name, the old was now called **monophonic sound** (*mono* for short). The use of multiple tracks to record sound from different directions to get a more realistic reproduction became all the rage at movie theaters in the 1930s. To market this new format for records used in the home, the manufacturers needed to come up with a way to do that inexpensively using a groove in plastic. Furthermore, there was a good financial incentive to do it in a way that was backward-compatible. That is, the customers should be able to play the new stereo records on their old mono machines and get acceptable mono results.

The solution was to include both side-to-side and up-and-down motions. The mono playback machines at the time were only sensitive to the side-to-side motion. The clever idea is that when creating the stereo groove, the extent of the side-to-side motion is used for the sum of the left (L) and right (R) signals ("L + R"), and the up-and-down motion is used to record their difference ("L – R"). The mono machines will only play back the sum of the two, which is appropriate. Adding and subtracting the signals from these two motions will allow you to separate the L and R signals. Note also that with this approach, both L and R are treated equally in that they rely equally on both the side-to-side and up-and-down motions.

Sensors to separately pick up the horizontal and vertical motions of the stylus might seem the obvious choice; however, some careful thought leads to the conclusion that the sum and difference of those motions correspond to motion at a 45° angle on either side of the vertical, one for L and one for R.[2] Thus, the L and R signals can each be picked up directly using identical motion sensors that are properly oriented.

Clever engineering was used in many other related entertainment technologies (radio, movies, the change from black-and-white to color television) to allow for backward compatibility – new offerings would still work on older machines – up until the more recent changes from analog to digital encoding. The drastic change in encoding format precludes backward compatibility. A digital audio recording (of any kind) will not produce a meaningful result if played on an old gramophone player.

There are two distinct types of electronic sensors that have been used for record playback. Older machines often used piezoelectric materials to convert vibrations into an electric signal, whereas newer devices use small moving magnets and/or coils. For the latter, there are designs with a fixed magnet and moving coils, fixed coils and moving magnet, and a piece of magnetic material moving near a fixed magnet and coils. The basic physics for the magnetic system is the same as was described earlier for the electric guitar pickup, only the vibrations here come from the stylus in contact with the record instead of a guitar string.

Piezoelectric Materials

A **piezoelectric** material is one that produces a voltage when it is bent. To play back a record, the piezoelectric material is physically connected to the stylus. As the stylus vibrates, the crystal vibrates and produces a corresponding electrical

---

[2] If the distance scale for up-and-down motion differs from the scale for the side-to-side motion, the appropriate angle will differ from 45°.

signal that can be amplified. Piezoelectric materials are an obvious choice for other types of vibration sensors as well, including some microphones.

In addition to their early use for record playback, piezoelectric sensors are often used as sound pickups from acoustic guitars, violins, and similar stringed instruments. For these acoustic instruments, the string makes very little sound; however, the string's vibrations create a force on the bridge of the instrument that is felt by the rest of the instrument. When used acoustically, the body of the instrument vibrates in response to that force, creating the sound you hear. By placing a piezoelectric sensor at or near the bridge, those same forces from the vibrations can be picked up electronically. Hence, the amplified electric signal can produce a sound more like that of the acoustic counterpart than, for example, that of the magnetically based electric guitar pickup.

Piezoelectric materials have other applications as well. Although the electric power produced is low, the voltage produced when these materials are struck can be quite high – high enough to produce a spark in some cases, which can then be used to ignite a flame. The piezoelectric effect also works the other way as well – if you apply a voltage, the material bends. This reverse process makes piezoelectric materials useful for low-power alarms and even small speakers. Piezoelectric devices are used for the fine positioning of objects or sensors, that is, as so-called *micro-positioners*. Micro-positioners are used in advanced devices such as atomic force and scanning tunneling microscopes. These microscopes can provide images of the surface of an object with atom-level resolution.

The discovery of piezoelectric materials in the late 1800s is generally credited to the French scientists Pierre and Jacque Curie. Not much more is ever said of Pierre's older brother Jacque (also known as Paul-Jacque). Pierre would later partner, both scientifically and through marriage, with the Polish-born and eventual two-time Nobel Prize winner Marie Skłodowska Curie.

While most materials are not piezoelectric, there are many common crystals that are. Examples include quartz, cane sugar (sucrose), and Rochelle salt.[3] For crystalline materials, the effect is strongest along certain crystalline directions. Recent developments have produced efficient ceramic piezoelectric materials. These noncrystalline solids are what are now commonly used in piezo alarms and speakers. Examples of these materials include barium titanate ($BaTiO_3$); the lead zirconium titanates ($Pb[Zr_xTi_{1-x}]O_3$, $0 < x < 1$), also known simply as *PZT*; and potassium niobate ($KNbO_3$). Dried bone has also been reported to exhibit piezoelectric properties.

Carefully cut quartz crystals play an important role in electronics. The crystal will have vibrational modes, similar to a plate or chime, that can be excited and detected electronically due to the piezoelectric properties. While the vibrational modes may have natural frequencies of many megahertz, they also have an extremely high quality factor ($Q \geq 1$ million). Hence, it is relatively easy to create

---

[3] Rochelle salt is also known as *potassium sodium tartrate*, $KNaC_4H_4O_6 \cdot 4H_2O$, which can be made using common kitchen ingredients.

an electronic frequency source, a "time base," using the vibration of carefully cut quartz crystals. Using commercially available crystals, it is relatively easy to achieve a frequency stability of 10 parts per million (ppm), and with some effort, one can do much better than that. To put that into perspective, an electronic timepiece that is fast by 1 ppm will gain only about 2.6 seconds in a month. Quartz crystal timing devices show up in radios, cell phones, computers, electronic keyboards, and electronic tuners used for musical instruments.

Magnetic Recording

As was already mentioned (see Chapter 17), magnetic materials exposed to a magnetic field become magnetized and may remain magnetized once the field is removed. This is the principle behind magnetic recording. To record a signal, the signal is sent as an electric current through an electromagnet. A piece of magnetic material is briefly exposed to that electromagnet, and that piece of material stores information about the field from the electromagnet experienced during that exposure. In the next instant, another piece of material is exposed, and so on. In practice, this is done by moving a continuous wire or tape of magnetic material past the electromagnet. To play back the recorded information, the magnetic wire or tape is moved past a playback coil (a nonenergized electromagnet), and due to Faraday induction, a voltage representing the signal is produced. The setup is shown schematically in Figure 19.2.

Modern digital recording devices will often use a separate **magnetoresistive** element for playback rather than a coil. A magnetoresistive material will change its electrical resistance in the presence of a magnetic field. That some materials, including iron and nickel, have this property was known in the mid-1800s,

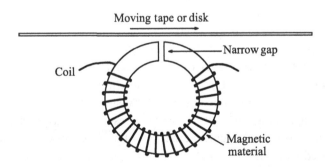

Figure 19.2 Recording and playback using a magnetic recording medium and an electromagnet. The magnetic material in the electromagnet has a small gap that is used to concentrate and control the location and extent of the magnetic fields produced and detected.

although the changes observed were very small. Highly engineered semiconductor material structures have since been developed where the changes can approach 100 percent.

The first magnetic recorders used a very thin steel wire to store and play back audio signals and are generally referred to as *wire recorders*. The first wire recorders date back to the late 1800s, almost the same time that Edison's phonograph was being commercialized. However, wire recorders did not have good commercial success until the mid-1900s, and then only briefly. These recorders had some technical problems, such as unpredictable twisting of the wire that could affect sound quality, wire breakage, and the inherent distortion due to the nonlinearity of magnetic materials. After considerable development, the quality of the wire reproduction could be comparable to the records of the time. For editing, or to fix a broken wire, wire ends were simply tied together. The knots caused an unavoidable short discontinuity in the signal as they passed by the playback electromagnet. There was no good way to use wires for multitrack (e.g., stereo) recordings.

Magnetic recording tapes and the machines to use them became affordable for home use starting in the 1950s. A recording tape is constructed by coating a thin plastic film with magnetic material, typically an iron oxide (γ-Fe_2O_3) powder, possibly with manganese (Mn), cobalt (Co), and/or chromium (Cr) included in the mix. The orientation of the tape can be well controlled, and the tape can be made wide enough so that multiple tracks can be recorded simultaneously on the same tape. Unlike plastic records, recording tape is easily erased and reused. Magnetic tapes were also developed and used for video recording and for computer data storage. Most computer storage eventually moved to disc-shaped magnetic media, where the entire recorded surface is always exposed, in order to speed up access.

Magnetically recorded information is stored in the local orientation and size of the magnetic domains in the medium (Figure 19.3). In an unmagnetized, or "blank," region, the domains will have orientations that produce no net magnetization (e.g., random orientations). To store the signal, the time-dependent magnetic field from the electromagnet organizes the domains as they pass by. This process is sufficient for digital ("on/off") signals, where, for example, the N–S direction represents the signal. Analog signals proved more challenging because the magnetizing process is quite nonlinear.

Figure 19.4 illustrates a certain "transfer function" – the relation between an input signal and the output, or final result – for a piece of magnetic material initially completely demagnetized and then briefly exposed to the magnetic field created by an input signal. In this case, small signals are suppressed, and large signals are clipped. Figure 19.5 illustrates what would happen for sinusoidal input signals of different amplitudes if recorded on a moving tape that had a transfer function as shown in Figure 19.4. The problem with saturation due to large signals is avoided by limiting the size of the input.

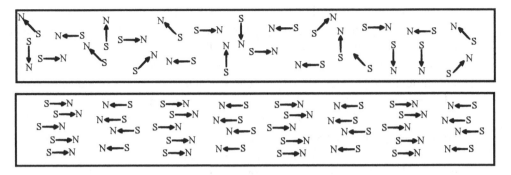

Figure 19.3 The magnetic domains are randomly oriented in a blank tape (*top*) but are oriented in a way to represent the original signal on a recorded tape (*bottom*).

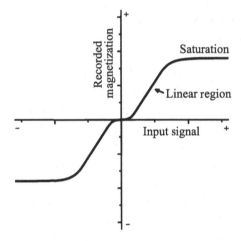

Figure 19.4 Representative transfer function for an unbiased magnetic recording.

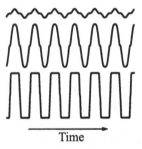

Figure 19.5 Sinusoidal input signals of different amplitudes as they would appear based on the transfer function shown in Figure 19.4.

There are two simple approaches used to deal with the magnetic-distortion problem for small signals. Both are generally referred to as *bias*. First, a constant offset to the input magnetic field can be added – that is, zero is redefined to be in the middle of the linear region. This is referred to as a *constant DC bias*, and it is easy to implement by adding a constant current to the signal or by putting a small permanent magnet next to the electromagnet. Recordings made using this first approach essentially waste more than half of the recording medium's range since you cannot cross zero. In addition, the entire tape becomes magnetized, which has disadvantages. The second approach, which is much less obvious, is to add a large sinusoidal (i.e., alternating) signal to the desired signal but at a frequency that is much too high to hear. This "AC bias" was successful and is used in modern analog recorders.

The AC signal added for normal audio recordings is very large in amplitude compared to the signal to be recorded, and it is at a typical frequency of 50–100 kHz. It may seem a bit strange to do this, but the interaction between the combined signal and the magnetic system, as the material passes the recording electromagnet, will result in a reasonably good representation of the desired signal. The large AC signal does not get recorded. As a word of caution, there are various explanations in the literature as to how this process works. Many of these explanations may seem quite reasonable on the surface but are fundamentally incorrect. In particular, the transfer function shown in Figure 19.4 is *not at all valid* for a large alternating signal – that function only arises from a brief exposure of an initially unmagnetized material to a *constant* field. Any explanation of the AC bias based on such a transfer function is therefore incorrect.

One of the first uses of an additional large AC signal for magnetic materials dates back to the early 1900s, when it was found that an AC signal that is gradually reduced in amplitude will leave the magnetic material in an unmagnetized state (i.e., "demagnetized"). That AC signal can be applied at frequencies well below 1 Hz and be equally effective. Speed is not important here, at least for a stationary magnetic material. The "high frequency" used for audio is only high as far as our hearing goes, and it is necessary because the magnetic tape is moving. For the magnetic systems used for recording, those high frequencies are still quite slow. *Fast* and *slow* are relative terms, and it is always important to ask: Fast (or slow) compared to what?

Magnetic systems can exhibit hysteresis – that is, the result depends on the history of the material, not just the current situation. Hysteresis arises in many systems when the system consists of many interacting components. In the case of magnetism, the magnetic domains interact with each other. For any particular applied magnetic field, the domains respond to that field plus the fields produced by their neighboring domains.

Magnetic systems may be difficult to visualize, so a simple analogy is perhaps useful. Consider a cylindrical container oriented so that its axis is horizontal. Inside this container is a small amount of sand (see Figure 19.6). If the container is gradually rotated, the sand will, at least for a while, move with the cylinder. If the cylinder is rotated even more, the pile of sand will start to move. If the cylinder is then brought back to its initial orientation, the center of the pile of sand will be found to be a bit to the side of its initial position. Where the sand ends up depends

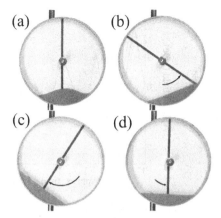

Figure 19.6 The initial slightly off-center pile of sand in (a) is "erased" through a series of rotations in (b)–(d).

on the final orientation of the cylinder and on the past history of the orientation of the cylinder. The system has hysteresis.

To center the sand pile about a particular spot using only slow rotations (e.g., no shaking), rotate the cylinder by large and equal amounts, both clockwise and counterclockwise, back and forth, with the desired spot down, on average. Then gradually reduce the amplitude of the back-and-forth rotations until there are none left. The pile of sand will end up close to the desired location. In fact, after a point, the pile will be in its final spot before the oscillations have been completely reduced to zero. The large rotations are the "bias" signal, and the desired spot represents the desired signal. Only the final resting position is recorded, not the rotations used to get it there.

For the magnetic system, a large alternating signal that is gradually reduced in amplitude is combined with the desired signal. For a magnetic recording, the various timescales are determined by the upper frequency to be recorded. Imagine a small piece of the recording tape that is to record the desired signal at a particular time. As that piece of tape passes over the recording electromagnet, it sees both the large AC bias and the desired signal. Initially, the AC bias is so large that any previous information on the tape is lost, and the desired signal is actually inconsequential. However, as that piece of the tape leaves the coil, the amplitude of the AC bias – and of the desired signal – will decrease due to distance. Since the system ends up "frozen" before the signals have gotten all of the way to zero, the desired signal is recorded as the tape departs.

To get this to work, the desired signal should be slow enough so that, as far as the piece of tape is concerned, the portion of the magnetic field produced by the signal looks almost constant in time, at least during the time the piece of tape is leaving the electromagnet. However, the AC bias should be fast enough so that, as it leaves, the piece of tape sees many alterations of the field that gradually decrease toward zero. The timescale, the dividing line here between what is fast and slow for the recording process, is set by how fast the piece of tape leaves the

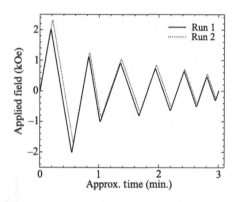

Figure 19.7 The time dependence of the uniform field applied to a stationary sample of recording tape for two runs – the first with zero offset and the second with a small positive offset. Such a field simulates the alternating field used during recording.

magnetic field produced by the electromagnet. To make a good recording, the latter, in turn, is set by the size and geometry of the electromagnet, the details of the tape, and the highest-frequency signal to be recorded. There will be an optimum amplitude for the bias field, which depends on the type of magnetic material used in the tape. Some recorders include a separate adjustment for the bias.

Since the process is already "slow" for the magnetic system, a magnetometer can be used to simulate the process to see what happens. Figure 19.7 shows the applied field used for two demonstration measurements for a small (stationary) piece of recording tape. In each case, the sample was prepared to have the same starting magnetization. For the first run (Run 1), a simple alternation about zero was used, which was gradually decreased in amplitude. For the second run (Run 2), a small positive offset was added, which was also gradually decreased in amplitude. That offset – zero for the first run and positive for the second – represents the signals to be recorded. Notice how small the difference is between the two applied signals.

The resulting magnetic behavior is often plotted as the magnetization in the sample as a function of the applied field. The time dependence is not evident in such a plot, but the sequence of events can be extracted. Figure 19.8 shows such plots for the two runs. Note that by the end of these runs, the applied fields have not decreased all the way to zero, but the recorded signals have already spiraled into their final values, and those values are quite different. Continuing the process beyond what is illustrated in Figure 19.8 would make little difference. The first case, Run 1 with zero offset, represents both the case where the signal happens to be zero and the process used for tape erasure, also referred to as *demagnetization*.

The bias field is used to create the spiral-like convergence to the final value but is not itself recorded. This biasing process is not necessary on playback. For playback, the tape moves past a magnetic-field sensor. A moving tape can use

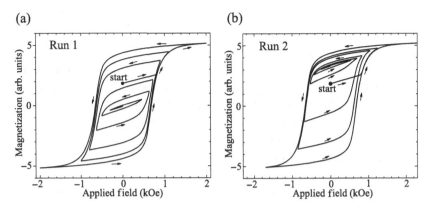

Figure 19.8 Magnetization of a piece of recording tape as the applied field is varied as shown in Figure 19.7. In (a), the tape is left demagnetized, whereas in (b), the final magnetization is positive.

a coil (an unenergized electromagnet) as a sensor. The coil sees a time-dependent magnetic field due to the stored magnetization as the material moves past. That time-dependent magnetic field creates an electric signal by way of Faraday induction, as was seen for the electric guitar pickup.

Optical Media

During optical recording and reproduction, the optical properties of a material (color, reflectance, transparency, etc.) are altered to represent the information to be stored. Those optical properties are then measured during playback. The two most common commercial uses of optical recording are the analog "sound on picture" used for movies and digital recording on compact discs (CDs), digital versatile discs (DVDs), and similar media.

The optical properties of a material can be altered in a number of different ways. The simplest are mechanical deformation (e.g., scratching or denting) and dyeing (or painting) the surface. Both are used. One method for dyeing makes use of light-sensitive materials, and then light is used to alter the optical properties. Such optical recording methods have been used for the production of sound in motion pictures and are currently used for user-writeable CDs and DVDs. The information on mass-produced CDs and DVDs is in the form of mechanical dents stamped into the surface – a very inexpensive process.

A detector of light – a **photosensor** – is required to read optically stored data. Aside from our eyes, perhaps the earliest devices used to measure light intensities were thermometers. Light is a wave, and waves carry energy. If the light is absorbed, that energy is converted to an alternate form, such as heat.

Adding heat will usually cause the temperature to rise.[4] William Herschel (1738–1822) discovered infrared radiation by sending sunlight through a prism, which spreads the light into a rainbow, and using a thermometer to compare the temperature rise from the different colors. He noticed that next to the red light was a region where the temperature would rise, but no light was visible. However, the ability of thermometers to measure changes at audio frequencies is quite limited – they are too slow to respond – and so this technology is not particularly useful for recorded music.

In the late 1800s, one of Hertz's many discoveries was the **photoelectric effect** – that shining light on metal plates can change their electrical properties. The full explanation of the effect was presented by Einstein in 1905 in one of his four "miracle year" papers, although the explanation was not fully verified experimentally for another decade. In 1921, Einstein was awarded the Nobel Prize for this work.[5] The basic idea is that the electrons in a metal that give rise to electrical conduction are relatively free to move around inside the metal; however, some energy is required to get them out of the metal.[6] Einstein used the idea that light comes in discrete packets of energy, now referred to as **photons**. Those photons can penetrate into the metal and, if they have enough energy, can knock electrons out. If there is an electric field present, those freed electrons can respond to that field, resulting in an electric current (Figure 19.9). Since it was well known that light was a wave, the existence of photons was greeted with considerable skepticism. This **wave–particle duality** – that something can behave both as a wave and as a particle – is one fact that led to the development of quantum mechanics.[7]

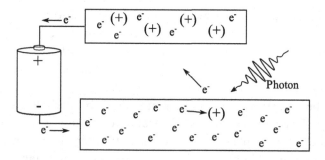

Figure 19.9 An electric current can result when a photon strikes an electron in a metal. If the photon has enough energy, the electron can be ejected from the metal and collected on a nearby positively charged electrode.

---

[4] There are a few conditions where adding heat does not cause a temperature rise. For example, adding heat to a glass of ice water can change the amount of ice present, but it does not change the temperature until all the ice is melted.

[5] This was the only Nobel Prize awarded to Einstein, despite his many other equally, if not more important discoveries.

[6] If this were not the case, the metal would disintegrate on its own.

[7] It was later discovered that electrons, normally thought of as particles, can also act as waves.

There are a number of photosensors based on semiconductor materials instead of metals. The basic principle of operation is similar to that of the photoelectric effect, except that the electrons are not ejected from the material. Instead, the electrons make a transition from one state within the material to another. For example, in a photoresistor, the electrons are (temporarily) knocked out of a state that does not conduct electricity into one that does. A semiconductor and a metal together can be used to create a device known as a *solar cell*, which ultimately converts the light energy into electrical energy. Such solar cells were known in the early 1800s. Modern solar cells are typically constructed with two semiconductor layers. In all these detectors, the principle is that photons come in and knock the electrons from their "normal" state to another, thus altering the electrical properties. The size of the effect depends mostly on the number of photons of sufficient energy that strike the material each second. All other factors kept constant, the number of photons per second is proportional to the intensity of the light.

Sound on Picture

Sound on picture was introduced in the early twentieth century in order to add synchronized sound to motion pictures. The sound information is included on the film alongside the images, which was an improvement over the use of synchronized gramophones. In essence, an image of the sound intensity is stored as a brightness on the film. The two primary competing technologies are referred to as **variable density** and **variable width**. For variable density, the sound is recorded as gray levels on the film. For variable width, the sound is recorded using a transparent trace, where the width of the trace controls how much light gets through. Both technologies can be played back using the same projector. A light source shines on one side of the film, and the intensity of the transmitted light is measured on the other (see Figure 19.10). A light sensor turns the intensity into an electrical signal that can be amplified and sent to a speaker.

The variable-density format is harder to produce, and reproduce, reliably. To get the correct shades of gray from the sound signal without over- or underexposing, over- or underdeveloping the film, or otherwise distorting the sound is comparatively much more difficult than with the fully exposed variable-width track. There is much more room for error with the on/off variable-width format – there is either enough light (or more) to fully expose the film, or none. As a consequence, the variable-density format did not last long. Creating a variable line width, which is then recorded on the film, can be accomplished using shutter arrangements, such as shown in Figure 19.11.

The audio signal that is to be recorded can be both positive and negative. Since there is no way to get "negative light," an offset must be added so that the most negative signal becomes positive. As a result, when the signal is zero, one-half of the maximum intensity must get through. When the signal is at the most negative possible, the light is just blocked, and when the most positive, the maximum

Figure 19.10
Sound-on-picture
information is read by
projecting light through
the sound track on the
film. A detector turns the
intensity of light into an
electrical signal.

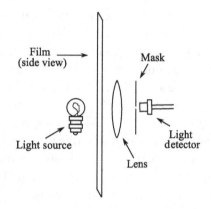

Figure 19.11 Light
through a narrow slit,
with a V-shaped shutter
that moves with the
sound amplitude, is
focused on the edge of
the film to create
a bilateral variable-width
recording.

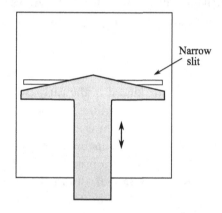

amount of light gets through. This is illustrated in Figure 19.12 for the variable-width format.

The variations in the amount of light getting through are sometimes mistakenly claimed to be variations in loudness. That is not accurate, of course, because loudness is closely related to power, proportional to amplitude squared, and not amplitude. The signal is loudest where there are the largest time-dependent *variations* in the amount of light getting through, positive and negative, from the average, not simply where the most light gets through.

The variable-width format adapts well to the use of stereo (and other multi-track) sounds by simply increasing the number of tracks. Those stereo movies, with left and right sound tracks, were backward-compatible so that they could be used with the older monophonic (single-track) projectors. The older projectors would see the total light intensity from both tracks, creating a sound based on the sum of the two. The newer projectors included two separate sensors to read the two separate tracks.

Since image projection and sound detection occur at different locations within the projector, the sound and images are offset from one another on the film.

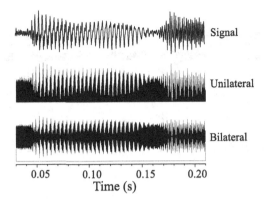

Figure 19.12 Example showing how the amplitude of a signal (*top*) is related to the variable-width recording. Both unilateral (*middle*) and bilateral (*bottom*) formats are shown. For this figure, the dark area of the recorded signal represents the transparent region on the film.

For most formats, they are about one second apart. The sound is read after the image projection, so the sound will precede the image on the film. Referring to Figure 19.13, the sound tracks represent the sound that will occur about one second after the adjacent images are shown on the screen.

Not all movie films used these formats. Magnetic recording tape can be attached in parallel to the film to include the sound. Newer digital formats are in common use for theater presentations. In those formats, the sound is represented as 0's and 1's and may include data-compression techniques. Also common now are entirely electronic movie formats, for both sound and image, that do not use film at all.

CDs and DVDs

With the development of high-speed electronics in the latter part of the twentieth century, many digital formats were developed. Digital magnetic tapes had a brief appearance, although the optical techniques, recorded on discs, were the commercial winner. **Compact discs** (CDs) and **digital versatile discs** (DVDs) are perhaps the most well known. Both developed from what were referred to as *laser discs*. The discs were initially available to consumers as read-only devices, although writable discs soon became available as well. Data are stored by modifying the optical properties along a spiral path around the disc. On these discs, the track goes from the inside out, in contrast to vinyl records, where the track goes outside in.[8]

Unlike vinyl records and the usual sound-on-picture formats, which are both considered analog formats, the digital formats use just two states to record the data. Optically, the two states might be stored using reflecting and nonreflecting

---

[8] For some dual-layer formats, the second layer starts from the outside and goes back into the center. Dual-layer systems are not considered here.

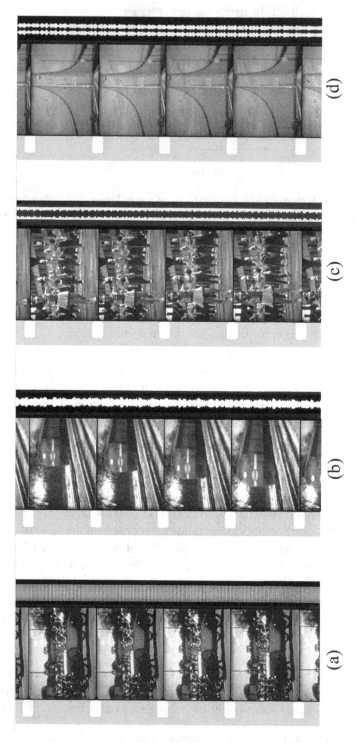

Figure 19.13 Examples of sound-on-picture formats. The sound information is on the right side. (a) Mono variable density; (b) mono variable width; (c) stereo, dual unilateral; and (d) stereo, dual bilateral.

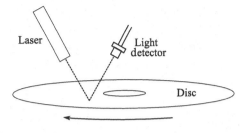

Figure 19.14
Signals on optical discs are read using the reflection of laser light off the surface. Changes in the detected signal correspond to 0's and 1's.

regions or, at least, more reflecting and less reflecting regions. The basic setup to read such a disc is shown in Figure 19.14. Some earlier devices used the variation of light transmission through a clear disc rather than reflection; however, reflection has proved to be easier to use. In modern discs, the *change* in the light intensity and the lack of a change when a change might be expected are considered the two states used to encode the information.

The size of the smallest spot of information (a "bit") on the disc is limited by the wave properties of the laser light. If the size of the spots on the disc is much smaller than the wavelength, diffraction effects become prominent. Hence, the smallest spots that can be used are comparable to the wavelength used. Audio CDs commonly use infrared light, video DVDs use red light, and the highest-capacity video discs now use blue light. As inexpensive diode lasers are developed for shorter and shorter wavelengths, the density of data on the disc can grow and grow. It is interesting that both the wave and particle nature of light come into play in these simple everyday devices – diffraction arises from the wave-like properties, and detection arises from the particle-like properties.

Digital Storage and Playback

Digital signals rely on a two-state system – call them 0's and 1's, "yes/no," or "on/off," or whatever you like. Digital signals are much, much easier to send, receive, and store than analog signals. In fact, the earliest forms of long-range communication were digital – examples include drums, smoke signals, and the telegraph. To be able to use such signals to, for example, store and/or play back a Puccini aria is, however, another matter. The transmission speed necessary for high-quality digital audio signals has only become available in the last few decades. Since then, the speed available with digital signals, storage, and transmission has grown approximately exponentially[9] to well beyond what is necessary for audio, and so digital recordings have virtually taken over.

---

[9] The growth is often described using "Moore's law" – that the number of transistors in an integrated circuit doubles roughly every two years.

There are many different formats and methods to store and play back digital music; however, the details rapidly become too specific to discuss here. Instead, consider two different extremes that can be used: those based on **base-2 binary** values and those based on on/off pulses (i.e., **1-bit encoding**) to encode and decode the signals.

1-Bit Encoding

Pulse-width encoding is a simple on/off scheme most often used only during playback, although it is possible to use it for storage as well. In the simplest scheme, the signal is represented by a rapid series of pulses with a fixed pulse spacing. The length of the pulse can vary from 0 to 100 percent of that pulse spacing. The concept is not unlike that of the variable-width optical tracks shown in Figure 19.13. The pulses are then sent through circuitry that outputs the average signal over a somewhat longer duration. The fraction of "on" time compared to the total time is referred to as the **duty cycle**. In order to produce a quality reproduction of the signal, the pulses must occur very rapidly compared to the highest frequencies in the signal. Another version of this scheme uses pulses of fixed length, but the time between pulses is varied. Again, the average of the pulsed signal is used. The former is referred to as *pulse-width modulation*, whereas the latter is referred to as *pulse-density modulation*. Both are illustrated in Figure 19.15.

The advantage of pulse-width and pulse-density encoding for playback is that only relatively inexpensive digital (i.e., on/off) electronics is required, and for sound reproduction, the time averaging can be done using a simple low-pass filter[10] or even using the mechanical inertia within the earphones or speakers used for playback. If the pulses occur very rapidly compared to human hearing, that high frequency will not be heard in any event. Highly efficient "Class D" amplifiers use this same idea but at much higher power levels.

Another simple scheme relies on pulses to represent *changes* in the signal. The idea is illustrated in Figure 19.16. To keep the electronics simple, only

Figure 19.15 Two schemes that represent a signal using the duty cycle of electric pulses. Pulse-width modulation is shown in (a), and pulse-density modulation is shown in (b).

[10] See Chapter 18 on RC time constants for examples.

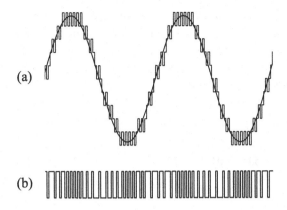

Figure 19.16 For delta–sigma encoding, the continuous signal in (a) is approximated with a signal that either increases or decreases by a fixed amount at fixed time intervals. The recorded signal (b) is a record of the direction of the changes.

two possibilities are included: the signal either increases or decreases by a fixed, small amount during each fixed (short) time interval. As shown, a rapid string of such changes can approximate the signal. The recorded signal is simply a record of the direction of the changes. The output signal is the running sum of the changes, sent through a low-pass filter. The pulses are generally sent at a rate much faster than even the highest-frequency component of the signal to be reproduced, and hence the demands on the filter are minimal. This 1-bit method is sometimes referred to as a "delta–sigma" (Δ–Σ) conversion. A delta is used to signify the use of changes in the signal, and the sigma represents the summation of those changes. This method can be used both for input and output. On input, during each fixed time interval, the digitally produced output signal is compared to the desired input signal to determine whether the next pulse should be an increase or decrease.

Base 2

Base-2 binary values are another convenient way to store numerical values using digital electronics. Recall that for base-10 values, digits range from 0 to 9, and depending on their position in the number, they are multiplied by an appropriate power of 10.

Example 19.1

The base-10 number $2{,}749 = 9 \times 10^0 + 4 \times 10^1 + 7 \times 10^2 + 2 \times 10^3$.

Base-2 values are formed in a similar manner, except only the digits 0 and 1 can be used (i.e., the digits that are less than 2), and instead of powers of 10, powers of 2 are used. Note that in both cases, the powers read from right to left and start with 0. Common errors include starting to count with 1 and/or reading values from left to right.

Example 19.2

The base-2 number $100111 = 1 \times 2^0 + 1 \times 2^1 + 1 \times 2^2 + 0 \times 2^3 + 0 \times 2^4 + 1 \times 2^5$
$$= 1 + 2 + 4 + 0 + 0 + 32 = 39 \text{ base } 10.$$

For both base 10 and base 2, if a decimal point is present, the digits to the right of the decimal have negative exponents.[11]

Examples 19.3

The base-10 number $23.14 = 2 \times 10^1 + 3 \times 10^0 + 1 \times 10^{-1} + 4 \times 10^{-2}$.
The base-2 number $101.11 = 1 \times 2^2 + 0 \times 2^1 + 1 \times 2^0 + 1 \times 2^{-1} + 1 \times 2^{-2}$
$$= 4 + 0 + 1 + 1/2 + 1/4 = 5.75 \text{ base } 10.$$

These examples also illustrate how to convert from base-2 values to base 10. To go the other way is somewhat more involved. There are several ways to think of the conversion, but one that works is to repeatedly subtract the largest power of 2 that is less than the remaining value, keeping track of which powers are used with a "1" and those that are not used with a "0." Once the remaining value is 0, the process is finished. It helps to start by writing down a list of the powers of 2.

Example 19.4

Convert 154 base 10 to base 2.
Powers of 2: 1, 2, 4, 8, 16, 32, 64, 128, 256
Exponents: 0, 1, 2, 3, 4, 5,   6,   7,   8
Subtract the largest that is less than or equal to 154: $154 - 128 = 26$.
Subtract the largest that is less or equal to 26: $26 - 16 = 10$.
Subtract the largest that is less or equal to 10: $10 - 8 = 2$.
Subtract the largest that is less or equal to 2: $2 - 2 = 0$.

---

[11] The term *decimal point* is often retained for bases other than 10.

The values subtracted were $128 = 2^7, 16 = 2^4, 8 = 2^3,$ and $2 = 2^1$. Put a 1 for the corresponding powers of 2, and use 0 for the rest. Remember that the lowest power of 2 is 0 and is recorded on the right.

Result = 10011010 base 2.

Electronic circuits to perform such a procedure will be similar, although all the subtractions are performed. If the result of the subtraction is negative, that result is discarded, effectively undoing the subtraction. There will be extra steps involved, but the extra electronics required to skip steps slows the process more than simply performing all of the subtractions.

Electronic representations will usually have a fixed number of digit positions available. Each is referred to as a *bit*. Thus, electronics built for 8-bit calculations would use eight 0's and 1's to represent base-2 values corresponding to the exponents 0 to 7. There will be 8 steps in the electronic conversion. Start with the highest power and work down to the lowest.

Example 19.5

Convert 154 base 10 to an 8-bit base-2 value:
Subtract 2^7: $154 - 128 = 26 \rightarrow$ not negative, so keep and record a "1."
Subtract 2^6: $26 - 64 = -38 \rightarrow$ negative, so skip and record a "0."
Subtract 2^5: $26 - 32 = -6 \rightarrow$ negative, so skip and record a "0."
Subtract 2^4: $26 - 16 = 10 \rightarrow$ not.negative, so keep and record a "1."
Subtract 2^3: $10 - 8 = 2 \rightarrow$ not negative, so keep and record a "1."
Subtract 2^2: $2 - 4 = -2 \rightarrow$ negative, so skip and record a "0."
Subtract 2^1: $2 - 2 = 0 \rightarrow$ not negative, so keep and record a "1."
Subtract 2^0: $0 - 1 = -1 \rightarrow$ negative, so skip and record a "0."
Result = recorded values (in order, left to right) = 10011010.

To convert a 16-bit value, start with 2^{15} and use 16 steps, and for 32-bit values, start with 2^{31} and perform 32 steps, and so forth.

To create a signal directly from base-2 values, a sum of signals (e.g., voltages), each of which is proportional to a corresponding power of 2, is used. If the corresponding bit is 1, the signal is added; if the bit is 0, then 0 is added instead. One efficient way to convert a signal to base 2 starts by turning a guess into a signal and then comparing that guess to the actual signal (e.g., a comparison of the two voltages). The process is the same as in the previous example, except that the value is converted to a signal (e.g., voltage) for the comparison. There are dedicated integrated circuits available that can make these conversions relatively quickly.

There are, of course, additional methods to convert to and from stored digital values and electrical signals. The interested reader is referred to a more specialized electronics text for more information about alternative methods.

Nyquist Folding

For all of the digital techniques, the signal is sampled at discrete time intervals, and the signal is assumed to vary smoothly between samples. When the signal does not vary smoothly between samples, the digital data will not produce an accurate representation of the original, continuous signal. This distortion is generally referred to as **aliasing**.

One important effect is referred to **Nyquist folding**. The effect is illustrated for a sinusoidal signal in Figure 19.17. There, a sinusoidal signal (*dotted*) is shown with a sampling frequency that is too low to accurately represent the sinusoid. If the sampling frequency is just a little too low, the signal may look distorted. If the sample frequency is very low, the result is a sinusoid, but at a much lower frequency. The frequency of the original sinusoid has been "folded" down to a lower frequency.

If the sampling frequency is f_s, then the sampled data will have a frequency no larger than $f_s/2$. To determine the resulting components of the recorded signal, take the spectrum of the original signal and, as necessary, add or subtract multiples of f_s to the frequency of each component until the result has all components in the range $-f_s/2$ to $+f_s/2$. Here, the negative frequencies act as positive frequencies but with an added phase shift (i.e., a minus sign is added to all of the sine terms in the Fourier series; see Chapter 9).[12]

Figure 19.17
Illustrating the effects when a continuous sinusoidal signal (*dotted*) is sampled a little too slowly (*open circles*) and much too slowly (*closed circles*).

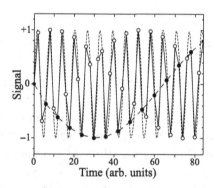

---

[12] In some movies, it is possible to see wheels on vehicles first go forward, but then appear to reverse direction. This is a visual version of Nyquist folding that arises since the movie consists of a series of visual samples that may not occur fast enough to avoid the "folding."

Two important strategies are used to avoid aliasing problems for audio recordings. First, the signal must be sampled at a frequency that is twice the maximum desired for the signal at hand. Since hearing ranges from about 20 Hz to 20 kHz, high-quality recordings typically sample at 40 kHz or higher. Second, the signal is filtered (see Chapter 18) to eliminate, to the extent possible, all Fourier series components that have a frequency higher than half of the sample rate. This filter is referred to as an **anti-aliasing filter**. It is not just the desired signal that is important since there may be considerable electrical noise (e.g., above 20 kHz), and without that filter, that noise could be moved down into the audio range. Without filtering, noise that cannot be heard in the original sound can appear in the recording.

Storage and Transmission Strategies

Storage and transmission of digital values will often include more bits than is necessary in order to better ensure accurate transmission. The simplest schemes simply help to detect errors. A scheme used for data on many CDs is to record data in 8-bit chunks ("bytes"), and each 8-bit chunk is encoded using a 14-bit code. Not all possible 14-bit codes are used. There are $2^{14} = 16,384$ unique 14-bit codes but only $2^8 = 256$ unique 8-bit codes, so most 14-bit codes are "illegal." While this may seem like overkill, the translation is cleverly designed so that if any single-bit error occurs within the 14-bit value, that error can be detected (the code does not correspond to a valid 8-bit value) *and* corrected. The scheme also has some other nice properties that ease the design requirements for the electronics.

While this discussion has moved away from physics for the moment, it is worth mentioning one more feature of digital data important for music – data compression. In order to use resources more efficiently, various data-compression techniques have been developed for both audio signals and images. There are two categories of compression: lossless and lossy. Lossless compression preserves the original (digital) information exactly but more efficiently deals with redundant information. For example, if the recorded data has 70 values in a row that are identical, rather than repeating the value 70 times, you store the value 70, the value that is repeated, and an indicator that there is a repeat. Lossy compression is very common for audio and images, and it relies on the fact that a good approximation to the data is often indistinguishable, by humans, from the original.

One common form of lossy compression relies on a cosine transform. To get into all the details of that compression goes beyond this presentation. However, the basic idea is that for short sections of the data, the cosine Fourier transform is computed, leading to a Fourier series representation based only on the cosine function (see Chapter 4). That is, a piece of the signal, $s(t)$, that starts at time $t = 0$ and ends at $t = T$ is written

$$s(t) = A_0 + A_1 \cos(\omega_0 t) + A_2 \cos(2\omega_0 t) + A_3 \cos(3\omega_0 t) + \dots, \qquad (19.1)$$

where $\omega_0 = 1/(4\pi T)$. If there are N data values in the range from $t = 0$ to T, then N values of the coefficients, A_i, are necessary to exactly match the data. However, many of those amplitudes, $\{A_i\}$, will be small and can be omitted without changing the perception of the result. The result is close enough without them. Refer back to Figure 4.2 for examples of approximations using such a Fourier synthesis. Efficient methods to compute the coefficients, along with fast electronics, allow this type of compression (and later, decompression) to be used in real time. For more information, look for articles about the "discrete cosine transform" (DCT) and "modified discrete cosine transform" (MDCT). Two-dimensional versions of these transforms are used for image compression.

Summary

Recordings of sound can use analog (continuous) or digital (on/off) values stored using a suitable medium. Early recordings used the physical modification of a medium (metal foil, wax, plastic) to store a representation of an analog signal. The recordings could be played back by using those modifications directly to reproduce the sound. As electronic technology developed, sensors, amplifiers, and speakers were used. Modern recording is mostly electronic.

Magnetic recordings store values by modifying the magnetization of a magnetic material. The nonlinear magnetization process can be made linear through the use of a large periodic biasing field, significantly larger than that of the signal to be recorded and at a frequency much too high to hear. The signal can be played back using Faraday induction and a pickup coil, followed by amplification.

Optical storage uses changes in a material's optical properties to store the signal. Earlier optical recording includes the analog sound-on-picture technology used by the motion picture industry. CDs and DVDs use changes in the optical properties of a plastic disc to store digital representations of the signal.

Digital data may be stored in a number of formats. One that is common uses a digital signal output that changes rapidly compared to the desired signal and is then electronically filtered to remove the high-frequency portion of the signal. Another uses base-2 binary representations of values. An appropriate sampling rate must be used for digitized signals to avoid aliasing and Nyquist folding effects. Digital signals can be compressed – stored more efficiently – using a number of techniques, some exact and some approximate (lossy).

ADDITIONAL READING

Berg, R. E., and D. G. Stork. *The Physics of Sound*, 2nd ed. Prentice Hall, 1995 [or 3rd ed., Pearson, 2005]. (See Chapter 7.)

Suits, B. H. *Electronics for Physicists: An Introduction*. Springer, 2020. (See Chapter 13.)

PROBLEMS

19.1 A long-playing (LP) vinyl record rotates at 33 1/3 rpm (rotations per minute). The music is recorded in a spiral path starting from the outside edge. A typical LP of 12 in. (30 cm) in diameter can contain about 20 to 25 minutes of recorded sound on each side. How long is the spiral track on each side?

19.2 Sixteen-millimeter movie film typically projects 24 frames (images) per second. The film is 16 mm wide, and each frame is 7.62 mm high (there are 40 frames per foot). If a 10-kHz sound is also recorded on the film, what is the spacing (on the film) between adjacent maxima of the recorded sound signal?

19.3 If a digital signal is represented with 8-bit base-2 binary values, what is the smallest change in the signal, expressed as a percentage of the maximum possible signal, that can be represented? If a resolution of no worse than 0.1 percent of the maximum is required, how many bits are required?

19.4 How old are you if your age in base 2 is 1,000,000 years? What is your current age in base 2?

19.5 Compare the fingerings for a trumpet with base-2 binary. How are they the same, and how are they different?

19.6 The energy for a photon of light depends on the frequency of the light: $E_{photon} = h\nu$, where h is Plank's constant and ν is the frequency in Hz. Look up the values necessary to compute how many photons per second strike a 1-cm^2 surface for red light with an intensity of 1 W/m^2 and then again for blue light. Describe in general how, for fixed intensity, the number of photons per second depends on frequency.

19.7 Consider a scheme where numbers are to be represented using 5 bits, exactly two of which are 1's and three are 0's. How many different numbers can be represented with such a scheme? How about a scheme with 6 bits, exactly three of which are 1's and three are 0's?

20   Electronics and Music

It is quite common to hear electronic effects and electronic instruments in today's music. Some electronic instruments have already been discussed, most notably the electric guitar. The goal here is to talk a bit more about how electronics can be used more generally in the creative process, at least where there is a connection to the physics already discussed in earlier chapters. What follows is a sampling of some electronic effects and electronic synthesis, with a particular emphasis placed on those with a physics connection. Many of these effects are now routinely accomplished using computers and software. As was the case in Chapter 19, it is not the goal here to explain how computers work. There are, however, many interesting effects that are, or can be, accomplished without computer-based techniques.

Electronic Effects

A large number of electronic effects have become popular. Interestingly, one of those, the various types of **distortion**, originally arose due to the limitations of the electronics. Distortion has now become part of the standard set of effects. As electric guitar players searched for more and more volume, they rapidly reached the limits of their amplifiers. One result was a **clipped signal** that is due to the maximum output of the electronics. Once the output signal reaches a maximum value, it can go no further, no matter what the input signal. Such an overdriven signal will sound distorted. The clipping can be "hard" or "soft." Hard clipping has a definite upper limit, like a ceiling in a room. The signal is undistorted as long as it stays below that level. Soft clipping has a more gradual limit – a padded ceiling. Above a certain level, the amplification becomes less and less until, finally, the maximum is reached. Both cases have a nonlinear relationship between input and output if the "ceiling" is reached. The guitar effects referred to as *overdrive* and *fuzz* are types of distortion.

(a)   (b)

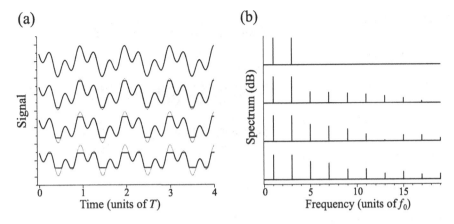

Figure 20.1 A sinusoidal wave with different levels of symmetric clipping is shown in (a), with corresponding spectra shown in (b).

Any nonlinear relationship between input and output will add overtones to the output that were not present in the input and can enhance some that were present. The changes to the overtones will, of course, change the sound. Additional overtones will be at frequencies that are multiples of the frequencies of the original overtones and also at sums and differences of multiples of the frequencies of the original overtones. An example, using a simple signal, is shown in Figure 20.1. The symmetric clipping in that example results in overtones only at odd multiples of the fundamental. Asymmetric clipping will produce overtones at all the multiples.

There are many other electronic effects based on signal modification within an amplifier. One is the **chorus effect**, where the signal is split into several copies, and then some small phase (or time) variation is added to one or more of the signals before the signals are recombined. This process simulates the imprecise match between different performers. There are several other effects based on splitting the signal, making alterations, and then recombining the signals.

An effect based on **automatic volume control**, to reduce static, turns off the signal when the level drops below a prescribed value. Another will compress the volume so that soft signals are made louder and louder signals are made softer. When applied to a signal that decays in time, this can create a **sustain effect**. Still other effects are various types of filters that allow and/or block certain frequencies, possibly based on signal amplitude. The list of effects based simply on electronic manipulation is extensive and is more about electronics than it is about physics.

A relatively simple effect that uses **time-dependent filtering** of the signal is the **talk box**. Here, a sound is sent through piping into someone's mouth. As they move their mouth, the normal vocal filtering action modifies the sound. If they talk or sing, but without producing any sound on their own, the incoming sound

Figure 20.2 A vocoder uses one signal, in this case, speech, to control another, in this case, the sound of a violin.

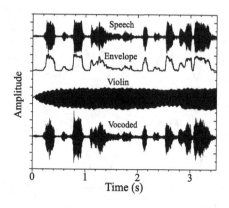

can be made to mimic speech. A more elaborate electronic version of this is the **vocoder**. A vocoder takes two audio inputs. One of the inputs is used to control the other. Typically, the control input is a speaking or singing voice. The vocoder electronics (or software) extracts, at least approximately, the filtering that had been applied (by the person's vocal tract) to get that signal and then applies those filters to a second signal. That second "carrier" signal might be from an electronic organ, guitar, violin, or other musical instrument. The process is illustrated in Figure 20.2, where the envelope from spoken words has been applied to the sound from a violin. The resulting sound is not as easy to understand as the original, but some words are understandable.

The basic design for the vocoder originated during the 1920s as a potential data-compression technique for early telephone systems. If only the relatively slowly changing "filter information" need be sent, then there may be cost savings. The carrier signal is reintroduced at the receiving end. The vocoder did not catch on for this purpose.

Several of the electronic effects used today were originally implemented using "electromechanical" devices. Two examples of such devices are the **spring-reverb** unit and the **motor-driven tremolo** unit.

To make a signal sound like it was produced in a reverberant room, reverberation can be added. This can be done in the recording studio in a number of ways. Early solutions included the use of a separate reverberant room. In this method, the signal is sent to speakers in a separate room, where microphones then pick up the sound and send the signal back. Long pipes can also be used to such effect. A speaker at one end of the pipe supplies the sound, and a microphone, either at the same or the opposite end, picks up the sound. A much more compact way to achieve a similar effect is the **spring reverb**, a popular choice for use within an electric guitar amplifier.

A spring-reverb unit uses a long, fairly weak (i.e., low spring constant) spring as an acoustic transmission line. The signal is applied to one end of the spring using an electromagnet and then picked up at the other end using Faraday's law – that is, a coil next to a moving magnet. That delayed signal is mixed with the original

Figure 20.3 A schematic of the inner workings of a spring-reverb unit. Springs can range from a few centimeters to almost a meter.

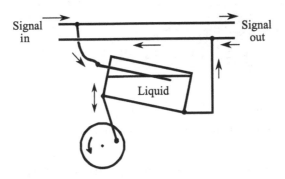

Figure 20.4 The inner workings of an electromechanical tremolo device. As the motor-driven wheel turns, the liquid makes contact with the inner wire. As the wire becomes increasingly submerged, more of the signal goes through the liquid, reducing how much makes it to the output.

signal to simulate reverberation. While the details may differ somewhat from unit to unit, the basic design is shown in Figure 20.3.

Tremolo – a periodic variation in volume, typically at a frequency of a few hertz – is a common effect. Tremolo is often confused with **vibrato**, which is a periodic variation in pitch. While there are numerous ways to produce the tremolo effect, one of the early versions was based on a metallic can containing an electrically conducting liquid. As the can moves, the liquid can cause an electrical connection across the signal lines. That connection will siphon off, so to speak, some of the electrical signal, causing a reduction in output volume. If the liquid were highly conductive, such as metallic mercury, the signal would simply be on/ off. To allow more control over the effect, a poorly conductive fluid is used, such as water or water with some additives. An example of the essential inner workings for such a device is shown in Figure 20.4. The first such device seems to have originated in the 1940s and is credited to Harry DeArmond, a pioneer in the development of the electric guitar pickup.

With a multi-speaker playback device, interesting effects can be designed based on filtering and signal processing. In particular, the addition of phase or time shifts to the signals is of use. For example, auditorium speakers above a stage can be fed slightly delayed signals that mimic signals from the stage that are reflected

Figure 20.5 The array of speakers at the Jay Pritzker Pavilion in Chicago can be used to simulate the reflections normally heard in an auditorium.

off the ceiling. As long as the volume from those speakers is not too high, the sounds will still seem to arise from the stage, although now with increased volume. That effect is due to the **precedence effect** (refer back to Chapter 15). Human hearing is such that the sound that is heard first tends to be what is used to locate the source, even if that sound is less intense than later reflections, or other reproductions, of that sound.

An extreme example of the use of delays to fool hearing is the Jay Pritzker Pavilion in Chicago (Figure 20.5). Here, a large array of speakers is used to mimic the reflections expected for an auditorium, but with no auditorium. The sounds produced on stage are fed to the sound system with appropriate delays to produce an auditorium-like acoustic experience. If the results of physics for an enclosed auditorium are appropriately simulated electronically, the audience will have a similar experience.

Other examples of effects based on signal delays and filtering include surround sound, expanded stereo, and various other "three-dimensional" audio effects. These are generally based on the physics of sound combined with psychoacoustics – the science of hearing perception. The finite speed of sound causes delays (i.e., phase shifts), and the environment causes various filtering actions. Both can be simulated electronically.

Making Music with Electronics

Originally developed in the 1930s, the **Hammond organ** was one of the earliest electromechanical musical instruments. A keyboard instrument, the Hammond

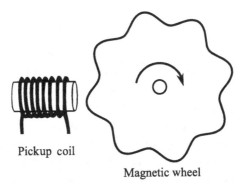

Pickup coil

Magnetic wheel

Figure 20.6 The basic principle used by the Hammond organ is a motor-driven wheel made of magnetic material and an electromagnet pickup, similar to an electric guitar pickup. The example wheel shown has eight "teeth."

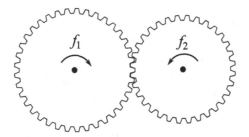

Figure 20.7 The gear on the left has 37 teeth, whereas the gear on the right has 32. Hence, their rotational frequencies are related by $f_2 = (37/32) f_1$.

organ uses tone wheels made from a magnetic material (i.e., iron or steel) next to a coil (Figure 20.6). The signal is generated by means of Faraday induction, such as previously described for electric guitars (Chapter 17). The number of "teeth" on the tone wheel and the wheel's spinning rate will determine the frequency of the signal. The shape of the teeth will affect the shape of the signal – that is, the overtone content. There are multiple tone wheels used for each key on the keyboard. When the corresponding key is pressed, which wheels contribute to the signal is determined by switches under the player's control. The signals are sent to an amplifier and output through a speaker system.

The frequency of the signals produced by a wheel with m "teeth" is m times the frequency of its rotation. The tone wheels are all driven by a common drive connected using gears. The notes are tuned by choosing the appropriate gear ratios. For example, as shown in Figure 20.7, if the driving gear has 37 teeth and is connected to the tone wheel using a gear with 32 teeth, then the wheel will spin at 37/32 times the rate of the common drive. More generally, if the common drive rotates at a frequency f_0, the tone wheel has m teeth, and the gear ratio is n_1/n_2, then the frequency of the signal produced, f, is given by

$$f = m\, f_0\, (n_1/n_2).\tag{20.1}$$

Hence, the frequencies of the signals produced by the Hammond organ are all related by integer ratios – that is, by rational numbers. To get a tuning that is close to equal temperament, relatively large integers are used. Due to practical limitations, using more than about 100 teeth for any gear becomes difficult.

The Hammond organ was marketed as a compact substitute for a pipe organ. In fact, many of the original design considerations were aimed more at making it behave as a suitable substitute rather than as an instrument in its own right.

Perhaps the earliest purely electronic instrument to endure past its initial conception was the **theremin**. The theremin was developed around 1919 by Léon Theremin, a Russian physicist, based on radio circuitry developed using what was then a new technology – vacuum-tube electronic components. The sound of the theremin, and its cousins, are often those sounds most associated with the phrase "electronic music." The theremin gained popularity particularly for the eerie sounds it could create. Examples designed to achieve this eerie effect can be heard in classic movies such as *The Wizard of Oz* (MGM, 1939) and the original version of *The Day the Earth Stood Still* (20th Century Fox, 1951), among others.

The theremin circuitry contains four electronic oscillators. Two of those oscillators are fixed in frequency, and the other two are controlled, somewhat, by the performer. The frequency of each oscillator will depend on a time constant, such as the RC time constants previously discussed (Chapter 18). In the case of a theremin, it is more likely that the time is determined by the combination of an inductor (L) and a capacitor (C). In any event, if the capacitance, C, is varied, the frequency of the oscillation will vary. For the theremin, there are typically two "antennas." They are best thought of not as antennas but each as one "plate" of a capacitor. The other plate for each is one of the player's hands. Humans are electric conductors, after all, and two conductors near each other form a capacitor. As the player moves their hand closer and farther from the "antenna," the spacing of the capacitor "plates" changes, and hence the capacitance changes.

The two antennas are used to control the pitch (traditionally a vertical antenna) and volume (traditionally a horizontal antenna) of the sound. See Figure 20.8. The use of one vertical and one horizontal antenna spaced on opposite sides of the instrument helps to reduce the capacitance between the two antennas, helping to keep the two functions separate. The inner workings of the two control signals are the same until the last stage of the electronics.

If a theremin is available, some simple observations can be made. First, if a nonconducting object, such as a piece of plastic, cardboard, or dry wood, is brought near the antennas, the effect is very much smaller than if a hand is used. There may be some effect since the object may have a nonzero dielectric constant, but the effect will not be nearly as large as the effect when a conductor is used. Secondly,

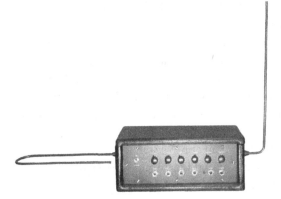

Figure 20.8 A theremin (with additional control functions available). Note the two "antennas" used by the player.

most theremins are adjusted so that as a hand comes close to the pitch antenna, the frequency of the sound goes up. This is in the opposite direction of what might be expected from the physics. As a hand gets closer, the capacitance gets larger, causing the time constant to get longer. A longer time constant should decrease the frequency. In addition, a computation of the capacitance, based on estimated areas and spacing, shows that the capacitance will be very small. That would mean that the other contributor to the time constant (R and/or L) would have to be impractically large to get the low notes. This is where the fixed oscillators come into play.

The key is to beat the signal from each variable oscillator with a fixed oscillator. To do that, the signals are electronically multiplied – a nonlinear process. The resulting signal will contain contributions at the sum and difference frequencies. This comes directly from the sinusoidal identities discussed earlier (Chapter 3), such as

$$\cos(\omega t)\cos(\omega_{\text{fixed}}t) = \frac{1}{2}\cos(\omega + \omega_{\text{fixed}}) + \frac{1}{2}\cos(\omega - \omega_{\text{fixed}}). \qquad (20.2)$$

If the frequency of the fixed oscillator is almost the same as that of the variable oscillator, the difference frequency, $(\omega - \omega_{\text{fixed}})$, can be small. If, in addition, the two individual frequencies are both well above hearing, whereas the difference frequency is within the audible range, then only the difference frequency will be heard. If the variable-frequency oscillator changes by, say, 1,000 Hz, then the difference frequency, the beat frequency, also changes by 1,000 Hz. The *changes* in frequency are preserved. For an oscillator operating at, say, 100 kHz, a 1,000-Hz change is only 1 percent. However, with this scheme, that small percentage change can result in a change of several octaves for the (audible) difference frequency. The small changes of the small capacitance, due to movement of the hand, are thus sufficient to produce audible pitches covering many octaves.

The fixed frequency oscillator within the theremin is normally adjusted to match the variable oscillator when the performer's hand is far away – there is typically an adjustment knob for this. When matched, the beat frequency is 0 Hz. As a hand is brought near, the difference between the two frequencies grows, and an audible tone can be generated. If the adjustment knob has sufficient range, the fixed oscillator can, instead, be adjusted to match when the performer's hand is nearby rather than far away, in which case moving the hand away creates an increase in frequency. Since only the difference frequency is heard, it is not possible to tell which of the oscillators is higher in frequency.

The signal that determines volume is treated similarly; however, it is converted to a frequency-dependent control signal that alters the amplitude of the output. It is normally adjusted so that the output signal (from the other pair of oscillators) is silenced as the performer's hand gets close.

In the language of radio electronics, the process of mixing two signals is called **heterodyning**. When two signals are adjusted so that the difference frequency is zero, that is referred to as **zero beating** the two signals. A typical radio receiver uses heterodyning. Inside the radio is an oscillator adjusted by the radio's frequency dial (or equivalent). The radio signals received by the antenna will include *all* the stations within range. By adjusting the internal oscillator, the signal from the desired station can be "heterodyned" to some single predefined frequency known as the *intermediate frequency* (i.f.). The electronics that follow will block signals from other stations, which are farther away from the i.f. frequency, while amplifying the desired signal. This process can cause problems if there are two strong radio signals that differ in frequency by the i.f. frequency. Those stations can heterodyne with each other, and the output will be some mix of both stations, possibly independent of the internal oscillator. In an environment with many radio transmitters, some coordination of the frequencies in use is necessary to avoid this **intermodulation** (or *intermod*, for short) problem.

There are many cousins to the theremin, differing largely in the interface between the performer and the electronics. For example, a similar-sounding instrument could use a rotary dial or resistive slide wire to alter a resistance, which then alters an RC time constant, which in turn controls an oscillation frequency. Such a device was used in the Beach Boys' recording of "Good Vibrations" (1966). In that case, a mechanical slide was rigged to turn the rotary dial of a signal generator. Such an interface is much easier to control (i.e., "play") than a theremin. Any old-fashioned audio-signal generator that includes volume and frequency knobs can be modified to create such instruments.

Electronic Synthesis

Effective **sound synthesis** is achieved by creating a model of the sound that can be implemented electronically. The basic functions to be simulated include the fundamental pitch, the overtone content, and how the volume of the sound

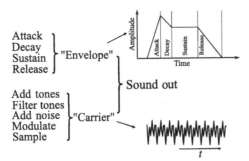

Figure 20.9 An illustration of the basic model used for sound synthesis intended for music.

changes from start to finish. It is convenient to consider these two as separate functions, although they may be integrated together in practice.

A simplified schematic of a simple synthesizer is shown in Figure 20.9. Like many physics models, it supplies the essential components of a sound signal. Its success suggests that in many cases, this model is close enough to be useful. In this model, a "**carrier**" tone is created, and its amplitude is controlled in time.

A simple form of amplitude control, briefly mentioned earlier (Chapter 9), that models the essential features of a musical sound is referred to as **ADSR**. *ADSR* stands for "attack, decay, sustain, release." The attack phase models the very beginning of the sound, as it turns on. There is then a transition period as the sound settles in, the "decay." Finally, the tone can be held for longer and shorter times, which is the sustain period. After that, the sound is turned off – the release. A time period, as well as starting and ending values for each of these periods, is specified. ADSR provides an "envelope" for the signal.

The carrier signal can be generated in a number of ways. A number of tones can be added together, a complex tone can be filtered to remove tones, noise can be added as a component, rapid modulation in frequency can be used to generate overtones, and sampling – a recording of an actual sound – may be used. In addition, any of those signals can be altered periodically in time to include tremolo, vibrato, glissando, or other effects.

Of course, the simple ADSR model can be extended in many ways to achieve much more complex effects. For example, the envelope may be made frequency dependent so that each overtone changes separately. With modern electronics, there are few limits on what can be done.

The first synthesizers to be produced as a commercial product were the **Moog synthesizers**, developed by Robert Moog in the 1960s. These devices relied heavily on voltage-controlled oscillators to produce pitches and voltage-controlled amplifiers and filters to control the envelope. An elaborate set of patch cords was used to interconnect the specific modules for each sound. The oscillators that control pitch were designed with an exponential behavior, just like the notes from a keyboard. Each octave, a factor of 2 in frequency, would correspond to the addition of a fixed voltage. For example, one might use 1 volt per octave – each time the control signal increases by 1 volt, the frequency goes up by a factor of 2. Various attack and decay times are controlled by signals that grow or shrink in time due to, for example, an RC time constant. The early popularity of the Moog synthesizer is undoubtedly related to the studio recordings of Wendy Carlos. These recordings, "Switched on Bach" and "The Well-Tempered Synthesizer," featured classical music performed using the synthesizer. For those recordings, each different sound was set up and recorded separately, and then they were all combined, to great effect.

Interfaces

This chapter is finished off with a brief discussion of electronic interfaces in common use – that is, the electronic connection between the player and the instrument. The "no-touch" capacitive interface, for the theremin, and the variable resistor were already mentioned. Many synthesizers will use a traditional keyboard that musicians find familiar. The simplest electronic keyboards are just an array of on/off switches configured to look like a piano keyboard. More sophisticated keyboards are force sensitive to mimic the behavior of a real piano – if you press more aggressively, the output sound is louder.

Virtually any electrical signal can be turned into a control signal. The signals from an electric guitar contain frequency and amplitude information that can be imposed upon other signals. Another popular interface is that of the electronic drum set (Figure 20.10). Here, the strike from a drumstick is picked up and used for control. In the simplest case, the drum strike provides an on/off signal, triggering another (electronically stored) sound. Of course, the intensity of the strike can be used as well. Piezoelectric sensors (Chapter 19) are commonly used as the detector for many electronic drum functions. More sophisticated drums may include multiple sensors, for example, to distinguish a hit near an edge from a hit near the center.

Wind instrument players have not been excluded. **Electronic wind instruments** (EWIs), or **wind controllers**, are played in a manner very similar to an instrument such as a clarinet or saxophone. An air-pressure sensor is used to measure the strength of the breath, and special switches are used to sense finger positions on the keys. A common finger sensor is resistive, where the electrical resistance is that of the player. One finger, often a thumb, is constantly held against one fixed electric connector. As the other fingers touch the keys, a very small electric current will flow through those fingers and out through the one fixed connection, for example, the thumb. The keys themselves need not move. Various other sensors and switches can be included for other effects.

Figure 20.10 A synthesized electronic drum set uses a layout similar to an acoustic drum set. Sensors detect the location and intensity of impacts and send that information to a computer, and the computer produces the preprogrammed sound associated with that sensor. (Photo by Guitarist Magazine/Future, Getty Images.)

With modern **computer animation**, virtual instruments can be created. Those are totally up to the imagination of the animator. Particularly elaborate animated musical performances can be conjured up where both the visual and audio portions of the performance are entirely constructed within a computer.

Concluding Remarks

This ends this journey into the physics behind music. Clearly, not every topic could be included, and those that are included cannot include great depth, but perhaps there was enough material to get started. And perhaps there were some project ideas along the way that will prove to be fruitful and enjoyable endeavors.

Summary

Many electronic effects are accomplished using computer programming. Some that do not use programming are created by devices that use simple mechanical and electrical properties to produce distortion, tremolo, and reverb. Electronic circuits can be used to produce effects such as sustain. Electronic signal delays can be used to produce a chorus effect and to emulate various sound environments.

Electronic musical instruments include the Hammond organ and the theremin and its relatives. The Hammond organ uses rotating magnetic materials and the Faraday effect to produce tones. The theremin is based on changes in capacitance, where the player's hand is part of the capacitor. To get signal frequencies throughout the audible range, the theremin creates beats between two high-frequency oscillators, one fixed in frequency and one with a frequency that depends on the player's hand position.

Electronics is also an important part of sound production in synthesizers. The tones, including those from models based on the attack, decay, sustain, and release (ADSR) of a sound, can be created in a number of different ways. A synthesizer's interface with the player includes switches arranged as a keyboard, as well as various other types of sensors used to control the sounds.

ADDITIONAL MATERIAL

Animusic. "Pipe Dream" [animated video]. Animusic, 2001.

Berg, R. E., and D. G. Stork. *The Physics of Sound*, 2nd ed. Prentice Hall, 1995 [or 3rd ed., Pearson, 2005]. (See Chapter 5.)

Carlos, W. *Switched-On Bach*. Columbia Records, 1968.

PROBLEMS

20.1 If a particular note from a Hammond-like organ uses the gear ratio of 37/32, as shown in Figure 20.7, what is the best gear ratio to use for the note one-half step higher based on equal temperament? Assume a common drive motor and the same number of teeth in the tone wheels, and limit the number of teeth in both gears to be less than 100.

20.2 If you are capable of *changing* an oscillator's frequency by up to 2 percent (e.g., by moving your hand as is done for a theremin), what is the minimum oscillator frequency that can be used to produce beats from 0 to 10 kHz?

20.3 A simple way to create electronic signals near the frequencies of an equal-tempered musical scale is to use a very high-frequency electronic oscillator, which is then divided electronically by integers. Both of these tasks are relatively easily accomplished with digital electronics. For example, if you start with a 2.0000-MHz signal and divide by 478, you get 4,184.1 Hz, which is four octaves (a factor of 16) above 261.5 Hz, middle C. What integer values, all smaller than 478, could be used for the rest of the chromatic scale, and what frequencies result? Compare to frequencies using the table in Appendix C.

20.4 Use an audio editor, or similar means, to synthesize the sound of an orchestral chime using the data from Figure 7.8. Those data show three prominent overtones at frequencies of 768, 1,129, and 1,547 Hz with amplitudes of −3.8 dB, −0.228 dB, and −1.814 dB, respectively. Create and add these three sinusoids together, and then use a rapid "fade-in" at the beginning and a very long "fade-out" to the end for the attack and decay, respectively. You may need to use the fades multiple times. Adjust the duration, attack, and decay to try to get a realistic chime sound.

20.5 Find an example of an electronic effect, either within performed music, in an audio editor, or available as hardware, and explain how it works.

20.6 Create your own electronic effect or music.

Appendices

APPENDIX A

Mathematics

I. Exponents and Standard International (SI) Unit Prefixes

Exponents are a shorthand way to indicate that a number is to be multiplied by itself. If the exponent is a simple integer, multiply the number by itself that number of times.

Example A.1

$$7.1^3 = 7.1 \times 7.1 \times 7.1.$$

If the exponent is zero, then the result is 1.0 (except for 0^0, which must be handled on a case-by-case basis). For negative integers, multiply the inverse.

Example A.2

$$6.3^{-2} = (1/6.3) \times (1/6.3).$$

When multiplying a number raised to a power by that *same* number raised to another power, you can *add* the exponents.

Examples A.3

$$8.3^2 \times 8.3^3 = 8.3^5 \text{ and } 4.5^2 \times 4.5^{-4} \times 4.5 = 4.5^{-1} = 1/4.5.$$

Fractional exponents can be defined as well. One fractional exponent that shows up often is ½, which corresponds to the square root. Using the rules described previously, $(2^{1/2}) \times (2^{1/2}) = 2^1 = 2$, so $2^{1/2}$ represents the number that, when squared, gives you 2. Hence, $2^{1/2} = \sqrt{2}$, even though you cannot really "multiply 2 by itself 1/2 times." This idea can be generalized to include all numbers as exponents, including irrational and imaginary numbers (see following discussion).

Powers of 10 are often expressed using exponents. For integer powers of 10, the exponent is the number of zeros you add to the right (for positive) or left (for negative) of the number 1 to get the value. Adding zeros in this way only works for powers of 10. When the exponent of 10 is positive, then put the zeros to the right.

Examples A.4

$$10^3 = 1{,}000 \text{ and } 10^6 = 1{,}000{,}000.$$

When the exponent of 10 is negative, then put the zeros on the left.

Examples A.5

$$10^{-3} = 1/1{,}000 = 0.001 \text{ and } 10^{-6} = 1/1{,}000{,}000 = 0.000001.$$

An exponent of 0 means do not add any zeros.

Example A.6

$$10^0 = 1.$$

For a large number such as 2,300,000, it is often convenient to write it as a smaller number multiplied by a power of 10.

Example A.7

$$2{,}300{,}000 = 2.3 \times 10^6.$$

This is known as *scientific notation*. The value written as 2.3×10^6 implies that the value is rounded off to two digits. The value 2.300×10^6 would be presumed to have four digits of accuracy, the last two of which just happen to be zeros.

The first would be said to have "two significant digits," and the second, "four significant digits." A calculator will treat the two values identically. In general, the final result from the calculator should not be written with more digits than the smallest number of significant digits used for any value during the calculation. Note that when applying this rule, a value that, by its nature, is an integer should be treated as having as many significant digits as your calculator can handle, even though extra zeros are not written down for integers.

Example A.8

If someone says they walk 2.5 km to work, that would be interpreted to mean it is approximately a 2.5-km walk, but that may actually be as far off as, say, 2.6 or 2.4 km. However, if they say they have 3 cats, it is generally interpreted to mean that they have exactly 3 cats.

Prefixes for units are often used as a convenient alternative way to indicate powers of 10. Refer to Table 1.1 for the most common of these prefixes.

Example A.9

$$2.1 \text{ millionths of a kilogram} = 2.1 \times 10^{-6} \text{ kg} = 2.1 \times 10^{-6} \times 10^3 \text{ g}$$
$$= 2.1 \times 10^{-3} \text{ g} = 0.0021 \text{ g} = 2.1 \text{ mg}.$$

On some calculator and computer displays, an E might be used to signify the exponent of 10. The E in this case should be read as "times 10 to the power." Hence, $2.1E - 3 = 2.1 \times 10^{-3} = 0.0021$. Note that the use of the E notation for this purpose should be reserved for calculator and computer displays and should not usually be used anywhere else.

II.   Some Mathematical Properties

Special Numbers

$$\pi = 3.14159265\ldots$$
$$e = 2.71828183\ldots$$

Radians and Degrees

There are 360° around a circle, and 2π radians around a circle.
To convert from radians to degrees, multiply by $180/\pi = 57.2957795\ldots$
To convert from degrees to radians, multiply by $\pi/180 = 0.01745329\ldots$

Identities for Sinusoidal Functions

$$\sin(-A) = -\sin(A)$$

$$\cos(-A) = \cos(A)$$

$$\sin A = \cos(A - 90^\circ)$$

$$\cos A = \sin(A + 90^\circ)$$

$$\sin^2(A) + \cos^2(A) = 1$$

$$\sin(A + B) = \sin A \cos B + \cos A \sin B$$

$$\sin(A - B) = \sin A \cos B - \cos A \sin B$$

$$\cos(A + B) = \cos A \cos B - \sin A \sin B$$

$$\cos(A - B) = \cos A \cos B + \sin A \sin B$$

$$\sin 2A + \sin 2B = 2 \sin(A + B)\cos(A - B)$$

$$\sin 2A - \sin 2B = 2 \sin(A - B)\cos(A + B)$$

$$\cos 2A + \cos 2B = 2 \cos(A + B)\cos(A - B)$$

$$\cos 2A - \cos 2B = 2 \sin(A + B)\sin(B - A)$$

$$\tan(A) = \sin(A)/\cos(A)$$

$$C \cos A + D \sin A = \sqrt{C^2 + D^2} \sin(A + \vartheta),$$
$$\text{where } \tan \vartheta = C/D.$$

If n is any integer (positive, negative, or zero), then

$$\sin(n\pi) = 0 \quad \text{and} \quad \cos\left(\left(n + \frac{1}{2}\right)\pi\right) = 0.$$

If n is any even integer (i.e., 0, ±2, ±4, etc.), then

$$\cos(n\pi) = 1 \quad \text{and} \quad \sin\left(\left(n + \frac{1}{2}\right)\pi\right) = 1.$$

If n is any odd integer (i.e., ±1, ±3, etc.), then

$$\cos(n\pi) = -1 \quad \text{and} \quad \sin\left(\left(n + \frac{1}{2}\right)\pi\right) = -1.$$

These last two relations can be combined so that if n is any integer,

$$\cos(n\pi) = (-1)^n \quad \text{and} \quad \sin\left(\left(n + \frac{1}{2}\right)\pi\right) = (-1)^n.$$

If A is in radians and $|A| \ll 1$, then

$$\sin A \approx A, \quad \cos A \approx 1 - A^2/2, \quad \tan(A) \approx \sin(A) \approx A,$$

and the smaller the angle, the better are these approximations. These come from "series expansions" or "series representations" of these functions,

$$\sin(A) = A - \frac{A^3}{2\cdot3} + \frac{A^5}{2\cdot3\cdot4\cdot5} - \frac{A^7}{2\cdot3\cdot4\cdot5\cdot6\cdot7} + \cdots$$

$$\cos(A) = 1 - \frac{A^2}{2} + \frac{A^4}{2\cdot3\cdot4} - \frac{A^6}{2\cdot3\cdot4\cdot5\cdot6} + \cdots,$$

valid only if A is in radians. Equality is achieved only in the limit that there are an infinite number of terms (with the exception of $A = 0$); however, a good approximation can usually be made using a relatively small number of terms if A is not too large. The notation for these particular series representations can be made more compact through the use of factorial notation, consisting of an exclamation point. For example, 5!, read as "five factorial," would represent the product of all the integers from 1 to 5.

Law of Cosines

For any triangle, the lengths of the sides (a, b, c) are related to the angle, θ, between sides of length a and b, by

$$a^2 + b^2 - 2ab\cos\theta = c^2.$$

If $\theta = 90°$, this reduces to the Pythagorean theorem for a right triangle,

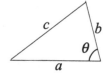

$$a^2 + b^2 = c^2.$$

Exponential Functions

The exponential function is given by

$$\exp(A) = e^A.$$

and is defined for all values of A.

Some properties of the exponential function include:

$$\exp(0) = 1$$

$$\frac{1}{\exp(A)} = \exp(-A)$$

$$\exp(A) \times \exp(B) = \exp(A + B)$$

$$\frac{\exp(A)}{\exp(B)} = \exp(A - B).$$

Logarithms

If $y = 10^x$, then, by definition, x is "the logarithm base 10 of y," or symbolically, $x = \log_{10}(y)$.

If $y = e^z$, then, by definition, z is the logarithm base e of y, or $z = \log_e(y) = \ln(y)$. Logarithms can be defined using any base greater than zero. Base 10 and base e are the most common. Base e logarithms are also referred to as *natural logarithms*.

Often, the "10" is left off when base 10 is used. Base e logarithms are often abbreviated "ln"; however, be aware that within many math tables, "log" with no subscript is used to refer to the natural logarithm. The extension of the logarithm to include complex numbers is signified by "Log."

The word *logarithm* is often shortened to *log*. So, for example, since $8 = 2^3$, one could say that "the log base 2 of 8 is 3," or symbolically, $\log_2 8 = 3$.

Since $10^x > 0$ for all x, the logarithm of a negative value is not defined (for real numbers).

Properties of logarithms (with z, A, and B all greater than 0):

$$\log_z A = \frac{\log_{10} A}{\log_{10} z} = \frac{\ln A}{\ln z},$$

$$\log(A \times B) = \log A + \log B,$$

$$\log(A^x) = x \log A,$$

$$\log(A/B) = \log A - \log B,$$

$$10^{\log_{10}(A)} = A,$$

$$z^{\log_z A} = A.$$

A useful expansion that can be used for approximation if $|x| \ll 1$ is

$$\log_e(1 - x) = -x - \frac{x^2}{2} - \frac{x^3}{3} - \frac{x^4}{4} - \cdots$$

$$\log_{10}(1 - x) = \frac{1}{\log_e(10)}\left(-x - \frac{x^2}{2} - \frac{x^3}{3} - \cdots\right).$$

Imaginary and Complex Numbers

An imaginary number can be written as a real number that defines its magnitude, multiplied by the square root of –1. The square root of –1 is designated with a lowercase i or, in some literature, with a lowercase j. Some identities are

$$i^2 = -1, \quad i^3 = -i, \quad i^4 = 1, \quad \text{and so forth.}$$

$$1/i = -i .$$

An imaginary value with magnitude B would be written iB.

A complex number has a real and an imaginary part. A complex number, C, where the magnitude of the real part is A and the magnitude of the imaginary part is B would be written $C = A + iB$. The magnitude of C is a real value and is written as $|C|$, where $|C| = \sqrt{A^2 + B^2}$. The phase factor, φ, associated with C is given by any (or all) of the following involving the arctangent, arcsine, and arccosine functions:

$$\varphi = \tan^{-1}(B/A) = \sin^{-1}(B/|C|) = \cos^{-1}(A/|C|).$$

Note that an electronic calculator may give an answer that is off by 180°, and the result will need to be corrected manually. For the arctangent, a correction is usually necessary when the real part is negative. The phase factor is particularly important when imaginary values are used to describe the relationship between two sinusoidal signals.

The relationship between the values A, B, $|C|$, and φ can be visualized using a right triangle with sides of length A and B; hypotenuse of length $|C|$; and with the phase angle, φ, opposite to the side of length B. As a shorthand, C can be written

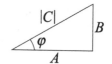

$$C = |C| \angle \varphi.$$

Example A.10

The complex number $C = 4 + 3i$ has real part 4, imaginary part 3, and magnitude and phase given by

$$|C| = \sqrt{4^2 + 3^2} = 5; \quad \varphi = \tan^{-1}(3/4) = 0.6435 \text{ radians} = 36.87°.$$

Multiplication and division of complex values are much easier when the values are expressed as a magnitude and phase. For multiplication, the magnitudes are multiplied, and the phases are added. For division, the magnitudes are divided, as usual, and the phase in the denominator is subtracted from the phase in the numerator. That is,

$$C_1 \times C_2 = |C_1| \times |C_2| \angle (\varphi_1 + \varphi_2); \quad \frac{C_1}{C_2} = \frac{|C_1|}{|C_2|} \angle (\varphi_1 - \varphi_2).$$

III. The Summation Symbol

An oversized, uppercase Greek sigma is often used to write summations involving many related terms in a compact form. The terms are generally related by an integer index, and that index may also serve as a numerical value within the terms that are to be added. Starting and ending values are supplied, and all integer values in between are included.

Example A.10

To add 10 values indicated by the variable b, indexed from 1 to 10, can be written

$$S = \sum_{k=1}^{10} b_k.$$

To write an expression corresponding to the sum of all the integers squared from 1 to 100,

$$S = \sum_{n=1}^{100} n^2 = 338{,}350.$$

Infinite summations can also be represented with this notation. For example,

$$\log_e(1-x) = -x - \frac{x^2}{2} - \frac{x^3}{3} - \frac{x^4}{4} - \ldots = -\sum_{n=1}^{\infty} \frac{x^n}{n}.$$

Sometimes one or more of the limits on the summation symbol may not be explicitly included but are implied from the context. Although not used in this text, an oversized uppercase Greek pi can be used in a similar way to write the product of, rather than the sum of, a series of terms.

APPENDIX B
Greek Alphabet

A α	Alpha		N ν	Nu
B β	Beta		Ξ ξ	Xi
Γ γ	Gamma		O o	Omicron
Δ ϱ	Delta		Π π	Pi
E ε	Epsilon		P ρ	Rho
Z ζ	Zeta		Σ σ	Sigma
H η	Eta		T τ	Tau
Θ θ	Theta		Y υ	Upsilon
I ι	Iota		Φ φ	Phi
K κ	Kappa		X χ	Chi
Λ λ	Lambda		Ψ ψ	Psi
M μ	Mu		Ω ω	Omega

APPENDIX C

Note Frequencies

The accompanying table lists equal-tempered frequencies, in hertz, based on the fixed value listed along the top row. For this notation, C4 is commonly referred to as *middle C*. A commonly used scale corresponds to A4 = 440 Hz.

Note	C4 = 256 Hz	A4 = 432 Hz	A4 = 438 Hz	A4 = 440 Hz	A4 = 442 Hz	A4 = 444 Hz
A3	215.27	216.00	219.00	220.00	221.00	222.00
A3#/B3♭	228.07	228.84	232.02	233.08	234.14	235.20
B3	241.63	242.45	245.82	246.94	248.06	249.19
C4	256.00	256.87	260.44	261.63	262.81	264.00
C4#/D4♭	271.22	272.14	275.92	277.18	278.44	279.70
D4	287.35	288.33	292.33	293.66	295.00	296.33
D4#/E4♭	304.44	305.47	309.71	311.13	312.54	313.96
E4	322.54	323.63	328.13	329.63	331.13	332.62
F4	341.72	342.88	347.64	349.23	350.82	352.40
F4#/G4♭	362.04	363.27	368.31	369.99	371.68	373.36
G4	383.57	384.87	390.21	392.00	393.78	395.56
G4#/A4♭	406.37	407.75	413.42	415.30	417.19	419.08
A4	430.54	432.00	438.00	440.00	442.00	444.00
A4#/B4♭	456.14	457.69	464.04	466.16	468.28	470.40
B4	483.26	484.90	491.64	493.88	496.13	498.37
C5	512.00	513.74	520.87	523.25	525.63	528.01
C5#/D5♭	542.45	544.29	551.85	554.37	556.89	559.40

To convert to other octaves, divide by 2 for each octave down, and multiply by 2 for each octave up.

Example A.11

To find C3, use the value for C4 divided by 2, and to find G7, multiply the value for G4 by $2 \times 2 \times 2 = 8$.

To use other fixed values, multiply the frequencies by the ratio of the new fixed value to the old fixed value.

Example A.12

To convert frequencies from the column A4 = 440 Hz to the scale with A4 = 446 Hz, multiply each frequency in the A4 = 440 Hz column by $446/440 = 1.013636$.

APPENDIX D

Answers to Selected Problems

1.1 10^{-8} cm

1.3 $9.46 \times 10^{+15}$ m

1.7 Hint: Think about the measurement processes used and the expectations of the participants.

2.1 Between $A^\sharp/B^\flat$ and B, a bit closer to B

2.3 33 Hz and 4,208 Hz

2.5 430.5 Hz

2.7 Carefully compare plots using linear and logarithmic scales. Also try using the inverse of the distances, instead of the distances, for the plot.

3.1 1.2 Hz, unison

3.3 14 Hz

3.5 $m = 19$, 265.6 Hz (about 1.2 percent higher)

3.7 469.8 Hz, 465.0 Hz, and 456.8 Hz

4.1 7.0 dB

5.1 In each case, the force on the car equals the force on the truck.

5.3 (b) g, down

5.5 Δk = factor of 4, or Δm = factor of $1/4$.

5.7 (a) 2,610

6.1 $A = 0.15$ m, $\lambda = 0.25$ m, $f = 10.5$ Hz.

6.3 10 cm

6.5 Comes from: $(1 - 1/2^{1/12}) \approx 1/18$

7.1 Translation and rotation

8.1 17 m and 1.7 cm

8.3 (b) 7.6 mm; (d) 3 m/s toward $-x$.

8.5 (a) $E = P \times T = \rho\omega^2 A^2 cT/2 = \rho\omega A^2 \lambda/2$.

(b) Note that the values represent a violin A string.

8.7 (a) cT; (b) $1/T$; (c) $f = 1/T, n = t/T$; (d) vt.

9.1 $\Delta f \approx 100$ Hz (± 50 Hz) for all. (a) About seven half steps.

11.1 1.3×10^{-14}

11.3 9 km

11.5 0.02 moles leave

12.1 1 percent at 0.244 radians, or about $14°$

12.3 About ± 1 mK

12.7 No, watching a video in reverse clearly looks improbable.
12.9 340 Hz

13.3 20 dB, -10 dB, 1,000 times
13.5 174 dB

14.1 5.3×10^6 and 1.3×10^6 acoustic ohms
14.3 2,200 Hz
14.5 37°

15.1 17 m
15.3 (a) 0.8 m
15.5 T_R

16.1 Near 440 Hz
16.3 85 Hz
16.5 1, 6, 3, 7, 9, 10, 5, 8, 4, 2

17.1 Never
17.3 (a) N will be up.

18.1 80–300 kV
18.3 2.54 times too large
18.5 (a) 5.7 V

19.1 About 500 m
19.3 0.4 percent, 10 bits

20.1 49/40 or 98/80
20.3 Values include 451, 379, 253, and 239.

Index

acoustic impedance, 233, 246
    specific, 246
adiabatic
    exponent, 221, 390
    process. *See* process, adiabatic
admittance, 242
ADSR, 166, 385
air, 198
air reed, 183
aliasing, 372
amu. *See* atomic mass unit
anechoic chamber, 280
anti-aliasing filter, 373
anti-node, 99
approximation, 217
architectural acoustics, 269
artificial harmonics, 106
atomic mass unit, u, 191, 193, 195
attack and release, 165
aural harmonic, 189
autocorrelation function, 62
Avogadro's constant, 193

Barkhausen effect, 318
beats, 34
bells, 137
    tubular. *See* chime, orchestral
Bernoulli effect, 181
Bessel function, 116, 129
Bohr model, 191
butterfly effect, 186

cantilever, 123
capacitance, 335
capacitor, 80, 335
    parallel plate, 335
carbon dating, 194
cartioid, 272
CD, 365
cents. *See* interval, cents
chaos, 186
chime, 117
    grandfather clock, 117, 120
    music box, 117

orchestral, 117
wind, 121
Chladni plate, 139
chord, 31
    barbershop, 41
    inversion, 41
    triad, 31
chorus effect, 377
circle of fifths, 29
clipping, 376
clock arithmetic, 305
Coanda effect, 181
cocktail party effect, 289
collision, elastic and inelastic, 202
complex number, 244, 396
condenser microphone, 344
conical pipe, 250
corner reflector, 289
correlation time, 56
cosine transform, 373

dalton (mass unit). *See* atomic
    mass unit
damping, 83
decibel, 53, 234
degeneracy, 131
degree of freedom, 199, 200,
    221
dielectric strength, 334
diffraction, 299
diffusion, 301
    active, 307
    primitive root, 306
    quadratic residue, 306
    random, 303
digital signals, 367
dipole, 271
dispersion, 156
distortion, 66
    effect, 376
drum head, 127
duty cycle, 368
DVD, 365
dyad, 20

echo, 280
effective absorbing area, 284
Einstein, Albert, 3, 195, 362
electret microphone, 345
electric breakdown, 334
electric field, 331
electric guitar pickup, 324
electrical conductivity, 333
electromagnet, 319
electromagnetic field, 332
electromotive force, 319
electronic wind instruments, 386
elementary charge, 331
energy density, 290
enharmonic, 21
exponential, 390
    decrease, 19, 340
    function, 394
    increase, 19

facade pipe, 185
Faraday's law of induction, 322
ferromagnetic, 316
fictitious force, 90
fifth, 20
finger holes, 254
formants, 264
Foucault pendulum, 90
Fourier series, 46
    phase constant, 46,
        51, 167
Fourier synthesis, 49
frequency, 16, 17
    angular, 46, 76
frequency domain, 45
friction, 174
    coefficient of, 176
    kinetic, 176
    static, 176
fundamental frequency, 48

gas law, ideal, 204
gauge pressure, 204
glockenspiel, 126

gramophone, 352
gravity, force of, 70

half step, 21
Hammond organ, 380
harmonic, 48, 103
    artificial, 106
harmonic oscillator, 67
harmonic series, 98
harmonic tuning. *See*
    temperament, just
Helmholtz resonator, 79
Herschel, William, 3, 362
hertz, definition, 16
heterodyne, 384
Hooke's law, 72
hum, 325
humbucker, 325
Huygens principle, 300
hyperbolic functions, 118
hysteresis, 358

ideal gas law, 204
ideal spring, 72
ideal string, 100
image source, 281
imaginary number. *See* number,
    imaginary
impedance, 242
    characteristic, 247
    input, 247
    matched, 248
impulse, 201
inductor, 80
inertia, 68
infrasound, 237
inharmonic, 117
interference, 55, 295
    constructive, 296
    destructive, 296
intermodulation, 384
interval, 20
    barbershop seventh, 42
    cents, 39
    fifth, 20, 29
    harmonic seventh, 42
    table, 36, 39
inverse square law, 278
irrational number. *See* number,
    irrational
isotopes, 194
isotropic source, 269

kalimba. *See* cantilever
keyboard, 21

law of reflection, 281
line source, 274
linear physics, 171
lip reed, 182
log scale, 23
logarithm, 23, 395
loudness, 236

magnet, 310
    electromagnet, 319
    permanent, 315
magnetic field, 312
    lines, 312
magnetic recording, 355
    AC bias, 358
    DC bias, 357
    tape, 356
magnetoresistor, 355
manometer, 204
marimba, 125
maximal-length sequence, 304
mbira. *See* cantilever
membrane, 127
microphone
    condenser, 344
    dynamic, 323
    electret, 345
    ribbon, 323
    sensitivity, dBV, 346
middle C, 22
missing fundamental, 60
mode locking, 185
modular arithmetic, 305
mole, 193
monochord, 32
monophonic, 352
mouth pressure, 209
music of the spheres, 3
music, definition of, 4
musical
    chord, 31
    fifth, 29
    note, 28
    scale, 28

natural abundance, 194
Newton, Isaac, 67
    laws of motion, 67
node, 99, 115

noise, 55
    colored, 56
    correlation time, 56
    pink, 56
    white, 56
nonlinear physics, 171
normal mode, 115
number
    complex, 244
    imaginary, 244, 396
    irrational, 36
    rational, 33
    real, 244
Nyquist folding, 372

ocarina, 79
octave, 20
Ohm's law, 243, 338
opera gong, 173
optical recording, 361
organ pipe, 184
    width scaling, 265
overtones, 48

paramagnetic, 318
parity, 304
partial, 48
pendulum, 77
    coupled, 133
    large amplitude, 173
period, 16
phantom power, 344
phonograph, 351
photoelectric effect, 362
photons, 362
photosensor, 361
physics, definition of, 6
piezoelectric materials, 353
pipe, 245
    cylindrical, 245, 249
    facade, 185
    organ, 184
pitch, 59
    number, 59
plane source in a baffle, 275
plate
    circular, 130
    square, 135
plucked string, 104
polarization
    longitudinal, 154
    transverse, 154

power, 156
Pythagorean theorem, 10, 298, 394
precedence effect, 289, 380
pressure, 203
    absolute, 204
primitive root sequence, 305, 306
process
    adiabatic, 214
    isobaric, 214
    isothermal, 214
    isovolumetric, 214
proximity effect, 278
pulse-width encoding, 368
pure tone, 48
Pythagoras of Samos, 1
Pythagorean comma, 39
Pythagorean temperament, 38
PZT, 354

quadratic residue sequence, 305, 306
quadrupole, 272
quality factor, Q, 86, 109, 265

random signals, 55
rational number. *See* number, rational
ray approximation, 280
RC time constant, 330
reciprocity, 276
reflection, 248, 295
reflection coefficient, 248, 265, 302
resistance, 338
resistor, 338
resonance, 67
    driven harmonic oscillator, 81
    sympathetic, 88
resonator, 126
retentivity, 317
reverberant energy, 291
reverberation, 280
reverberation time, 281

ripple tank, 299
root mean square (rms), 200

sabin, 285
scale
    major and minor, 29
    musical, 28
    organ pipes, 265
    Shepard, 60
    table, 29, 399
schematic diagram, 339
Schroeder diffuser, 305
scientific notation, 9, 391
series expansion, 217
Shepard's scale, 60
signal-to-noise ratio, 292
sinusoidal functions, 34, 393
sound
    complex, 48
    composite, 48
    speed. *See* speed of sound
sound on picture, 363
sound synthesis, 384
soundboard, 276
speaker enclosure, 80
spectrograph, 166
spectrometer, 50, 59
spectrum, 50
speed of sound, 212
    temperature dependence, 225, 226
spring constant, 72
spring reverb, 378
standing wave, 101
states of matter, 197
stereophonic, 352
strike tone, 138
string
    ideal, 100
    piano, 124
    plucked, 104
    under tension, 94
superposition, 136
sustain effect, 325, 377
synthesizer, 385

temperament, 37
    equal, 38
    just, 37
    meantone, 39
    Pythagorean, 38
temperature, 199
    absolute, 200
theremin, 382
timbre, 48
transformer, 326
transmission coefficient, 248, 265
transverse, 154
    motion, 95
traveling wave, 143
    sinusoidal, 148
tremolo, 127, 184, 379
tremulant, 184
tuning fork, 78, 274

ultrasound, 237
uncertainty principle, 161
units
    derived, 8
    prefixes, 9, 392
    standard international, 7

vibraphone, 127
vibrato, 184
vinyl record, 352
vocal folds, 263
vocoder, 378

wave, 101, 143, 230, 295
    peak-to-peak amplitude, 102
    standing, 101, 151
    traveling, 143
wavelength, 102, 148
wave number, 148
wire recorder, 356
wolf tone, 137

xylophone, 125

zero beat, 384

Printed in the United States
by Baker & Taylor Publisher Services